Probability for Data Scientists

1st Edition

Probability for Data Scientists

1st Edition

Juana Sánchez

University of California, Los Angeles

cognella®
SAN DIEGO

Bassim Hamadeh, CEO and Publisher
Mieka Portier, Acquisitions Editor
Tony Paese, Project Editor
Sean Adams, Production Editor
Jess Estrella, Senior Graphic Designer
Alexa Lucido, Licensing Associate
Susana Christie, Developmental Editor
Natalie Piccotti, Senior Marketing Manager
Kassie Graves, Vice President of Editorial
Jamie Giganti, Director of Academic Publishing

Printed in the United States of America.

cognella® ACADEMIC PUBLISHING
3970 Sorrento Valley Blvd., Ste. 500, San Diego, CA 92121

Contents

Detailed Contents

To my mother Juana and the memory of my father Andrés,
with love, admiration and gratitude.

The enlightened individual had learned to ask not "Is it so?" but rather "What is the probability that it is so?"

Ross, 2010

In investigating the position in space of certain objects, "What is the probability that the object is in a given region?" is a more appropriate question than "Is the object in the given region?"

Parzen, 1960

Preface

Probability is the mathematical term for chance. Much of statistics, data science and machine learning theory and practice rests on the concept of probability. The reason is that any conclusion concerning a population based on a random sample from that population is subject to uncertainty due to variability. It is probability theory what enables one to proceed from mere description of data to inferences about populations. The conclusion of a statistical data analysis is often stated in terms of probability. Understanding probability is thus necessary to succeed as a statistician and data scientist in artificial intelligence, machine learning or similar endeavors.

This book contains a mathematically sound but elementary introduction to the theory and applications of probability. The book has been divided in two parts. Part I contains the basic definitions, theorems, and methods in the context of discrete sample spaces, which makes it accessible to readers with a good background in high school algebra and a little ability in the reading and manipulation of mathematical symbols. Part II contains the corresponding ideas in the continuous case, and is accessible to readers with a working knowledge of the univariate and multivariate differential and integral calculus, and mastery of Part I. The book is designed as a textbook for a one-quarter or one-semester introductory course, and can be adapted to the needs of undergraduate students with diverse interests and backgrounds, but it is detailed enough to be used as a self-learning tool by physics and life scientists, engineers, mathematicians, statisticians, data scientists and others that have the necessary preparation. The text aims at helping the reader become fluent in formulating probability problems mathematically so that they can be attacked by routine methods, in whatever applied field the reader resides. In many of these fields of application, books on chance quickly jump to the most advanced probability methods used in research without the proper apprenticeship period. Probability is not to be learned as a cookbook, because then the reader will have no idea how to start when encountering an unfamiliar problem in their field of application. Numerous examples throughout the text show the reader how apparently very different problems in remotely related contexts can be approached with the same methodology, and how probability studies mathematical models of random physical, chemical, social and biological phenomena that are contextually unrelated but use the same probability methods. For example, the law of large Numbers is the foundation of social media, fire, earthquake and automobile insurance, and gambling, to name a few.

Having those who have to deal with data, data science or statistics in mind, the main goal of this book is to convey the importance of knowing about the many (the probability distribution for random behavior) in order to predict individual behavior. The second learning goal is to appreciate the principle of substitution, which allows the manipulation of basic probabilities about the many to obtain more complex and powerful predictions. Lastly, the book intends to make the reader aware of the fact that probability is a fundamental concept in Statistics and Data Science, where statistical tests of hypothesis and predictions involve the calculation of probabilities.

In part I, Chapters 1 to 6 review the origin of the mathematical study of probability, the main concepts in modern probability theory, univariate and bivariate discrete probability models and the multinomial distribution. Chapters 7–10 make up Part II. Sections that are too specialized and more advanced are indicated and the author recommends passing them without loss of continuity, or refers the reader to other sections of the book where they will be explained in detail. To enhance the teaching and self-learning value of the book, all chapters and many sections within chapters start with a challenging question to encourage readers to assess their prior conceptions of chance problems. The reader should try to answer that question and discuss it with peers. At the end of each chapter, the reader should go back to that question and compare initial thoughts with thoughts after studying the chapter. Exercises at the end of most sections of the book and at the end of each chapter give the reader an opportunity to apply the methods and reasoning process that constitutes probability topic by topic. Some of them invite research and broader considerations. Because random numbers are used in many ways associated with computers nowadays, including the adaptive algorithms used by social media to modify behavior, computer games, generation of synthetic data for testing theories, and decision making in many fields, every chapter contains guided exercises with the software R that involve random numbers.

Relevant references for further analysis found throughout the book will allow the reader to continue training in the more advanced way of approaching probability after they finish this book. There are so many fields of engineering and the physical, natural, and social sciences to which probability theory has been applied that it is not possible to cite all of them. Probability is also at the heart of modern financial and actuarial mathematics, thus exercises in health care and insurance are also included.

The book is intended as a tribute to all those who have made an effort to make probability theory accessible to a wide audience and those that are more specialized. Consequently, the reader will find many examples and exercises from a wide array of sources. I am deeply indebted to them. By bringing many of these authors to the reader's attention I wish to direct enquiries to sources with correct information and give students a sense of the depth and breadth of thinking probabilistically and of how they can move to more difficult aspects of the theory. If I have missed acknowledging or have misquoted some author, I hope the author will bring this to my attention, and I apologize in advance.

In studying this book, the reader must make an effort to talk about what is or is not understood with peers. Sharing results of experiments, chatting with colleagues about recent discoveries, learning a new technique from friends are common experiences for working scientists and is necessary for anyone wishing to apply probability theory. Probability literacy is a necessity. The success of data scientists in the application of probability is the product of multidisciplinary teams. Explaining a problem to others quite often helps see the solution of the problem.

The book title says "for data scientists," and indeed most of the examples of the book as well as many of the exercises and case studies, although adapted for beginners, come from interdisciplinary contexts that use scientific methods and processes such as probability modeling to extract knowledge from data. Fields such as genetics, computational biology, engineering, quality control, marketing, to name a few rely on the logic of probability to make sense of data. Genetic microarrays, medical imaging, satellite imaging, internet traffic involve large quantities of data, in some cases streaming data (available as it is produced) analyzed in real time. Where there is data there should be a good grasp of probability theory to make sense of the data.

I am indebted to my students at UCLA who, throughout the years, with their questions and their enthusiasm for the subject, have helped me improve my lecture notes on which this book is based. I am also thankful of the supportive teaching environment that my colleagues of 21 years at UCLA's Statistics Department have provided. I also take this opportunity to gratefully acknowledge my debt to the Affordable Course Materials Initiative (ACMI) of the UCLA Library, in particular Tony Leponte and Elizabeth Cheney (the latter currently at CSUN), for their help compiling resources for students of probability. I am most grateful to Alberto Candel for contributing very interesting resources and suggestions. I offer my sincere gratitude to Senior Acquisitions Editor Mieka Portier, Project Editor Tony Paese, Developmental Editor Susana Christie, Production Editor Sean Adams and Associate Production Editor Abbey Hastings for their constant guidance, encouragement, and careful scrutiny of the work done. Thanks also to all those at Cognella who have helped in the publication process and have helped improve the first notes considerably.

Juana Sánchez
University of California, Los Angeles
June 2019

Part I

Probability in Discrete Sample Spaces

When the mathematical theory of probability started in the seventeenth century, discrete sample spaces were the only spaces that could be handled with available mathematical methods at the time. It is then natural to try to start understanding probability by examining experiments with discrete sample spaces. These types of experiments lend themselves to all the scrutiny pertinent to continuous sample spaces without the additional concepts and conventions needed to handle the continuous case. Consequently, the reader can learn the main subjects of probability theory without the mathematical background hurdles.

This part of the book contains topics that are accessible to readers with a good background in high school algebra and a little ability in the reading and manipulation of mathematical symbols. Supplementary sidebars with review of some of the mathematics, and references to good sources to review the necessary mathematics, make the navigation smoother. Reference is also made to continuous sample spaces when pertinent, but those will be studied thoroughly in Part II of the book.

Numerous references to authors, web sites, and other supplementary materials at the accessible level of Part I can be found throughout the chapters. The reader should be aware that notation varies by authors, and vocabulary for the same thing is different across the disciplines, but the probability theory method may be exactly the same in all of them.

Probability theory is not a bag of different tricks to solve problems but a very condensed set of a few methods to solve a bag of very different and contextually unrelated problems. When doing problems, the reader should try to see what is the common methodology in them. For example, a problem that asks to compute an expectation for some finance random variable will read to the reader as different from a problem that asks to compute an expectation for a biology variable. However, both the biology and the finance problem will use the same method to compute the expectation.

Chapter 1

An Overview of the Origins of the Mathematical Theory of Probability

> One way to understand the roots of a subject is to examine how its originators thought about it.
> **(Diaconis and Skyrms 2018)**

▶ Look at Table 1.1 carefully

Table 1.1

(1,1) Sum = 2	(1,2) Sum = 3	(1,3) Sum = 4	(1,4) Sum = 5	(1,5) Sum = 6	(1,6) Sum = 7
(2,1) Sum = 3	(2,2) Sum = 4	(2,3) Sum = 5	(2,4) Sum = 6	(2,5) Sum = 7	(2,6) Sum = 8
(3,1) Sum = 4	(3,2) Sum = 5	(3,3) Sum = 6	(3,4) Sum = 7	(3,5) Sum = 8	(3,6) Sum = 9
(4,1) Sum = 5	(4,2) Sum = 6	(4,3) Sum = 7	(4,4) Sum = 8	(4,5) Sum = 9	(4,6) Sum = 10
(5,1) Sum = 6	(5,2) Sum = 7	(5,3) Sum = 8	(5,4) Sum = 9	(5,5) Sum = 10	(5,6) Sum = 11
(6,1) Sum = 7	(6,2) Sum = 8	(6,3) Sum = 9	(6,4) Sum = 10	(6,5) Sum = 11	(6,6) Sum = 12

What do you think this table represents? What could it be used for? What kind of things can you predict with it? Ask someone else the same questions and compare your thoughts. Are you uncertain about your guess?

1.1 Measuring uncertainty

How often do you think about uncertainty? Have you ever tried to measure your uncertainty about the outcome of some action you are planning to take in some way? For example, when you were debating whether a prescribed medicine for a cold would lead to recovery? Neglecting all possible influence of diet, stress, and financial problems, perhaps you found online information claiming that 80% of all of those taking this medicine in the past year got cured, and then you adopted this 80% as the measure of your uncertainty about the outcome that would ensue if you take the medicine for your cold. Certainly, some individuals that took the medicine recovered, and some did not, and you have no idea whether you will be among the former; taking the medicine does not always lead to the same outcome. Taking a medicine for a cold is a *random or chance experiment*. If the information online had said that 80% that took this medicine in the past year died, certainly the decision you made would perhaps have been different.

> A probability is a number that gives a precise estimate of how certain we are about something. **(Everitt 1999)**

1.1.1 Where do probabilities come from?

Another question about the hypothetical example given is how did the online source get to the 80% figure about the effectiveness of the drug? Where do probabilities come from? Are they based on *data* (the relative proportion of *many* people that recovered in the past after taking the medicine)? Are they based on some *model* that assumes that figure based on the chemical composition of the drug or some other factor? Or is it totally *subjective*, based on the pharmaceutical company's opinion? This chapter will discuss all these approaches and other names given to them.

Example 1.1.1 Distinction between model, data-based and subjective probability

When faced with a six-sided die, we are all inclined to believe that there is equal chance of getting any of the numbers when we toss it. The *model* we usually have in mind is shown in Table 1.2.

Table 1.2 A model for the toss of a die

Number in the dice	**1**	**2**	**3**	**4**	**5**	**6**
chance	1/6	1/6	1/6	1/6	1/6	1/6

However, we do not know that the die is physically fair, or that this model holds. A way to find out is with *data*, the other approach to calculating probabilities. To obtain data, you should complete the experiment proposed in Table 1.3, using what you think is a fair six sided die. Roll first 10 times and stop. Compute the number of 6's you would be expecting to get, based on the model, with 10 rolls and look at how many you really got. Then roll 40 more and stop. Now you will have accumulated 50 rolls. Count how many of those 50 rolls are 6 (include the ones in the first 10 and the ones in the last 40 rolls). Continue calculating how many you would have expected and so on, stopping at the number of rolls indicated on the left column. Complete Table 1.3.

Table 1.3 Data obtained by an experiment that consists of rolling a real six-sided die to observe what the proportion of sixes converges to. We do not know if the die is fair or not.

(1) Roll up to this number of rolls	(2) Expected number of sixes (based on model)	(3) Observed # of sixes	(4) Observed # minus expected #	(5) Expected proportion	(6) Observed proportion of 6's = (3)/(1)	Observed proportion – expected proportion
10	(1/6)10			1/6		
50	(1/6)50			1/6		
100	(1/6)100			1/6		
200	(1/6)(200)			1/6		
300	(1/6)(300)			1/6		
400	(1/6)(400)			1/6		
500	(1/6)500			1/6		
600	(1/6)600			1/6		
700	(1/6)700			1/6		
800	(1/6)800			1/6		
900	(1/6)900			1/6		
1000	(1/6)1000			1/6		

If your experiment is successful, the proportion of sixes that you get in column (6) will be closer and closer to 1/6 (in column 5) as the number of tosses increases if the model given in Table 1.2 is indeed a good model for the physical die you are using. But if you had tossed a loaded die, the results you put in Table 1.3 will contradict the model in Table 1.2. Thus, although we are not able to predict whether a single roll will give us the number 6 or not, we are able to predict that a large number of rolls will give a 6 with a very stable proportion of 1/6 if the die is fair, or other proportion if not.

It is common in data science to compare a probability model with data collected randomly. If the model is correct, a large amount of collected data (by experimentation, like you will do to complete Table 1.3) will support the model. If the model is incorrect, the data will not support it. To do the comparison of models to reality, statisticians and data scientists collect a lot of data when they can.

Returning to the medicine example at the beginning of this chapter, and by analogy with the die experiment, not much can be said by anyone about a particular individual in a large population that took the medicine we were talking about but, thanks to probability theory, there is more certainty about the combined behavior of all of the individuals, and methods to measure uncertainty. In other words, whether using an assumed model or the data they collect, data scientists may not be able to say whether an individual will be cured by taking the medicine, like you cannot say that you will be cured, but they may be able to say that there is a high chance (80%) of a person getting cured because 80% of all the people that took it were cured (assuming a lot of data and given that the data are random). As Venn put it many years ago,

> Let me assume that I am told that some cows ruminate; I cannot infer logically from this that any particular cow does so, though I should feel some way removed from absolute disbelief, or even indifference to assent, upon the subject; but if I saw a herd of cows I should feel more sure that some of them were ruminant than I did of the single cow, and my assurance would increase with the numbers of the herd about which I had to form an opinion. Here then we have a class of things as to the individuals of which we feel quite in uncertainty, whilst as we embrace larger numbers in our assertions we attach greater weight to our inferences. It is with such classes of things and such inferences that the science of Probability is concerned. (Venn 1888)

The calculus of probability makes possible statistics and gives statistics a foundation. Data scientists and statisticians think of probability models as models representing the population's random behavior. They constantly search in samples of data for what those probability models are. Because their data may not be the whole population, they may even use probability further to attach some error to their estimates. Probability is at the core of the search engines such as Google or Yahoo that we use every day to gather information. The goal of social media is to treat you like the average in the population, presenting to you what the summary of the combined behavior of many is, using the past behavior of other users. The way they try to predict your behavior as an individual is by them knowing about everybody prior to you approaching social media. Your behavior in turn leads them to update their algorithms about everyone. Probability theory also guides population genetics and genetic testing, medical diagnoses, language processing, surveillance, quality control, climate change research, social networks, psychology of people, and behavior of agents in video games, to name a few areas. Probability theory is the background behind all scientific and social endeavors.

> Students must obtain some knowledge of probability and must be able to tie this concept to real scientific investigations if they are to understand science and the world around them.
> **(Scheaffer 1995)**
>
> Probabilistic reasoning is a plain necessity in the modern world.
> **(Weaver 1963)**

1.1.2 Exercises

Exercise 1. You are given a new twelve-sided die by the host of a party you are attending. You are told that this die will be used to play a game after dinner in which you will lose $100 if the number is less than 6 and win $100 if the number is larger than or equal to 7. You are uncertain about the legitimacy of the die. What if the die is not fair? You do not want to insult your host, so you decide to check secretly while the host is in the kitchen preparing dinner. How would you decrease your uncertainty about the die?

Exercise 2. You are uncertain about the outcome of taking your significant other to a new restaurant to celebrate your birthday. Your significant other has never been to this restaurant and the invitation has to be a complete surprise (but not a complete failure). How do you decrease your uncertainty about the restaurant's quality?

Exercise 3. Suppose you are an economist who has been teaching in an economics department for quite some time. Someone asks you to choose between the following two things and earn $1,000 if you get it right: (a) Predict whether a new hire, Shakir, in the reception office of an economics department at a university will leave the job after a year (if you predict yes, and the person leaves, you get the $1,000); (b) Predict whether there will be some (not needing to give names) new hires among the 100 new hires in the reception offices of many economics departments across the US who will leave the job after a year. Do you choose (a) or (b)? Why?

Exercise 4. An individual 45 years old chooses to live in a neighborhood that has cheap housing but not a good safety and hygienic record. The individual is perfectly healthy, works hard, has a new car, has a very clean house, and has never been harmed or inconvenienced by anybody in the neighborhood. This individual is pretty much a mirror image of another individual of the same age who lives in a very fancy gated neighborhood with lots of security surveillance, who has the same health, the same car, the same job, and the same safety record. An insurance company offers a life insurance to both. But the premium of the first individual is much higher than that of the second individual. What explains that? Try to tie your response to what we have discussed in this Section 1.1.

Exercise 5. Brian Tarran (2015) interviewed Dan Bouk, a historian who wrote a book about how people see themselves as a statistical individual—one that that is understood and interpreted as the statistical whole, meaning as the average of everybody else (for example, a middle age individuals thinks there is 40% chance of death by heart attack, 20% chance of being hit by a car, etc.). Think about the things you think about yourself, and think hard about where those thoughts come from. How much is it based on data that you have seen on people your age? List three or four things that you believe about yourself based on something you have read about people your age (for example: risks, health items).

Exercise 6. Comment on what Jaron Lanier (2018) says in his recently published book:

> Behavior modification, especially the modern kind implemented with gadgets like smartphones, is a statistical effect, meaning it's real but not comprehensively reliable; over a population, the effect is more or less predictable, but for each individual it's impossible to say. (Lanier 2018)

1.2 When mathematics met probability

The mathematical theory of probability is relatively young. A reasonable place to start to connect formally with the calculus of probability is by placing ourselves in the 17th century, along with the pioneers. This section contains a few simple questions asked and solved during that period to get you started thinking about the origins of the mathematic measurement of chance. They are questions raised by observation that you can answer yourself by observing repeated particular outcomes in the rolls of dice, which may be bought at many convenience stores. Those are simple questions that initiated the development of the calculus of probability centuries ago. The roots of what you are about to learn in this book are in how gamblers and mathematicians answered those questions.

> Although probability theory today has about as much to do with games of chance as geometry has to do with land surveying, the first paradoxes arose from popular games of chance.
> **(Szekely 1986)**

1.2.1 It all started with long (repeated) observations (experiments) that did not conform with our intuition

When it comes to relative frequencies at which events occur, our intuition (you may call it our a priori "model") often does not conform to repeated observation. It is with this clash that mathematical probability started (a clash would occur, for example, if Table 1.2 in this chapter was contradicted by the relative frequency results that you will get in the last row of column 6 of Table 1.3). These clashes still happen now (Stigler 2015). The reader is encouraged to look at Side Box 1.1 for a definition of relative frequency.

Box 1.1

Relative frequency in long observations

A relative frequency is the proportion of times that something occurs. For example, if your quiz grades throughout a quarter are: 8, 10, 4, 4, 5, 10, 9, 4, then you got a grade of 4 at 37.5% of the time or 3/8 of the time. That is the relative frequency.

Event of interest: getting a 4
Count how many times you took a quiz the outcome is "favorable" (is a 4): 3 times
Long observation length: 8 quizzes
Relative frequency: 3/8

Assuming your performance does not change between this quarter and the next, it can be estimated that the probability that any of your quizzes will be 4 in the future is 3/8 or 37.5% or 0.375. Probability can be expressed in various forms: as fractions, percentages or decimal fractions.

The discrepancy between observation and intuition (or a-priori model) is still very prevalent nowadays. For example, if you record the first digit of every number you encounter (except phone numbers, address numbers, social security numbers, lottery numbers or numbers with an assigned maximum or minimum), intuition (our a-priori model) tells us that each of the numbers 1 to 9 are equally likely to be the first digit. However, long observation of many first digits in many numbers contradicts that intuition. Smaller first digits are more frequent than larger ones. This law is known as Benford's or first digit's law after the physicist Frank Benford who rediscovered it. Data that have that nature follow Benford's law. See Box 1.2.

Box 1.2

Using probability to detect fraud.

Repeated observations of many genuine large sets of numbers support Bedford's law. Hence data given to us on first digits of many numbers that do not satisfy Bedford's law could be indication that the numbers are fraudulent. Thanks to awareness of Benford's law tax accounting fraud can be detected. If you are curious about this, you may see this use and other uses of the law by yourself by trying some fun activities that use this first digit law in the NUMB3RS activity "We're Number 1" (Erickson 2006). Detecting fraud is one of the very extensive uses of Bedford's law in the last 20 years (Browne 1998; Hill 1999). But not everybody agrees with this last statement. Take, for example, William Goodman (Goodman 2016). This author says that "without an error term it is too imprecise to say that a data set 'does not conform' to Benford's law. By how much does it have to differ from expected value to not conform?" (Goodman 2016).

It turns out that probability theory also helps data scientists determine that error probabilistically. Chapter 9 in this book talks about the theorems that allow data scientists to attach errors to their estimates. Data scientists try to match data from populations with probability models of populations, but they have designed additional tools (beyond the scope of this book) to be able to use probability to measure also errors.

The history of probability is plagued since its beginning with examples where empirical facts did not present relative frequencies that were expected based on intuition (an a priori model). *In fact, the modern probability theory that you are going to study in this book is the result of efforts by gamblers, mathematicians, social scientists, engineers and other scientists to create a framework for thinking about the frequency of empirical facts so that we do not rely solely on intuition or a priori models. When using a mathematical probability approach to think about reality, we are bound to make less mistakes in our predictions.*

Making decisions based on long observations (when we can) or based on models supported by long observations, pays in data science, public policy, and our daily lives. Nowadays, the terms "evidence-based decision making" are very popular in many circles. For example, knowing the usual frequency of SIDS (Sudden Infant Death Syndrome) deaths in each county in a given state (possibly measured as deaths per hundred thousand) may help raise a flag in an anomalous year that has an unusually large frequency.

1.2.2 Exercises

Exercise 1. If you have never done this problem in a class or reading you may have done on your own, test your intuition by writing down all the possible outcomes of tossing three coins and enumerating the probability of those outcomes. Do not look for the answer anywhere. You want to write your own thoughts on the matter to assess your intuition. Look at your outcomes and probabilities up and down, add the probabilities, see if it all makes sense. If you have taken probability before and this is not the first time you do a problem like this, think how you would have answered before you took probability.

Exercise 2. Test your intuition by thinking about this problem: If you roll a die three times, what is the probability of getting at least one six? Again, do not look anywhere for an answer. This question is just for you to assess your intuition or a-priori model.

Exercise 3. A student of probability was asked to record the first digit of every number encountered throughout a week. If the student bought a coffee for $3.45 the student would record 3; if the student arrived to class at 10:05, the student would record 1, and so on. Phone numbers, zip codes and student id numbers were not allowed. Then the student was asked to write a table with the relative frequency of each first digit recorded. This student produced a perfectly uniform table, which said that each number was equally likely to happen: relative frequency of 1 was 1/9, relative frequency of 2, 1/9, and so on. Do you think this student used observed data to do this homework?

Exercise 4. Do the student activity found in Erikson (2006)

1.2.3 Historical empirical facts that puzzled gamblers and mathematicians alike in the seventeenth century

Consider a game that consists of rolling three supposedly fair six-sided dice like those in Figure 1.1, of different color each, and observing the value of the sum of the numbers. For example, if you get (3,4,5) the sum of the three numbers is 12. If you had to bet on a sum of 9 or 10, which one would you choose? 10 or 9? Would you be indifferent? Explain your reasoning to someone you know and is willing to lend a friendly ear. Ask your friends what they think.

If the three-dice game sounds too complicated, consider an easier game: rolling two fair six-sided dice like those in Figure 1.2, of different color, to find the value of the sum of the points. If you had to bet on 8 or 7, which one would you choose? 7 or 8? Would you be indifferent? Can you explain your reasoning to someone?

1.2.4 Experiments to reconcile facts and intuition. Maybe the model is wrong

Dice players experiment when they play the same game many times. It is *experimentation* what led gamblers of the seventeenth century to question their intuition (or models) and mathematics of games of chance. Experimentation is done with physical devices. We experiment to see if data support the a-priori model we have or to just discover some model.

Figure 1.1 Rolling three six-sided symmetric dice.

Figure 1.2 Rolling two six-sided symmetric dice.

> The observation of many games like those dice games just mentioned made dice players in the sixteenth and seventeenth century consider that there was a difference between the relative frequencies, whether practically significant or not, and ask for an explanation. If, playing with three dice, 9 and 10 points may each be obtained in 6 different ways, they thought, why was there a difference between the relative frequencies observed? Similarly, if playing with two dice, 7 and 8 each may be obtained in 3 different ways, why was there a difference in the relative frequencies observed? **(Apostol 1969)**

We could replicate the experience of the dice players playing the games of Section 1.2.3 by conducting an experiment with fair dice bought in some store. Equally likely numbers in a single six-sided die, for example, is a reasonable model assumption if the information we possess about the die is that it is symmetric or fair, and we do not possess any other information. The observations and concerns of gamblers were based on that assumption. *If the dice used were fair, why were the frequencies observed in those games different from what they expected based on their model?*

In the case of the game consisting of rolling two dice, a *repetition* of the experiment would consist of rolling two dice and recording the two numbers as a pair, for example (3,2), and then, separately, the sum of the pair, respectively 5. Repetition of trials, say m times, and recording how many trials gave a sum of 8 and how many of 7 out of the m trials would give an approximation to the frequencies of 8 and 7. The number of repetitions, m, would have to be large. Exercise 1 in Section 1.8 invites you to do that.

A trial of the experiment that would help us estimate the frequency of 9 and 10 for the sum of the points in the roll of three dice would consist of rolling three dice and recording the sum. Repetition of trials m times and recording the proportion of the m trials giving 9 or 10 would give us the approximation sought.

Box 1.3

Difference between experiment and simulation. Steps of a simulation.

The repetition of a physical activity like dice rolling many times under the same conditions while observing the relative frequency of a particular event of interest is called an experiment. Experimentation is a way to find whether a real die is unfair.

Simulation is different. When we simulate, we assume that our model is correct and produce data from that model. That is why simulation is done with computers.

The steps of a simulation are:

a. Determine the probability model to use, for example, a fair die (numbers 1 to 6 each with the same probability of 1/6)
b. Define what a trial consists of, for example, roll a die twice
c. Determine what to record at each trial, for example, we will record the sum of the numbers
d. Repeat a), b), c) many times, say 10000
e. Calculate what you are looking for, for example, what proportion of the 10000 trials gave us a sum equal to 7.

Repeating a trial many times requires patience, and lots of time, but it is worth doing. To achieve an accurate approximation requires many trials. For that reason, software is often used to conduct many trials of a simulation. Section 1.7 introduces the free software R and gives R code to conduct the simulation in Chapter Exercise 1.

Example 1.2.1

These days, applets created for the purpose of simulating, under known assumptions, can be found on many web sites. For example, a dice tossing applet that you can find at http://www.randomservices.org/random/apps/DiceExperiment.html allows you to do the simulations needed to determine how to answer the questions posed by gamblers that occupy our attention in this section 1.2. For example, by setting $n = 2$ (number of dice), options "fair" and $Y = \text{sum}$, and $\text{stop} = 100$ (number of trials), you will see the computer tossing two dice and showing to you what numbers come up, and you will see their sum. You will see that a sum equal to 7 appears more often than a sum of 8, even though the differences between the relative frequencies are small. You can then do the analysis with $n = 3$ to see what you discover about the question posed at the beginning of Section 1.2.3. If you are curious, you can explore further to see if the conclusions are different when the die is not fair.

1.2.5 Exercises

Exercise 1. We mentioned at the beginning of this chapter that the probability of an outcome could be found by observing many times the experimental outcome and counting how many of the many times observed the outcome occurred. But we also said we could just subjectively make up the probability. Still, we could have a mental model of the probability not based on observation but some other knowledge. In which of these three categories would you place the simulation approach we are talking about? How could you have figured out the answer with a model? What kind of model?

Exercise 2. "Forensics sports analytics" uses probability reasoning to help identify and eliminate corruption within the sports sector (Paulden 2016). Chris Gray (2015), a tennis follower, wrote an article where he presented a version of the widely used (in tennis) IID probability model for a player, player A, winning a tennis game. He gave the following model which depends on the probability of player A winning a point on serve (denoted by p, and assumed constant)

$$P(A\ winning) = \frac{p^4(-8p^3 + 28p^2 - 34p + 15)}{p^2 + (1-p)^2}$$

Paulden (2016) talks about an alternative version of this model, the O'Malley tennis formulae. Gray's and O'Malley's models are based on assumptions about the game, but they are also filled with probabilities that were obtained from past data on many players. How do you think you could validate either of the models mentioned by these authors? Use concepts seen in Sections 1.1 and 1.2 of this chapter to answer.

Exercise 3. Think of a situation where you had a very clear model of how often something that interests you would happen and your model clashed with the evidence you obtained from repeated observations.

1.2.6 The Law of large numbers and the frequentist definition of probability

Empirical observation by experimentation of the play of the game a large number of times, under the same conditions, was common in the seventeenth century. The relative frequency of an event, calculated from observations under the same circumstances, was believed by everyone to be more accurate if a large number of observations is taken. But it was not until the following century that this practice brought up the following question: *Does the probability that the estimate obtained with an experiment is close to the truth increase with the number of observations?*

Mathematicians in the eighteenth century, in particular Jacob Bernoulli, sought a theoretical counterpart to that empirical question, showing that the probability that the estimate is close to the truth increases with the number of trials. This theoretical counterpart is the theorem known as the *Law of Large Numbers*, a theorem studied in Chapter 9 of this book.

Defining probability of an event E as the long-run frequency of the event in a large number of trials, m, is known as the *frequentist definition of probability of an event*.

$$P(E) = \lim_{m \to \infty} \frac{\text{number of occurrences of the event}}{m}.$$

The law of large numbers gave legitimacy to using repeated experimentation to arrive at the probabilities and to the frequentist definition of probability. This law guides the day-to-day practice of statisticians by legitimizing the collection of large amounts of random data to obtain relative frequencies that are close to the true probabilities of events.

Example 1.2.2

I rolled a die 1,000,000 times and found that I got 400,000 times the number 6. According to the frequentist definition of probability, this means that we estimate the probability of a 6 to be 0.4. Because we simulated 1,000,000 rolls, we are almost convinced we are very close to the true probability and can conclude that the die is not fair. By the law of large Numbers, we give high probability to the fact that

$$\frac{400000}{1000000} - P(6).$$

is 0. P(6) means the true probability of 6, which based on our experimentation is very close to 0.4.

Statisticians, data scientists, insurance companies, and managers of social media make wise use of the law of large Numbers in designing their methods to analyze data and their policies and resources. The relative frequency with which something happens to a large number of subjects, is a good approximation to the true probability that this something happens to an individual.

1.2.7 Exercises

Exercise 1. Comment on the following statement: "I cannot predict one fair coin toss, but I can predict quite accurately that the proportion of heads in 1,000 tosses of a fair coin will be close to the theoretical probability of 1/2 assumed by the equally likely outcomes model."

1.3 Classical definition of probability. How gamblers and mathematicians in the seventeenth century reconciled observation with intuition.

Back in the seventeenth century, it was clear by repeated experimentation (gambling) that there was a difference in frequencies that did not conform to intuition. The law of large Numbers then made it clear that the relative frequencies obtained in repeated experimentation should be trusted. How to reconcile observation with the model gamblers believed in? How to translate that discrepancy into mathematics? What was wrong with the gamblers' model? Between 1613 and 1623 Galileo Galilei gave an explanation in Sopra le Scoperte dei Dadi (On a discovery concerning dice).

> *Galileo took crucial steps in the development of the calculus of chance. For the game with the three dice, Galileo lists all three-partitions of the number 9 and 10. For 9, there are 6 partitions: 1/3/5, 1/2/6, 1/4/4, 2/2/5, 2/3/4, 3/3/3. But this is not what we should count, Galileo claims. Each of those partitions covers several possibilities, depending on which die exhibits the numbers. What we must count is the number of permutations of each partition. For three different numbers there are 6 permutations, for example. For the partitions given, we have the following 25 outcomes (out of 216): (1,3,5), (1,5,3), (3,1,5), (3,5,1), (5,1,3), (5,3,1), (1,2,6), (1,6,2), (2,1,6), (2,6,1), (6,1,2), (6,2,1), (1,4,4), (4,1,4), (4,4,1), (2,2,5), (2,5,2), (5,2,2), (2,3,4), (2,4,3), (3,2,4), (3,4,2), (4,2,3), (4,3,2), (3,3,3). Repeating the process for a sum of 10 points, we can show that there are 27 different dice-throws (out of 216). In that way Gallileo proved "that the sum of 10 points can be made up by 27 different dice-throws (out of 216), but the sum of points 9 by 25 out of 216 only." His method and result are the same as Cardano's. Galileo takes for granted that the solution should be obtained by enumerating all the equally possible outcomes and counting the number of favorable ones.* (Hald 1990)

This implicitly assumes independence of the rolls, that all 216 possible outcomes are equally probable.

Although limited to this special case, Cardano and Galileo provided a theoretical counterpart to the observed phenomena by modeling the situation.

> *In spite of the simplicity of the dice problem, several great mathematicians failed to solve it because they forgot about the order of the cast. (This mistake is made quite frequently, even today.)* (Szekely 1986)

Chapters 2 and 3 of this book further discuss the role that the independence assumption makes in the calculation of probabilities.

We have seen that gamblers observed a difference between relative frequencies, whether significant or not, asked for an explanation and got an explanation from mathematicians. The explanation just described is a precursor of the concepts of sample space, events and random variables, three fundamental concepts of modern probability theory introduced in Chapter 2.

Galileo's solution for the dice problem implicitly used what we call now the *classical definition of probability of an event E, namely if E is an event,*

$$\text{Probability}(E) = \frac{\text{Number of favorable cases}}{\text{Total Number of logically possible cases}}.$$

Finding the probability entailed knowing all the logically possible cases and being able to count the ones that were favorable. Implicitly, this assumed that all outcomes were equally likely and implicitly assumes independence. *The mistake of the gamblers was that they were not counting all the logically possible cases.*

Using the classical definition of probability properly, i.e., counting all the outcomes that matter, helped solve mathematically the puzzle of the gamblers, i.e., it helped reconcile intuition with long observations.

Example 1.3.1

In the case of the two dice, let's go back to Table 1.1 to see that there are 36 logically possible outcomes that we enumerated there.

If we call the case of a 7 "favorable," the number of favorable outcomes where the sum is 7 is 6 out of 36, so the classical probability is 6/36 whereas the number of favorable outcomes where the sum is 8 is 5, making the classical probability 5/36. A not very significant difference, yet a difference that helps explain the gamblers' observed difference. Denoting probability by P,

$$P(\text{"}sum\ of\ 2\ dice\ is\ 7\text{"}) = \frac{6}{36}, \quad P(\text{"}sum\ of\ two\ dice\ is\ 8\text{"}) = \frac{5}{36}$$

Example 1.3.2

In the case of the three dice, let's go back to our earlier discussion to see that there are 216 logically possible outcomes that we enumerated there.

If we call the case of a 9 "favorable," the number of favorable outcomes where the sum is 9 is 25 out of 216, making the probability of 9 to be 25/216 whereas the number of favorable outcomes where the sum is 10 is 27, making the probability of 10 to be 27/216. A not very significant difference, yet a difference that helps explain the gamblers' observed difference.

$$P(\text{"}sum\ of\ 3\ dice\ is\ 9\text{"}) = \frac{25}{216}, \quad P(\text{"}sum\ of\ 3\ dice\ is\ 10\text{"}) = \frac{27}{216}$$

1.3.1 The status of probability studies before Kolmogorov

Not all probabilities are as simple to calculate as the ones described in the previous section. Sometimes it is necessary to combine the probabilities of two or more events or two or more outcomes. Continued efforts to reconcile observation with mathematical theory during the seventeenth century lead to solving more complex problems by using rules that govern the way that probabilities can be combined. Complex problems require rules to combine probabilities. We learn all those rules, which apply to any definition of probability, in Chapter 3, and use them throughout the book.

> "A gambler's dispute in 1654 led to the creation of a mathematical theory of probability by two famous French mathematicians, Blaise Pascal and Pierre de Fermat. Antoine Gombaud, Chevalier de Méré, a French nobleman with an interest in gaming and gambling questions, called Pascal's attention to an apparent contradiction concerning a popular dice game. The game consisted in throwing a pair of dice 24 times; the problem was to decide whether or not to bet even money (lose or win the same amount of money) on the occurrence of at least one "double six" during the 24 throws. A seemingly well-established gambling rule led de Méré to believe that betting on a double six in 24 throws would be profitable, but his own calculations indicated just the opposite"
> **(Apostol 1969).**

Using rules that we learn in Chapter 3, we would support de Méré's calculation as follows:

$$P(\textit{at least one } (6,6) \textit{ in } 24 \textit{ throws}) = 1 - P(\textit{no } (6,6) \textit{ in } 24 \textit{ throws}) = 1-(35/36)^{24} = 0.4914639.$$

Alternatively, you could get the same answer by looking at Table 1.1 to find the probability of (6,6) and then using the complement rule and product rule for independent events presented in Chapter 3 of this book.

This result indicates that the probability of getting at least one (6,6) is less that 0.5; it is more favorable that there will be no (6,6) pairs in 24 throws.

1.3.2 Kolmogorov Axioms of Probability and modern probability

The reliance on the mathematics of equally likely cases and the assumption of independence dominated the study of chance phenomena until the early nineteenth century. By the early 19th century, mathematical probability was mainly defined as the classical definition of probability, which applies only if all outcomes are equally likely, and for a finite number of outcomes or an infinitely large number of countable outcomes. Although this mathematical solution helped model properly the evidence from long observations, it suffered from circularity and did not help solve continuous problems.

To address that situation, attempts at defining probability differently were made, which gave rise to the *subjective definition of probability*. The disputes were resolved when Kolmogorov put probability in a solid mathematical foundation, thus initiating the *modern approach to probability*, which embeds all the definitions of probability mentioned so far in this chapter (classical, subjective and frequentist).

We start our study of the modern approach to probability in Chapter 2.

In the early twentieth century, Kolmogorov gave probability an axiomatic foundation, thus making it mathematically possible to tackle the uncountable, hence what cannot be approached with the classical definition of probability. Probability is a function P defined on sets of the larger set containing all logically possible outcomes of an experiment, S, such that this function satisfies Kolmogorov's axioms, which are:

- **Axiom 1.** The probability of the biggest set, the *sample space S*, containing all possible outcomes of an experiment, is 1.
- **Axiom 2.** The probability of an *event* is a number between 0 and 1.
- **Axiom 3.** If there are events that cannot happen simultaneously (are mutually exclusive), the probability that at least one of them happens is the sum of their probabilities.

Measure theory is a theory of sets. *Probability is a measure defined on sets. What is remarkable is that the frequentist, the classical, and the subjective definitions of probability satisfy the axioms.* The assumption of the existence of a set function P, defined on the events of a sample space S, and satisfying Axioms 1,2,3, constitutes the modern mathematical approach to probability theory. Any function P satisfying the axioms is a probability function. With those axioms, it is straightforward to prove the most important properties of probability, which we do in Chapter 3.

Because P is a function defined on events, and events are, mathematically speaking, sets, it is necessary to use the algebra of sets when studying probability. Chapter 2 guides your review of the algebra of sets. The axiomatic approach allows us to talk about probability defined in continuous sample spaces, and probability models defined on continuous random variables, which we do in Chapters 7 and 8. But discrete sample spaces and discrete random variables equally fall under the umbrella of the axiomatic approach. We study those in Chapters 2 to 6.

1.4 Probability modeling in data science

By probability modeling in data science we mean the act of using probability theory to model what we are interested in measuring. The conclusions that we reach will be as valid as the model is. Laplace (1749–1827) used to say that the most important questions of life are indeed for the most part only problems of probability. In most of these problems, we build models to describe conditions of uncertainty and provide tools to make decisions or draw conclusions on the basis of such models.

> Not only are probabilistic methods needed to deal with noisy measurements, but many of the underlying phenomena, including the dynamic evolution of the internet and the Web, are themselves probabilistic in nature. As in the systems studied in statistical mechanics, regularities may emerge from the more or less random interactions of myriad of small factors. Aggregation can only be captured probabilistically.
> **(Baldi et al. 2003)**

During the last decades, probability laws for classification, for social networks, internet traffic, the human genome, biological systems, the environment and many other interests of society in the 21st century have been sought.

With the proliferation of the world wide web (the Web) and internet usage, probabilistic modeling has become essential to understand these networks.

Spam filtering, for example, has made it possible for computer users to read their email without having to worry as much as they used to about spam mail (Goodman and Heckerman 2004). Spam filters are mostly based on the principles of conditional probability and Bayes theorem, which is covered in Chapter 3 of this book, and subsequent chapters. See http://paulgraham.com/bayeslinks.html for a brief survey of the topic. The increasingly popular field known as Machine Learning makes extensive use of the probability calculations that we will be learning in this book, and more advanced ones.

Conditional probability and Bayes theorem are used in classification of items where a system has already learned the probabilities.

Example 1.4.1

Suppose there are two classes of email, good email and spam email. We let the random variable $Y = 1$ if the email is good, and $Y = 2$ if the email is spam. Let W represent a new email message. Our decision is to classify a new email message W which contains the word "urgent" into class 1, good email, if

$$P(Y = 1)P(W \mid Y = 1) > P(Y = 2)P(W \mid Y = 2)$$

Otherwise, the email W is classified as spam email and rejected by the server. Why we use this decision rule given will become very clear to you after you study chapter 3. The conditional probabilities of $P(W \mid Y = 1)$ and $P(W \mid Y = 2)$ and the prior probabilities $P(Y = 1)$ and $P(Y = 2)$ are known and are based on past observations of the frequency of good and spam messages and the contents of good messages and spam messages.

Another area of machine learning where probability plays a very important role is text processing. Indexing, scoring and categorization of text documents is required by search engines such as Google http://www.stat.ucla.edu/~jsanchez/oid03/csstats/cs-stats.html.

The areas of application of probability mentioned should give you an idea of possible career paths that can be pursued with sound skills in probability reasoning like those you will acquire by studying this book. There are many other career paths that will become transparent as you study the book. Actuarial science, the science of insurance, for example, can not be pursued without first passing the first exam, for which this book prepares you well. At http://q38101.questionwritertracker.com/EQERFHHR/ry.com you will find sample exams.

> Engineering and computer science cannot survive without probability modeling.
> **(Carlton and Devore 2017)**

Most data science problems involve more than one variable and more than one events. A book on probability for data scientists would be incomplete if it did not include the study of probability of more than one random variable. This book, and in particular chapters 6 and 8 will give the necessary foundation to prepare yourself for the use of probability theory in multivariate problems.

Before we conclude, you should read the data science application case that follows to appreciate how a simple discovery like the solution of the dice problems helped model a very relevant problem in Physics (see Figure 1.3).

> The dice problem has some links with 19th and 20th century microphysics. Suppose that we play with particles instead of dice. Each face of the die represents a phase cell on which the particles appear randomly and which characterizes the state of the particles. Here dice is equivalent to the Maxwell-Boltzmann model of particles. In this model (used mostly for gas molecules) every particle has the same chance of reaching any cell, so in a list of equally probable events, the order must be taken into account, just as in the dice problem. There is another model in which the particles are indistinguishable, and for this reason the order must be left out of consideration when counting the equally possible outcomes. This model is named after Bose and Einstein. Using this terminology the point of the (dice paradox studied in this chapter), is that dice are not of the Bose-Einstein but of Maxwell-Boltzmann type. It is worth mentioning that none of these models are correct for bound electrons because in this case, only one particle may occupy any cell. In dice-language it means that after having thrown a 6 with one of the dice, we can not get another 6 on the other dice. This is the Fermi-Dirac model. Now the question is which model is correct in a certain situation. (Beside these three models, there are many others not mentioned here.) Generally we can not choose any of the models only on the basis of pure logic. In most cases it is experience or observation that settles the question. But in the case of dice, it is obvious that the Maxwell-Boltzmann model is the correct one and at this moment that is all we need.
> **(Szekely 1986, 3–4)**

1.5 Probability is not just about games of chance and balls in urns

We have talked a lot about dice in this chapter. That is because the mathematical theory of probability had its origin in questions that grew out of games of chance. The reader will find more dice and even balls and urns in this book and in almost every probability theory book that comes to the reader's attention, but not because probability theory is about them.

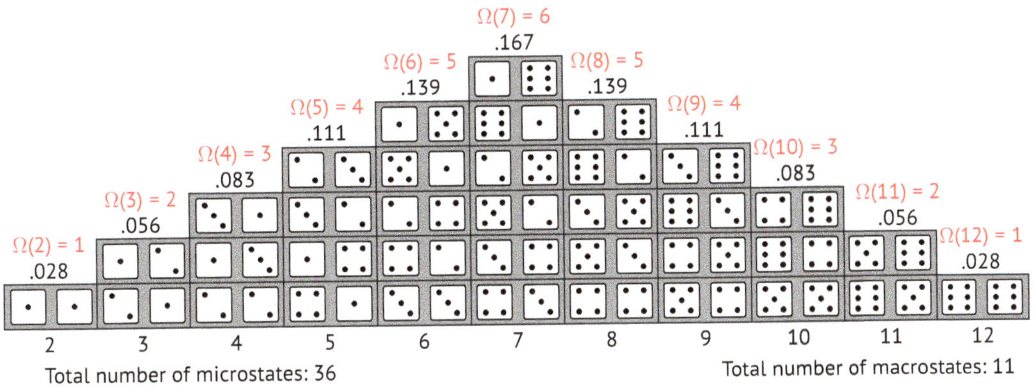

Figure 1.3 A simple six-sided die model helps clarify a rather complicated physics concept.

Source: http://hyperphysics.phy-astr.gsu.edu/hbase/Therm/entrop2.html.

> The early experts in probability theory were forever talking about drawing colored balls out of "urns." This was not because people are really interested in jars or boxes full of a mixed-up lot of colored balls, but because those urns full of balls could often be designed so that they served as useful and illuminating models of important real situations. In fact, the urns and balls are not themselves supposed real. They are fictitious and idealized urns and balls, so that the probability of drawing out any one ball is just the same as for any other.
> **(Weaver 1963, 73)**

Example 1.5.1

In India in 2012, the probability of dying before age 15 was 22%. The parents of 5 children are worried that dying before age 15 could happen to their children. One can think of a box with 100 balls, 22 of which are red and 78 of which are green. What is the probability of drawing, in succession, 5 red balls with replacement? Would this box model simulate well the real situation of dying before age 15, even though it is a box with balls? Friedman, Pisani, and Purves, authors of an introductory statistics book, introduced probability using box models like this (Friedman, Pisani, and Purves 1998).

The reader should be warned that science books use different names for the same concepts that we talk about in this book. A book in physics, another in psychology, another in linguistics, for example, may be using the same "rolling two dice" experiment model that you saw in this chapter yet each of them uses different names for the total number of outcomes, for the number of sets, for the sum and such concepts that are very standard in the probability theory books. Physics, Probability and Linguistics require the background that you are going to learn in this book to solve their seemingly unrelated problems. The fact is not that probability theory consists of a bag of an endless number of tricks to solve problems as it may appear to the beginner, rather probability theory is what an endless number of real problems have in common. The reader will be well served by focusing in mastering the methods that probability theory provides in order to be prepared to apply the same method to a wide array of dissimilar problems that require the same method.

1.6 Mini quiz

Question 1. You are playing with three fair six sided dice. You are interested in the sum of the points. Which is more favorable: 9 or 10? That is, if you had to bet on 9 or 10, which one would you choose?

a. 9
b. 10
c. either one

Question 2. You are playing with two fair six sided dice. You are interested in the sum of the points. Which is more favorable? 7 or 8? That is, if you had to bet on 7 or 8, which one would you choose? 7 or 8?

a. 7
b. 8
c. either one

Question 3. Which of the following is most likely?

a. at least one six when 6 six-sided dice are rolled
b. at least two sixes when 12 six-sided dice are rolled
c. at least three sixes when 18 six-sided dice are rolled

Question 4. Where do probabilities come from? Circle all that applies.

a. models
b. data
c. subjective opinion
d. all of the above
e. none of the above

Question 5. The classical definition of probability has some limitations. Which of the following are some limitations?

a. It cannot be used when the outcomes are not equally likely.
b. It can only be used when there are finite or infinite countable outcomes.
c. It does not satisfy Kolmogorov's axioms.
d. We could not double-check it with long observations.

Question 6. In the context of rolling 3 six-sided dice, what is the most important factor contributing to obtaining the correct answer to the probability of the sum being 14, for example, without having to do long observations?

a. counting not only the favorable partitions but also the number of permutations of each partition.
b. using the law of large numbers
c. use your subjective opinion
d. Taking into account that the number of possible outcomes is: any of the numbers from 3 to 18, that is, there are 16 outcomes. One of those outcomes is favorable, 14. So the probability 1/16 will be the correct probability.

Question 7. The dice model that reconciled observations with the intuition of seventeenth-century gamblers is similar to what model for particles in physics?

a. Fermi-Dirac's
b. Bose-Einstein's
c. Maxwell-Boltzmann's
d. Jaynes'

Question 8. Use the classical definition of probability to find the probability that in two rolls of a four-sided die the sum is 5.

a. 1/5
b. 1/4
c. 1/3
d. 1/8

Question 9. The law of large Numbers (LLN) added only what to the belief that more observations obviously give more accurate estimates of the chances?

a. The LLN showed that the probability that the estimate is close to the truth increases with the number of trials.
b. The LLN tells us that we can be more certain that long observations give us accurate estimates the more the observations made.
c. The LLN legitimizes the frequentist definition of probability.
d. All of the above.

Question 10. Kolmogorov made it possible to

a. calculate probabilities of outcomes that can take any value in an interval of the real line
b. use the same rules of probability that are consistent with axioms in both the discrete and continuous outcomes scenario
c. none of the above
d. (a and b)

Box 1.4

R and Rstudio

R code is code that is understood by the software R. It is widely used by data scientists in their day to day data analysis routines. It is also used to generate random numbers that allow us to simulate many random phenomena.

We can simulate many rolls of three dice and compute the probability of the event of interest in seconds using R.

R is a free open source software that can be downloaded into any computer. Rstudio is an interface that makes working with R much easier. To use it, R must be installed. R can be downloaded from

https://cran.r-project.org/

and Rstudio can be downloaded from

https://www.rstudio.com/

In the RStudio website, at the address https://www.rstudio.com/online-learning/ you will find tutorials on how to get started typing and practicing basic R code. The reader is also encouraged to visit the following address, which has introduction to R coding https://stats.idre.ucla.edu/r/

For example, if I wanted to roll a fair six-sided die with R 5 times, I would type in the R console

```
sample (6, size = 5,prob = c(rep(1/6,6)), replace =
```

This gives R the order to sample 5 number from 1 to 6, where each number has probability 1/6, and that is true for each number (guaranteed by typing replace = T).

1.7 R code

1.7.1 Simulating roll of three dice

The reader should read Side Box 1.4 before starting the simulation with R.

To do a simulation with software to estimate the proportion of times the sum of three fair six-sided die is 10 or 9, we may use the following R code. Type the code in the Editor window of RStudio, then execute it line by line by placing the cursor on the line and clicking on Run.

```
#This line is a comment. R does not do anything with it
n=1000 # number of trials (change this number for exercise 1)
sum.of.3.rolls=numeric(0) # storage space opens
for(i in 1:n) { # this is a loop to fill the storage space
trial=sample(1:6, 3, prob=c(rep(1/6, 6)), replace=T) #with rolls
sum.of.3.rolls[i] = sum(trial) #then calculate the sum of rolls
```

```
} #This ends the loop after 1000 trials
sum(sum.of.3.rolls==10) # Count how many times you got a sum=10
sum(sum.of.3.rolls==9) # Count how many times you got a sum=9
```

1.7.2 Simulating roll of two dice

To do a simulation to estimate the proportion of times the sum of two fair six-sided die is 7 or 8, we may use the following R code.

```
n=1000 # number of trials (change this number for exercise 1)
sum.of.2.rolls=numeric(0)
for(i in 1:n) {
trial=sample(1:6, 2, prob=c(rep(1/6, 6)), replace=T)
sum.of.2.rolls[i] = sum(trial)
}
sum(sum.of.2.rolls==8)/n # Find relative frequency of 8
sum(sum.of.2.rolls==7)/n # Find relative frequency of 7
```

1.8 Chapter Exercises

Exercise 1. You will do a simulation in this problem. A trial of this simulation consists of rolling two fair six-sided dice of different color. The number in both is recorded as a pair (a,b), where *a* is the first roll and *b* is the second. For example, you could obtain (3,2), where 3 is the number on the first die, and 2 is the number on the second die.

You will do 125 trials, by hand or using software. If you use the R code given in section 1.7 you could do many more trials. Alternatively, you may use the applet introduced in Example 1.2.1.

a. At each trial, record the sum of the two numbers. For example, if the outcome is (3,2) the sum is 5. The sum is called a random variable, because until you actually know its value the value is not known, it is determined by chance. We will call this random variable Y.

   Y = the sum of the two numbers in the two rolls.

   Table 1.4 below illustrates the process. For someone to double check your numbers they need to see what they are. So a table of some of the trials is always recommended. Record on Table 1.4 some of your trials.

Table 1.4

Trial Number	(a, b)	$Y = a + b$
1		
2		
3		
4		
.....		
.....		
125		
Total number of trials:		**With $Y = 7$:** **With $Y = 8$:**

b. Based on the results recorded on Table 1.4, what proportion of the trials gave you a sum equal to 7 and what proportion gave you a sum equal to 8? Compare with the result you would get using the applet introduced in Example 1.2.1, run 10000 times. Explain the difference using the frequentist definition of probability introduced in Section 1.2.6.

c. If you used the classical definition of probability introduced in Section 1.3, what would be the probability that the sum of the two dice is 7? What assumption would you have to make?

Exercise 2. As we have seen in this chapter, Galileo, and Cardano before him, suggested that in order to educate our intuition about dice games of their time discussed in this chapter, we should start by considering all the possible outcomes of the games. For example, the game with the two dice has the outcomes and the corresponding values of the random variable representing the sum indicated in Table 1.1.

If we assume that all numbers of one dice are equally likely to appear (the dice is assumed to be fair), then the solution for how frequently each outcome appears is given by counting the number of times it appears (the number of favorable cases). The classical probability would be that number divided by 36.

a. Write a table indicating in one column the value of the sum of the faces of two dice and in the second column the number of times the sum appears divided by 36. Is 7 or 8 more frequent?

b. Is there a mathematical formula that would model the value of the sum of two dice? Why did you write the formula you wrote? Talk to friends about it.

c. Create a table with all the possible outcomes of the roll of three dice and the value of the sum associated with each outcome. In that table, you write (a,b,c), where $a =$ the number in the first roll, $b =$ the number in the second roll and $c =$ the number in the third roll. The sum $= a + b + c$. Then write separately another table that has in the first column, the value of the sum, and in the other, the relative frequency of the sum. Is 9 or 10 more frequent?

Exercise 3. Suppose the prior probabilities that an email message is spam ($y = 1$) is $P(y = 1) = 0.4$, and the prior probability that it is not spam ($y = 2$) is $P(y = 2) = 0.6$. Also suppose that the conditional probabilities for a new email message, w, containing the word urgent are $P(w \mid y = 1) = 0.5$ and $P(w \mid y = 2) = 0.3$. Into what class should you classify the new example? Show the work.

Exercise 4. Suppose two players, A and B, toss a fair coin in turn. The winner is the first player to throw a head. Do both players have an equal chance of winning the game? You may investigate this question doing a simulation.

The probability model is a fair coin. A trial of the simulation consists of a game. For example, A starts and gets a head in the first toss. Another example, A starts and gets a tail in the first toss, B gets a tail in the second toss, and A gets a head on the third toss.

Repeat the trials 100 times recording whether A or B wins. At the end, compute the relative frequency of A winning and the relative frequency of B winning. Then answer the question asked.

Exercise 5. Suppose you are playing a game that involves flipping two balanced coins simultaneously. To win the game you must obtain "heads" on both coins. What is your classical probability of winning the game? Explain.

Exercise 6. Esha and Sarah decide to play a dice rolling game. They take turns rolling two fair dice and calculating the difference (larger number minus the smaller number) of the numbers rolled. If the difference is 0, 1, or 2, Esha wins, and if the difference is 3, 4 or 5, Sarah wins. Is this game fair? Explain your thinking.

Exercise 7. What is the proportion of three-letter words used in sports reporting? Write down a thoughtful guess. Then design an experiment to find out.

Exercise 8. The molecule DNA determines the structure not only of cells, but of entire organism as well. Every species is different due to the differences in DNA. Even though DNA has the same structure for every living thing, the major differences arise from the sequence of compounds in the DNA molecule. The four base molecules that form the structure of DNA are adenine, guanine, cytosine, and thymine, often referred as A, G, C, and T for short. The entire DNA sequence is formed of millions of such base molecules, so there is a lot of different combinations, and hence, lots of different species of organisms.

Research what a palindrome is and come up with a strategy to conclude whether palindromes are randomly placed in DNA or not.

Exercise 9. What does the forecast "60% chance of rain today" mean? Do you think the forecaster has erred if there is no rain today?

Exercise 10. Use the classical definition of probability to calculate the probability that the maximum in the roll of two fair six-sided dice is less than 4.

1.9 Chapter References

Apostol, Tom M. 1969. *Calculus*, Volume II (2nd edition). John Wiley Sons.

Baldi, Pierre, Paolo Frasconi, and Padhraic Smyth. 2003. *Modeling the Internet and the Web. Probabilistic Methods and Algorithms*. Wiley.

Browne, Malcolm W. 1998. "Following Benford's Law, or Looking out for No. 1." *New York Times*, Aug. 4, 1998. http://www.nytimes.com/1998/08/04/science/following-benford-s-law-or-looking-out-for-no-1.html

Carlton, Mathew A., and Jay L. Devore, 2017. *Probability with Applications in Engineering, Science and Technology*, Second Edition. Springer Verlag.

Diaconis, Persi, and Brian Skyrms. 2018. *Great Ideas about Chance*. Princeton University Press.

Erickson, Kathy. 2006. *NUMB3RS Activity: We're Number 1!* Texas Instruments Incorporated, 2006. https://education.ti.com/~/media/D5C7B917672241EEBD40601EE2165014

Everitt, Brian S. 1999. *Chance Rules*. New York: Springer Verlag.

Freedman, David, Robert Pisani, and Roger Purves. 1998. *Statistics*. Third Edition. W.W. Norton and Company.

Goodman, Joshua, and David Heckerman. 2004. "Fighting Spam with Statistics." *Significance* 1, no. 2 (June): 69–72. https://rss.onlinelibrary.wiley.com/doi/10.1111/j.1740-9713.2004.021.x

Goodman, William. 2016. "The promises and pitfalls of Benford's law." *Significance* 13, no. 3 (June): 38–41.

Gray, Chris. 2015. "Game, set and starts." *Significance*. (February): 28–31.

Hald, Anders. 1990. *A History of Probability and Statistics and Their Applications before 1750*. John Wiley & Sons.

Hill, Theodore P. 1999. "The Difficulty of Faking Data." *Chance* 12, no. 3: 27–31.

Lanier, Jaron. 2018. *Ten Arguments For Deleting your Social Media Accounts Right Now*. New York: Henry Holt and Company.

Paulden, Tim. 2016. "Smashing the Racket." *Significance* 13, no. 3 (June): 16–21.

Scheaffer, Richard L. 1995. *Introduction to Probability and Its Applications*, Second Edition. Duxbury Press.

Stigler, Stephen M. 2015. "Is probability easier now than in 1560?" *Significance* 12, no. 6 (December): 42–43.

Szekely, Gabor J. 1986. *Paradoxes in Probability Theory and Mathematical Statistics*. D. Reidel Publishing Company.

Tarran, Brian. 2015. "The idea of using statistics to think about individuals is quite strange." *Significance* 12, no. 6 (December): 16–19.

Venn, John. 1888. *The Logic of Chance*. London, Macmillan and Co.

Weaver, Warren. 1963. *Lady Luck: The Theory of Probability*. Dover Publications, Inc. N.Y.

Chapter 2

Building Blocks of Modern Probability Modeling

Experiment, Sample Space, Events

- Look at Figure 2.1.

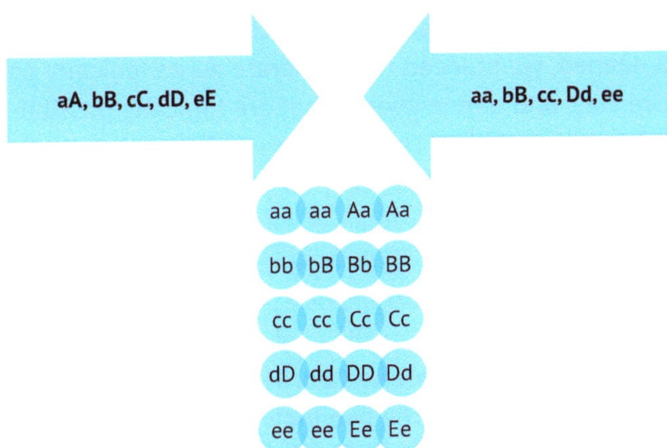

Figure 2.1

What is this figure representing? What phenomenon could it help us understand?

> *The concept of probability occupies an important role in the decision-making process, whether the problem is one faced in business, in engineering, in government, in sciences, or just in one's own everyday life. Most decisions are made in the face of uncertainty.*
> **(Ramachandran and Tsokos 2015)**

2.1 Learning the vocabulary of probability: experiments, sample spaces, and events.

Probability theory assigns technical definitions to words we commonly use to mean other things in everyday life.

In this chapter, we introduce the most important definitions relevant to probability modeling of a random experiment. A probability model requires an experiment that defines a sample space, S, and a collection of events which are subsets of S, to which probabilities can be assigned. We talk about the sample space and events and their representation in this chapter, and introduce probability in Chapter 3.

A most basic definition is that of a *random experiment,* that is an experiment whose outcome is uncertain. The term experiment is used in a wider sense than the usual notion of a controlled laboratory experiment to find the cure of a disease, for example, or the tossing of coins or rolls of dice. It can mean either a naturally occurring phenomenon (e.g., daily measurement of river discharge, or counting hourly the number of visitors to a particular web site), a scientific experiment (e.g., measuring blood pressure of patients), or a sampling experiment (e.g., drawing a random sample of students from a large university and recording their GPAs). Throughout this book, the reader will encounter numerous experiments. Once an experiment is well defined, we proceed to enumerate all its logically possible outcomes in the most informative way, and define events that are logically possible. Only when this is done, we can talk about probability. This section serves as a preliminary introduction of the main concepts. Later sections in this chapter talk in more detail about each of the concepts.

Denny and Gaines, in their book *Chance in Biology*, introduce fringeheads—fish that live in the rocky substratum of the ocean. The authors describe how when an intruder fringehead approaches the living shelter of another fringehead, the two individuals enter into a ritual of mouth wrestling with their sharp teeth interlocked. This is a mechanism to establish dominance, the authors add, and the larger of the two individuals wins the battle and takes over the shelter, leaving the other homeless. Fringeheads are poor judges of size, thus they are incapable of accurately evaluating the size of another individual until they begin to wrestle. When they enter into this ritual they do not know what their luck will be. As the authors claim, since they cannot predict the result of the wrestling experiments with complete certainty before the fringehead leaves the shelter to defend it, but have some notion of how frequently they have succeeded in the past, these wrestling matches are random experiments. Every time a fringehead repeats the experiment of defending its home, there is one of two possible outcomes (an outcome is a specific output of the experiment). The set of all possible logical outcomes of an experiment is called a sample space for that experiment. In the case of the fringehead wrestling, there are only two possible elementary logical outcomes, success (*s*) or failure (*f*), and these together form the sample space. We say that $S = \{s, f\}$ This S is a discrete finite sample space, to put it more technically. (Denny and Gaines 2000, 14)

Individual outcomes of an experiment are *elementary events*. For example, in the fringehead example, *s* is an elementary event, and *f* is another elementary event. Elementary events can be joined in compound events. The largest compound event is the sample space itself. The compound events are denoted by capital letters. Some elementary events are defined from elementary events in other sample spaces, for example observing *two* fringeheads defending their shelter, with sample space

$$S = S_1 \times S_2 = \{f,s\}\times\{f,s\} = \{ff,\ fs,\ sf,\ ss\},\ \text{where } S_i = \{s,f\},\ i = 1,2.$$

An experiment, the logical outcomes of the experiment, and hence the sample space vary depending on the problem being studied.

Example 2.1.1

For physicists, for example, a random experiment may consist of observing the number of photons measured by a detector (Cĭrca 2016), with sample space set $S = \{0,1,2,\ldots.\}$, the set of positive integers. An elementary event in the photon experiment is a single integer number. A compound event is, for example, the event A that the detector sees more than 10 photons, with $A = \{11, 12, \ldots..\}$.

Example 2.1.2

In genetics, consider another example explained by the Random Project (Siegrist 1997).

> In ordinary sexual reproduction, the genetic material of a child is a random combination of the genetic material of the parents. Thus, the birth of a child is a random experiment with respect to outcomes such as eye color, hair type, and many other physical traits. We are often particularly interested in the random transmission of traits and the random transmission of genetic disorders.
>
> For example, let's consider an overly simplified model of an inherited trait that has two possible states (phenotypes), say a pea plant whose pods are either green or yellow. The term "allele" refers to alternate forms of a particular gene, so we are assuming that there is a gene that determines pod color, with two alleles: g for green and y for yellow. A pea plant has two alleles for the trait (one from each parent), so the possible genotypes are
>
> gg, alleles for green pods from each parent.
> gy, an allele for green pods from one parent and an allele for yellow pods from the other (we usually cannot observe which parent contributed which allele).
> yg, an allele for green pods from one parent and an allele for yellow pods from the other (in reverse.
> yy, alleles for yellow pods from each parent.
>
> Thus the sample space for the mating of gy, gy genotypes is:
>
> $$S = \{gg,\ gy, yg, yy\}.$$
>
> The genotypes gg and yy are called homozygous because the two alleles are the same, while the genotype gy and yg is called heterozygous because the two alleles

are different. The event heterozygous is then $H = \{gy, yg\}$, and the event that the child is homozygous *is* $M = \{gg, yy\}$. Typically, one of the alleles of the inherited trait is dominant and the other recessive. Thus, for example, if g is the dominant allele for pod color, then a plant with genotype gg or gy has green pods, while a plant with genotype yy has yellow pods. Genes are passed from parent to child in a random manner, so each new plant is a random experiment with respect to pod color.

Pod color in peas was actually one of the first examples of an inherited trait studied by Gregor Mendel, who is considered the father of modern genetics. Mendel also studied the color of the flowers (yellow or purple), the length of the stems (short or long), and the texture of the seeds (round or wrinkled). (Siegrist 1997)

Example 2.1.3

Selecting a sample of three persons from a group of six people to form a committee of three people is an experiment that may result in the choice of any one of the $\binom{6}{3}$ or 20 committees, while selecting a treasurer, a captain and a typist out of the group of six is an experiment that results in $6 \times 5 \times 4 = 120$ outcomes.

Example 2.1.4

Playing a lottery where you must select five numbers from 49 is an experiment that has $\binom{49}{5} = 1{,}906{,}884$ possible outcomes.

Box 2.1

Math Tidbit

$$\binom{6}{3} = \frac{6!}{3!3!} = \frac{6\times5\times4\times3\times2\times1}{3\times2\times1\times3\times2\times1}.$$

In general, if n and k are nonnegative integers,

$$\binom{n}{k} = \frac{n!}{k!(n-k)!} = \frac{n\times(n-1)\times(n-2)..\times..\times1}{k\times(k-1).....\times1\ (n-k)\times(n-k-1)\times....\times1}.$$

See http://www.randomservices.org/random/foundations/Structures.html for more counting formulas. Chapter 4 in this book will revisit these formulas again.

2.1.1 Exercises

Exercise 1. Consider the experiment of monitoring the credit card activity of an individual to detect whether fraud is committed. Write the sample space of this experiment.

Exercise 2. Consider the following experiments and list their most informative sample space, i.e., the sample space consisting of all the possible orders in which the outcomes may appear.

- Sampling three students at random to determine whether they have bought Football Season tickets.
- Planting three tomato seeds and checking whether they germinate or not.
- Tossing a coin three times and checking whether head or tail appears.

Exercise 3. Consider the following experiment and list the outcomes of the sample space: observing wolves in the wilderness until the first wounded wolf appears.

Exercise 4. Consider the following experiment and list the outcomes of its sample space: screening people for malaria until the first 3 persons with malaria are found.

Exercise 5. This problem is inspired by Mosteller et al. (1961, 5). Suppose parents are classified on the basis of one pair of genes, and that *d* represents a dominant gene and *r* represents a recessive gene. Then a parent with genes *dd* is pure dominant, *dr* is hybrid, and *rr* is pure recessive. The pure dominant and the hybrid are alike in appearance. Offspring receive one gene from each parent, and are classified the same way. Write the sample space for the mating of *dr* with *rr*.

Exercise 6. A Chess club is debating which two people in the club should be supported to attend the world championship. They decide to select the two people at random by placing their name in a box and drawing the two names without replacement. The people in this club are:

Alison, Rosa, Hasmik, Jeonwong, Qing, Julie, and Edelweiss.

List the sample space of this sampling experiment.

2.2 Sets

In talking more formally from now on about the sample space S and events, we will need the concept of *set*. The mathematical theory of probability is now most effectively formulated by using the terminology and notation of sets. *Events are sets*. The foundation of probability in set theory was laid in 1933 by the Russian probabilist A. Kolmogorov.

Example 2.2.1

Consider the set H of numbers that may result from the toss of a 6-sided die. We may list it as:

$$H = \{1, 2, 3, 4, 5, 6\}.$$

Definition 2.2.1

A set V is a subset of a set A, denoted by $V \subseteq A$, if each element of V is also an element of A. The null set $\varnothing$ is a subset of every set.

Definition 2.2.3

Two sets A and B are equal ($A = B$), if and only if they have exactly the same elements. If one of the sets has an element not in the other, they are unequal and we write $A \neq B$.

We can also specify this set by describing it, instead of listing it:

$$H = \{y : y \text{ is a number from 1 to 6}\},$$

where y is a place-holder. We use braces when we are listing the elements of a set or specifying its properties.

Example 2.2.2
Let $A = \{$"*an odd number less than 7*"$\}$.
A is a subset of H, in Example 2.2.1.

Example 2.2.3
The sets $A = \{2,4,6\}$, $B = \{4,6,2\}$ are equal but the following sets: $W = \{5,1,2\}$ and $T = \{3,1,4\}$ are not equal.

The rest of this chapter makes extensive use of sets and their properties. Readers who need a refresher in the theory of sets and Venn diagrams may benefit from studying first the lesson on sets found at http://www.randomservices.org/random/foundations/Sets.html, a resource of the Random Project (Siegrist 1997), and then coming back to this chapter, where we use the theory under the assumption that the reader understands the concept of a set. The rest of the chapter relies on definitions given in that resource and is devoted to naming and illustrating sets as used in Probability Theory.

2.2.1 Exercises

Exercise 1. Do all the computational exercises at the end of the following lesson on sets that you should review before you go on studying this chapter:

http://www.randomservices.org/random/foundations/Sets.html

Exercise 2. Consider the set of movies showing in the movie theaters of the town where you live. List their names. Then create a subset containing the drama movies. If your town has no movie theaters, look at the nearest town.

Exercise 3. An academic department has 13 members. Their names are Abbott, Cicirelli, Cuellar, Liu, Pham, Mason, Danielian, Abe, Martinez, Mojica, Naseri, Engle, and Zaplan. From this set, list the subset containing the names that start with M.

Exercise 4. One of the many tasks in machine learning is to classify objects or individuals into subsets with common characteristics. What subsets of the periodic table could we make?

Exercise 5. (This exercise is based on an activity by Weber, Even, and Weaver (2015). Two people are playing a game of "Odd or Even" with two six-sided fair die. A trial of the game occurs when two cubes are rolled once and points are assigned. The game consists of many trials, the number of which are decided upon by Player A and B before the first trial. Consider the following rules.

Odd or Even Rules:
Roll two standard number cubes.
If the sum is 6 or an odd number, then Player A scores a point.
If the sum is 6 or an even number, then Player B scores a point.
Repeat the above steps an agreed number of times.
Whoever has the most points at the end of the game wins.

Suppose you play a game consisting of many trials. Regarding the points earned at the end of the game, what is the set of possible outcomes? List the outcomes.

2.3 The sample space

Definition 2.3.1

The sample space of an experiment is the set S of all logically possible outcomes of the experiment.

We now introduce the most basic set of probability theory, the sample space. In this chapter, we restrict ourselves to the consideration of finite sample spaces.

When an experiment is performed, it results in one and only one member of the set S.

The sample space defines the experiment. The sample space serves as the universal set for all questions concerned with the experiment. All outcomes logically possible in the experiment must be listed in the sample space.

Box 2.2

Sample Spaces

A sample space is *finite* if an integer can be assigned to each possible element. The outcomes of a single wrestling match of the fringehead is a finite discrete sample space.

A sample space is *countably infinite* if the elements can be counted, i.e., can be put in one-to-one correspondence with the positive integers. The number of wrestling matches a fringehead will enter to get its first loss is a discrete sample space which has an infinite number of logically possible outcomes, assuming the fringehead is immortal.

A sample space is *uncountably infinite* if it cannot be put in one-to-one correspondence with the positive integers. The time it takes to complete a standard homework is an uncountably infinite sample space.

Finite and countable infinite sample spaces are also called discrete sample spaces. Uncountably infinite sample spaces are also called continuous sample spaces. Part I of this book is about the former, Part II about the latter.

Being a set, S could have a finite number of logically possible outcomes or an infinite one. Experiments with finite number of outcomes or countably infinite outcomes are called *discrete sample spaces*. Experiments with outcomes any of the real numbers have *continuous sample spaces*. We will focus on the finite discrete sample spaces in this chapter.

The sample space is our universe of discourse, it has to represent the experiment properly. For example, consider the experiment that consists of observing what happens to drivers passing a section of a busy highway. The following sample space:

$$S = \{\textit{the squirrel ate the grapes}, \textit{the squirrel did not eat the grapes}\}$$

does not seem to be a good description of the possible outcomes of the random phenomenon that interests us. Instead,

$$S = \{\textit{flat tire}, \textit{bumper to bumper}, \textit{no incident}, \textit{fatal accident}, \textit{minor crash}\}$$

might represent our interest in this phenomenon much more accurately.

Example 2.3.1

"Receiving a letter grade after completing your intro probability class at a public American University" like that where the author works is a 7-outcome experiment. The sample space of the logically possible outcomes of the experiment is

$$\text{S} = \{\text{A, B, C, D, F, P, NP}\}.$$

Example 2.3.2

Tossing three coins—a dime, a quarter, and a penny—in a row and keeping track of the sequence of heads and tails that results is an 8-outcome experiment. Thus the sample space is the set

$$\text{S} = \{\text{HHH, HHT, HTH, HTT, THH, THT, TTH, TTT}\},$$

which has eight elements and provides a list that represents the logically possible outcomes of one toss, if we understand that the first letter in a pair designates the outcome for the dime, the second letter that for the quarter, and the third letter that for the penny. Thus HTH means that the dime fell heads, the quarter fell tails and the penny fell heads. Every logically possible outcome of the experiment corresponds to exactly one element of the set S.

Sometimes, when the sample space is not too big, we could use trees as a tool to find out what to put in the list of elements of S. For example, in the case of the three coins experiment, Figure 2.2 shows that the way to construct the sample space for an experiment like this is to first consider the possible outcomes for the first coin, then the ones for the second and then the ones for the third.

Notice that the methodology can also be used to represent the sample space for similar experiments in different contexts. Example 2.3.3 is basically the same experiment but in a totally different context.

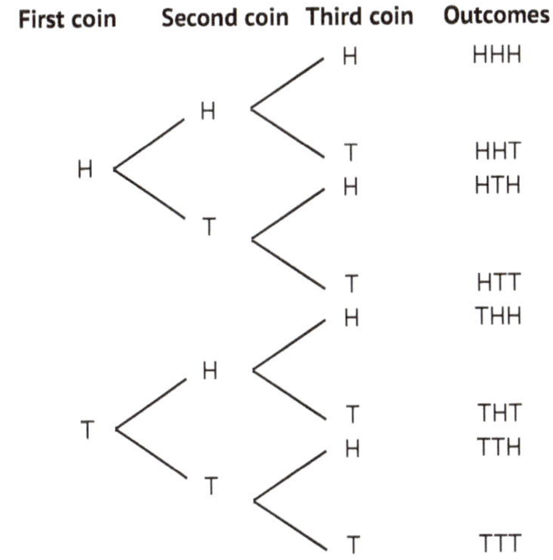

Figure 2.2 Tree representation of an experiment consisting of tossing a dime, a quarter and a penny.

Example 2.3.3.

Capoeira is an Afro-Brazilian martial art that combines elements of dance, acrobatics, and music. It was developed in Brazil at the beginning of the sixteenth century. It currently has United Nations Cultural Heritage status and is practiced in many parts of the world. Drawing a random sample of three youngsters from the young population in three different neighborhoods a, b, and c of Rio de Janeiro, Brazil, to see if they practice capoeira or not is an 8-outcome experiment. Practicing capoeira is denoted a success (s) and not practicing a failure (f). The listing of the 8-outcomes sample space is:

$$S = \{sss, ssf, sfs, sff, fss, fsf, ffs, fff\},$$

where, for example, ssf denotes an outcome where the individuals from neighborhoods a and b practice capoeira, and the one from neighborhood c does not.

Example 2.3.4

Consider the experiment of observing SAT scores for a student randomly chosen among those that have taken the SAT. Note: SAT is a standardized test for college admissions. Scores are multiples of 10, and therefore discrete numbers. There are three sections: reading, math, and writing, each section with positive scores between 200 and 800. So the total possible score is between 600 and 2400. We may describe the sample space, instead of listing all its elements, since it is a very large set.

$$S = \{score \mid 600 \leq score \leq 2400 \text{ and } score \text{ is multiple of } 10\}.$$

2.3.1 A note of caution

Although there is more than one way to represent a sample space, we prefer to use the representation that allows us to answer the largest number of questions, that is the most detailed one indicating the order (see Section 1.3 to understand why).

For example, in the experiment of example 2.3.2, we said that S is

$$S = \{\text{HHH, HHT, HTH, HTT, THH, THT, TTH, TTT}\}.$$

Many of you are probably thinking: "why not

$$S = \{\text{one head, two heads, three heads}\}?"$$

This last representation lets us talk about the number of heads, but does not let us talk about the order of heads and tails or which coin is head, while the first sample space with the 8-outcome representation allows us to address any question, for example, the probability that the dime is Heads. The reader will see that because of that, in this book, when listing the outcomes of the sample space we always choose the sample space representation that allows us to answer all questions about any events in the sample space.

2.3.2 Exercises

Exercise 1. Five students, numbered 1,2,3,4,5 to keep their identity secret, are competing for the "best data scientist of the year" and the "best mathematical statistician of the year" awards, offered by the undergraduate student association in their school. All five students have applied for the two awards. A student can get at most one award. What are logically possible outcomes of this experiment? Give the most informative listing of the sample space. Explain the notation you use.

Exercise 2. A foreign-exchange student from a Mediterranean country arrives at the US for the first time and is mesmerized by the amount of cakes offered for dessert at the orientation party, a total of five different cakes labeled a,b,c,d,e. The student decides to try three of them chosen at random (and not eat anything else). What is the set of possible logical choices? That is, list the sample space S.

Exercise 3. As an international student coming to the United States, there are three different student visas that a foreign student could be issued: F1 Visa, J1 Visa or M1 Visa. Descriptions of these visas can be seen at https://www.internationalstudent.com/study_usa/preparation/student-visa/. If we select three international students at random from the population of international students at a particular point in time to observe their visa status, what would be the sample space? List it. To simplify the notation, we use the letters F, J, and M, respectively, for the types of visas.

Exercise 4. Sometimes, companies downsize by laying off the older workers. Consider an experiment that consists of keeping track of layoffs in a major company that is under the radar of the Equal Employment Opportunity Commission until three employees older than 40 are laid off. List some outcomes of the sample space of this experiment, at least six members of S.

Exercise 5. (This exercise is inspired by a related problem on pages 34–35 of Khilyuk, Chilingar, and Rieke (2005), but uses the more recent EPA standards.) The Environmental Protection Agency (EPA) in the United States evaluates air quality on the basis of the Air Quality Index (AQI), which classifies air quality into five major categories: Good (AQI 0–50), Moderate (AQI 51–100), Unhealthy for Sensitive Groups (AQI 101–150), Unhealthy (AQI 151–200), Very Unhealthy (AQI 201–300), Hazardous (PSI 301–500). The following document,

https://www3.epa.gov/airnow/aqi-technical-assistance-document-sept2018.pdf

contains on pages 4 and 5 pollutant-specific sub-indices corresponding to those categories. A specification of the sample space based only on the air quality categories given above without specifying the pollutant compositions, is not very helpful. Air quality is identified with the worst category of any particular contaminant. For example, carbon monoxide larger than 40 ppm would result in AQI being hazardous, even though the other pollutants are at the Good level. Indicate what would be a more appropriate and informative listing of the sample space under this classification of AQI system.

2.4 Events

> It is to be emphasized that in studying a random phenomenon our interest is in the events that can occur (or, more precisely, in the probabilities or degree of uncertainty with which they can occur). The sample space is of interest not for the sake of its members, but for the sake of its subsets, which are the events. (Parzen 1960, 12)

The largest event is the sample space. In every experiment, some outcome in the sample space must happen, logically speaking, with certainty.

A subset or event may be represented by listing all its elements, as in Example 2.3.2, or by describing it with algebraic equations and inequalities, as we did in Example 2.3.4. The listing is preferred if there is a small number of outcomes in the event, but a compact mathematical description is preferred when the number of outcomes in *S* is very large. The mathematical language of events is the same as that of set theory. We denote events by capital letters, for example, A, B, C,. ...

Definition 2.4.1

Let a sample space S be given containing all logically possible outcomes of an experiment. An *event* is a subset of a sample space S of an experiment. We say that event A occurs if the outcome of the experiment corresponds to an element of the subset A.

We say that an event has occurred if an outcome in the event occurs.

The most basic events associated with an experiment are those that correspond to an elementary outcome- for example, winning a mouth wrestle in the fringehead experiment, or getting three heads in the coin tossing experiment. We call these elementary events, or simple events. Events can also include several outcomes-for example, getting two heads in the tosses of three coins.

Example 2.4.1

Think of the experiment in Example 2.3.2.

$$S = \{HHH, HHT, HTH, HTT, THH, THT, TTH, TTT\}.$$

We may be interested in the event described verbally as A = {"two heads occur"} and list it as:

$$A = \{HHT, HTH, TTH\}.$$

We recognize A as a subset of the sample space S. The subset A is the mathematical counterpart of the event "two heads."

We could define many other events in S, for example:

- B = {"Number of tails is 1"} = {HHT,HTH, THH}
- C = {"The first toss is a head"} = {HHH, HHT, HTH, HTT}
- D = {"The number of heads is larger than the number of tails."} = {HHT, HTH, THH, HHH}

Each event described above is given precise mathematical meaning by the corresponding set.

Box 2.3

Events and sets

Sets are the main building blocks of probability theory. The sample space *S* is a set, each of the outcomes in S is a simple set, a set with one outcome or elementary events, and a bigger event is a set of outcomes of S. If the outcome of an experiment is contained in the collection of outcomes in the event, then we say that the event has occurred. Thus an event can occur in several ways. Event A occurs if either of the outcomes in it happens.

Example 2.4.2

A clinical study can screen patients until it finds one with a disease, but budget allows only at most four patients screened. We will list the sample space S, the event E that "a patient with the disease is found" and the event B that "the patient with the disease is found in at most 3 attempts."

Let *s* denote that the disease is present and let *f* denote that no disease is present.

$$S = \{s, fs, ffs, fffs, ffff\},$$

where *ffs* means that three individuals are screened, the first two without the disease and the third with the disease.

$$E = \{All\ outcomes\ in\ S\ except\ outcome\ ffff\},$$
$$B = \{s, fs, ffs\}.$$

Example 2.4.3

An experiment consists of searching the internet for an email address of the history teacher you had in high school 25 years ago. The sample space of this experiment is:

S = {address found, address not found}.

A subset of S is A = {address found}.

Example 2.4.4

The experiment is to assign seats to four students with the same academic background. It is known that two of the students speak Japanese fluently and two don't. We must assign the students to four chairs.

The set of all possible seating arrangements based on whether the student speaks Japanese or not may be described as follows:

$$S = \{(x_1, x_2, x_3, x_4) \text{ such that } x_i = 0 \text{ means no Japanese}, x_i = 1 \text{ means Japanese spoken}\}$$
$$= \{(0110), (0101), (1010), (0011), (1001), (1100)\}.$$

Let the event A represent the seating arrangements with the Japanese-speaking students not sitting together.

$$A = \{(1001), (1010), (0101)\}.$$

2.5 Event operations

Set operations allow us to obtain new sets from subsets of the sample space. We consider in this section some of the most important set operations as applied to events.

Consider two events A and B.

The *union of events A, B,* is the event C consisting of outcomes that are in at least one of the events:

$C = A \cup B = \{s_i \in S : s_i \text{ in } A \text{ or } s_i \text{ in } B \text{ or both}\}$, where s_i is an outcome in S, $i = 1, 2, \ldots, N$, and

N is the number of elements in S.

The *intersection of A and B* is the event E consisting of elements of S that belong to both A and B:

$$E = A \cap B = \{s_i \in S : s_i \text{ in } A \text{ and } s_i \text{ in } B\}.$$

It should be noted that many writers denote the intersection of two events A and B by AB instead of $A \cap B$.

The *complement of an event A* is the event A^c consisting of all elements of S that are not in A.

$$A^c = \{s_i \in S : s_i \text{ not in } A\}.$$

Box 2.4

A note on "or" and "and"

When we use the expression "A or B" in referring to events, the meaning is never in doubt because we always use the inclusive "or" of everyday English. That is, "A or B" means "A or B or both." When we ask "A or B" we mean

$$A \cup B.$$

When we use the expression "A and B" we mean

$$A \cap B.$$

Definition 2.5.1

Two events A and B are *disjoint* if they are *mutually exclusive*, i.e., if

$$A \cap B = \varnothing.$$

Generalizing to more than two events, consider events $E_1, E_2, \ldots, E_n$. The events E_i, E_j are pairwise disjoint if

$$E_i \cap E_j = \varnothing \text{ for all } i, j, i \neq j.$$

The *impossible event* $\varnothing$ is the empty set in set theory. One important property of the impossible event is that it is the complement of the certain event S. Clearly, $S^c = \varnothing$, for it is impossible for S not to occur. A second important property of the impossible event is that it is equal to the intersection of any event A and its complement A^c:

$$A \cap A^c = \varnothing.$$

Furthermore,

$$A \cap \varnothing = \varnothing; \quad A \cup \varnothing = A.$$

Example 2.5.1

Consider the sample space depicted in Figure 2.3 consisting of all the pairs of numbers that you can get when you roll a red die and a white die (you do not know ahead of time whether the dice are fair or not). We will associate with each outcome the sum of the points (as we did in Table 1.1 of Chapter 1).

Let A be the event that the sum is smaller than or equal to 5. Let B the event that the sum is larger than 3 but smaller than 7. Then,

$$A = \{(1,1), (1,2), (1,3), (1,4), (2,1), (2,2), (2,3), (3,1), (3,2), (4,1)\},$$

$$B = \{(1,3), (1,4), (1,5), (2,2), (2,3), (2,4), (3,1), (3,2), (3,3), (4,1), (4,2), (5,1)\},$$

$$C = A \cup B$$
$$= \{(1,1), (1,2), (1,3), (1,4), (2,1), (2,2), (2,3), (3,1), (3,2), (4,1), (1,5), (2,4), (3,3), (4,2), (5,1)\},$$

$E = A \cap B = \{(1,3), (1,4), (2,2), (2,3), (3,1), (3,2), (4,1)\}$, and
$A^c = \{(a,b) \text{ in } S : (a+b) > 5\}$.

Example 2.5.2

Suppose that key elections for the position of president of a country are held. There are three voting districts, I, II, III. The winner of the election has to have won the majority of votes in two of the three voting districts. Assume that there are two candidates, A and B, for the position. The sample space for the outcome of the election can be seen in Figure 2.4.

$$S = \{AAA, AAB, ABA, ABB, BAA, BAB, BBA, BBB\},$$

where, for example, ABB means that candidate A won in district I, and B won in districts II and III.

Let W be the event that A wins the election,

$$W = \{AAA, AAB, ABA, BAA\}.$$

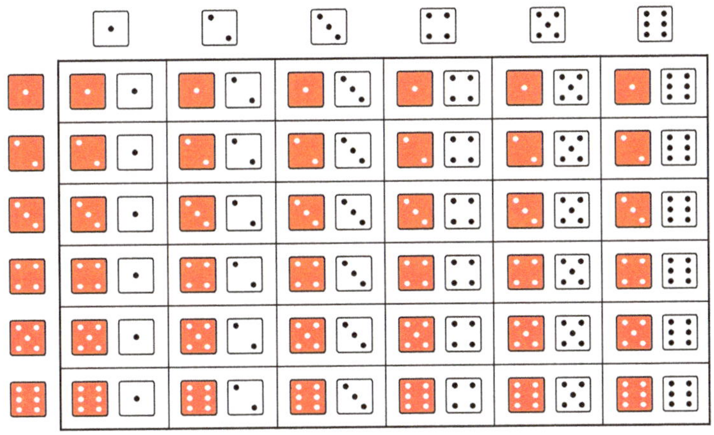

Figure 2.3 The 36 outcomes of the roll of two six-sided dice.

Source: http://dsearls.org/courses/M120Concepts/ClassNotes/Probability/440_theo.htm; Source: https://1.bp.blog-spot.com/_P75nQNhmdtE/TIkdopwHQiI/AAAAAAAAA88/J2qd7TUfpXA/s1600/backgammon-dice-probability.gif.

The complement of W is "B wins the election,"

$$W^c = \{ABB, BAB, BBA, BBB\}.$$

Let T be the event that the first two districts, I and II, voted for the same candidate. Then,

$$T = \{AAA, AAB, BBA, BBB\}.$$

The event

$$C = W \cap T = \{AAA, AAB\}.$$

The reader may want to mark these events in Figure 2.4 using Venn diagrams.

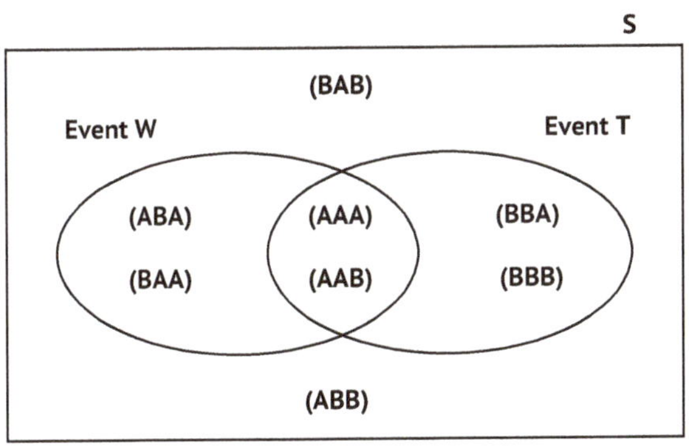

Figure 2.4 Venn diagrams for election example 2.5.2.

In particular, if the union of pairwise disjoint events $E_1, E_2, \ldots, E_n$ is S, then the collection of these events forms a partition of S. A partition of S requires: (a) that all the events forming the partition are pairwise disjoint; (b) that the union of all those events is S. The two conditions must be checked. An example of a partition can be seen in Figure 2.5. The dots indicate that many other mutually exclusive events could be included in the picture.

Definition 2.5.2

A *partition* of a set A is a subdivision of the set into subsets that are disjoint and exhaustive, i.e., every element of A must belong to one and only one of the subsets. Thus $E_1, E_2, \ldots, E_n$ is a partition of A if $E_i \cap E_j = \varnothing$ *if* $i \neq j$ and $E_1 \cup E_2 \cup \ldots \cup E_n = A$.

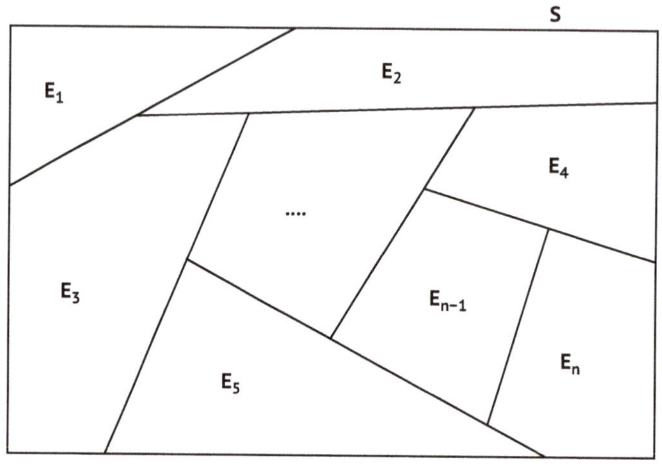

Figure 2.5 A partition of the sample space.

Partitions are very useful, allowing us to divide the sample space into small, non-overlapping pieces, as in a puzzle. Visualizing partitions often helps doing the proofs of the main theorems of probability in Chapter 3.

Example 2.5.3

The possible partitions of the sample space S = {1, 2, 3, 4} are:

(*i*) [{1}, {2,3,4}]; (*ii*) [{2}, {1,3,4}]; (*iii*) [{3}, {1,2,4}]; (*iv*) [{4}, {1,2,3}]; (*v*) [{1,2}, {3,4}];
(*vi*) [{1,3}, {2,4}]; (*vii*) [{1,4}, {2,3}]; (*viii*) [{2,4}, {1}, {3}]; (*ix*) [{3,4}, {1}, {2}]; (*x*) [{1,2,3,4}];
(*xi*) [{1}, {2}, {3}, {4}]; (*xii*) [{1,2}, {3}, {4}]; (*xiii*) [{1,3}, {2}, {4}];
(*xiv*) [{1,4}, {2}, {3}]; (*xv*) [{2,3}, {1}, {4}].

Example 2.5.4

Let us consider the logical possibilities for the next three games in which England plays Russia in a FIFA World Cup. We can list the possibilities in terms of the winner of each game:

$$S = \{EEE,\ EEU,\ EUE,\ EUU,\ UEE,\ UEU,\ UUE,\ UUU\},$$

where E denotes England and U denotes Russia. The outcome or simple event EUU means that England wins the first game and Russia wins the next two games.

A partition of the sample space is made by the sets A = {"England wins two games"}, B = {"Russia wins two games"}, C = {"England wins three games"}, and D = {"Russia wins 3 games"}. Can you list the outcomes contained in each of these events?

We can construct many other events using the basic set operations of union, intersection, and complement. Visit the Venn diagram app to see what some of those other events are. The app can be found at this url: http://www.randomservices.org/random/apps/VennGame.html.

Example 2.5.5

Visit the traffic light applet to visualize this problem. It can be run if your browser has Java and you will find it at http://statweb.stanford.edu/~susan/surprise/Car.html.

Set the applet to run at low speed and the number of lights to three. Then think about the following problem:

> Driving a customer, a taxi driver passes through a sequence of three intersections with traffic lights. At each light, she either stops, s, or continues, c. The sample space is
>
> $$S = \{sss, ssc, scs, scc, css, csc, ccs, ccc\},$$
>
> where, for example, csc denotes "continues through first light, stops at second light, and continues through third light. Let event A denote that the taxi driver stops at the first light,
>
> $$A = \{scc, scs, ssc, sss\}.$$
>
> And let event B be the event that the taxi driver stops at the third light,
>
> $$B = \{css, ccs, scs, sss\}.$$
>
> The event C = {"stops at the first or the third light but not both"} is:
>
> $$C = (A \cap B^c) \cup (B \cap A^c) = \{scc, ssc, css, ccs\}.$$
>
> The event D = {"does not stop at the third or the first lights") is:
>
> $$D = \{csc, ccc\}.$$

Box 2.5

Verbal description of events

Events may be defined verbally, and it is important to be able to express them in terms of the event operations. For example, consider two events, E and F. The event C that exactly one of the events will occur is equal to the event $C = (E \cap F^c) \cup (E^c \cap F)$. The event W that none of the events will occur is equal to the event $W = E^c \cap F^c$. The event T that at most one (that is, one or less) events happen is $T = (E \cap F)^c$.

As you keep reading this book, pay attention to this verbal translation of the mathematical expressions and to alternative mathematical expressions for the same event, because this is crucial for communicating verbal material to wide nontechnical audiences.

2.6 Algebra of events

The algebra of events is the algebra of sets. This algebra tells us the relations among sets that are obtained by set operations. The following are important relations (of sets in general) that help simplify calculations. For any three events, A, B, and C, defined on a sample space S,

Commutativity:

$$A \cup B = B \cup A \text{ and } A \cap B = B \cap A.$$

Associativity:

$$A \cup (B \cup C) = (A \cup B) \cup C \text{ and } A \cap (B \cap C) = (A \cap B) \cap C.$$

Distributive Laws:

$$A \cap (B \cup C) = (A \cap B) \cup (A \cap C) \text{ and } A \cup (B \cap C) = (A \cup B) \cap (A \cup C).$$

De Morgan's Laws is a law about the relation between the three basic operations.

$$(A \cup B)^c = A^C \cap B^c \text{ and } (A \cap B)^c = A^C \cup B^c.$$

The proof of De Morgan's laws can be found in several sources. See Ross (2010) or Siegrist (1997).

All of the properties mentioned can be generalized to a number $n > 2$ of events. Let $E_1, E_2, \ldots\ldots, E_n$ be events defined in the sample space S. Then,

$$(\cap_{i=1}^{n} E_i)^c = \cup_{i=1}^{n} E_i^c.$$

$$(\cup_{i=1}^{n} E_i)^c = \cap_{i=1}^{n} E_i^c.$$

In prose form: the complement of the intersection of n events is the union of the complements of these n events. And the complement of the union of n events equals the intersection of the complements of these events.

2.6.1 Exercises

Exercise 1. If we pull a card from a deck like that in Figure 2.6 and consider the events A = spade, B = heart, are these events mutually exclusive? Why?

If we pull a card from a deck and consider the events A = spade and B = ace, are these events mutually exclusive? Why?

Exercise 2. Consider the sample space of Figure 2.7 with elements representing the ages at which one can apply for an annual free ticket to an attraction park, i.e.,

$$S = \{1, 2, 3, 4, 5, 6, 7, 8, 9, 10, 11, 12, 13, 14, 15, 16, 17, 18, 19, 20\}.$$

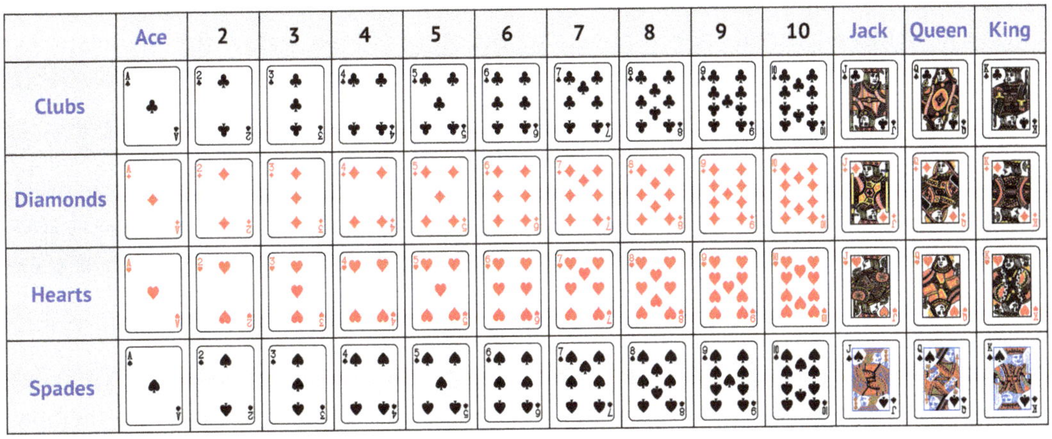

Figure 2.6 Deck of cards.

The event A contains all ages that are multiples of three, while event B contains all ages that are multiples of five. (i) Identify in the Venn diagrams the events *A, B*. (ii) List the elements in the events

$A \cap B$, $A \cap B^c$, $A^c \cap B$ *and* $(A \cup B)^c$, and determine whether these events form a partition of S.

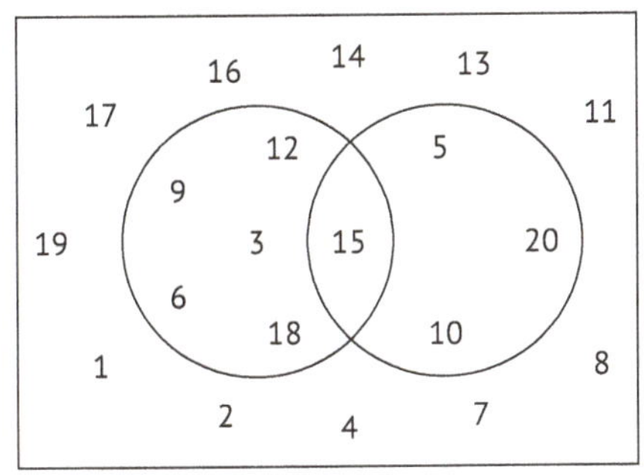

Figure 2.7 Ages of free-ticket-eligible individuals.

Exercise 3. Four components are connected to form a system as shown in Figure 2.8. The subsystem 1–2 will function if both of the individual components function. The subsystem 3–4 functions if both of the individual 3–4 components function. For the entire system to function, at least one of the two subsystems must function. (i) List the outcomes in the sample space. (ii) Let *A* be the event that the system does not work, and list the elements of *A*. (iii) Let *B* be the event that the system works, and list its elements. What is the relation between events *A* and *B*?

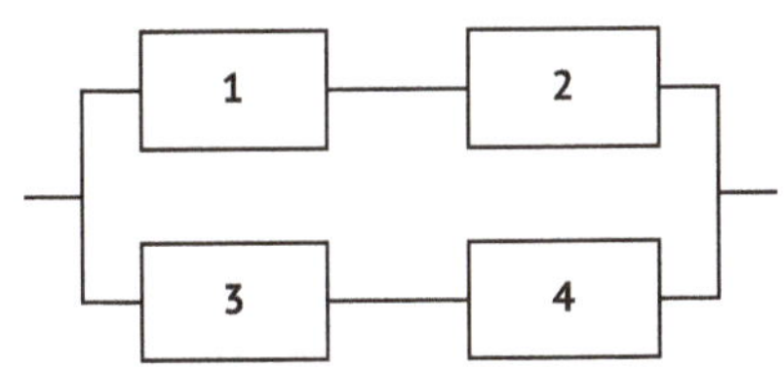

Figure 2.8 System with 4 components.

Exercise 4. Here is some information about the 16,000 participants in the Big Data Annual Competition organized by a statistical society.

- 50% are data scientists
- 25% have a car
- 60% of those with a car drive to school or to work
- 40% are second time participants
- 80% are from New Jersey
- 15% are well known experts in algorithms
- 10% are from New York
- 5% are from California

(i) How many participants are from outside the state of New Jersey? (ii) How many participants drive to school or work? (iii) How many participants drive to school or work and have a car?

Exercise 5. If the outcome of an experiment is the order of finish in a car race among 4 cars having post positions 1,2,3,4, then how many outcomes are there in the sample space and how many outcomes are there in the event E consisting of outcomes in which car 3 wins the race?

Exercise 6. A series of 3 jobs arrive at a computing center with 3 processors and could end up in any of the processors. List the members of the sample space and then list the members of the event that all processors are occupied.

Box 2.6

Classification, clustering and partitions

In data science, classification is a technique that first partitions observed data into a set of mutually exclusive classes. First, subjects observed are represented by a tuple describing their characteristics, i.e., (educated, old, rich, famous) or (not educated, old, rich, not famous). Each tuple is then linked to a predefined class, called the class label attribute. For example, (educated, old, rich, famous) subjects are classified as "excellent" credit rating. The observed tuples and the class label given to them is called the *training set*. This training set is used to teach the algorithm how to classify new objects. Since the class label of each training tuple is known, this first step is called *supervised learning,* because we know to which class the tuple belongs. This process identifies the logically possible outcomes. In the second step, the model created in the first step is used to identify customers whose class label is unknown. Thus, classification uses data to come up with a definition of the sample space and the prior known partition of the sample space.

On the other hand, cluster analysis looks at new objects and tries to determine whether the tuples representing them suggest some sample space and partition. Cluster analysis algorithms are called unsupervised learning.

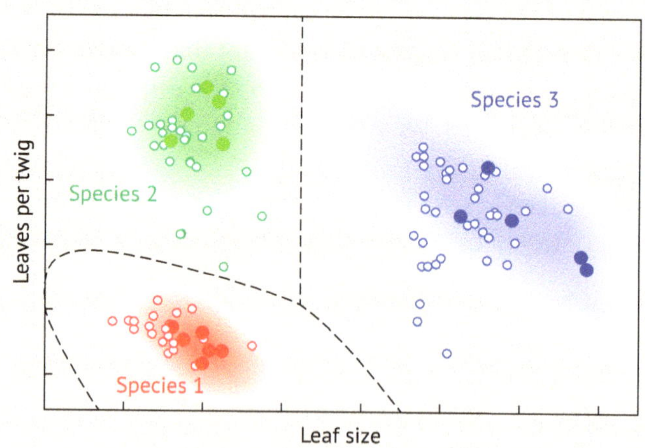

Figure 2.9 Supervised learning partitions the sample space into mutually exclusive sets or classes. This is known as supervised machine learning classification.

Source: https://astrobites.org/wp-content/uploads/2015/04/image22.jpg.

2.7 Probability of events

The third building block of probability theory is a probability function defined on the sample space, mapping to the real numbers. All together, a sample space, the events defined in a sample space and a probability function form a probability space. With a probability space well defined, we can approach probability problems of any complexity.

Chapter 3 is dedicated to this third building block. To understand the material in Chapter 3, it is important to first feel proficient in all the material seen in Chapter 2.

2.8 Mini quiz

Question 1. The building blocks of probability theory are (select all that apply):

a. the sample space of the experiment
b. events in the sample space
c. a probability function defined on the sample space.
d. random experiments

Question 2. A partition of the sample space must satisfy which of the following?

a. The sets in the partition must be disjoint
b. The union of the sets must be equal to the sample space
c. The complement of the union of the sets comprising the partition is not empty
d. The intersection of the sets in the partition is not empty

Question 3. Consider two events, A and B, in a sample space. The event $(A \cap B^c) \cup (B \cap A^c)$ represents the event

a. only A or only B happens, but not both
b. A and B happen
c. Neither A nor B happens
d. A or B happens

Question 4. Which of the following is an experiment?

a. Observing whether a fuse is defective or not
b. Observing the duration of time from start to finish of rain in a particular place

Question 5. Consider two events A and B in a sample space S. The event $A \cap B$ is not empty. The event

$$(A \cap B^c) \cup (A \cap B) \cup (B \cap A^c)$$

contains the same elements as which event?

a. $(A \cap B^c) \cup (A \cap B)$
b. $(A \cup B)^c$
c. B
d. A^c
e. $A \cup B$

Question 6. Two six-sided dice are rolled. Let A be the event that the sum is less than nine, and let B be the event that the first number rolled is five. Events A and B are (select all that applies)

a. mutually exclusive
b. complements of each other
c. equal
d. not mutually exclusive

Question 7. "Has teeth" and "has feathers" are

a. empty events
b. mutually exclusive events
c. complement events of each other
d. a union of events

Question 8. The event $(A \cap B^c) \cup (A^c \cap B)$ contains

a. outcomes that are in both A and B happen
b. outcomes that are only in A or only in B happen
c. outcomes that are in A or B happen
d. outcomes that are in neither A nor B happen

Question 9. (This problem is inspired by Pfeiffer (1965, 25).) A certain type of rocket is known to fail for one or two reasons: (1) failure of the rocket engine because the fuel does not burn evenly or (2) failure of the guidance system. Let the experiment consist of the firing of a rocket of this type. We let A be the event that the rocket fails because of engine malfunction and B the event the rocket fails because of guidance failure. The event F of a failure of the rocket is thus given by: $F = A \cup B$. Consider the following three events:

W = (engine fails) or (engine operates and guidance fails)
T = (guidance fails) or (guidance operates and engine fails)
R = (engine operates and guidance fails), or both fail, or (engine fails and guidance operates)

Each of the events W, T, and R are (choose all that apply)

a. equal to F
b. equal to $A \cap B^c$
c. a partition of F
d. a partition of $A \cup B$

Question 10. (This problem is inspired by Johnson (2000, chapter 7).) A tract of land in the Alabama Piedmont contains a number of dead shortleaf pine trees, some of which had been killed by the littleleaf disease, some by the southern pine beetle, and some by the joint attack by both agents. If one of these dead trees were selected blindly, one could not say that it was killed by either the disease or the beetle because

a. we should say "or" which is what we use for union
b. the events "killed by disease" and "killed by beetle" are disjoint
c. the events "killed by disease" and "killed by beetle" are not disjoint
d. the event "killed by disease" is the complement of "killed by beetle"

2.9 R code

Being aware of the many possible outcomes that can arise in an experiment can be illustrated with the matching problem.

An online dating service research team has matched individuals A1, A2, A3, and A4 with individuals B1, B2, B3, and B4, respectively. If they call for a date, the number after the letter determines who they are matched with. That is, the research team thinks that A1 matches well with B1, etc. The assistant who responds to date requests does not use the information provided by the research team. Instead, the assistant assigns a date to the A individuals at random. That is, for example, A1 is given a randomly chosen person from the B pool of candidates. This is like putting four numbers in an urn, and drawing one number at random, without replacement. In R, the activity done for the four in the A pool is:

```
sample(1:4, 4, replace=F) # one trial of the matching experiment.
```

If we had gotten `3,1,4, and 2` as a result, this would mean that A1 gets B3, A2 gets B1, A3 gets B4, and A4 gets B2.

Listing the possible outcomes of this experiment could take a lot of space if done by hand. You could do many trials at once by using a for loop in R. For example, suppose you want three trials. Then you would use the following code:

```
trials=matrix(rep(0,12),ncol=4)
for(i in 1:3){ # put each of the three trials in a row
trials[i,]=sample(1:4, 4, replace=F)
}
trials
```

Suppose we got the following rows, which are edited:

A1 A2 A3 A4

1 4 3 2 #A1 will be happy, as it matches

2 4 1 3 # nobody will be happy

1 2 3 4 # all of them will be happy

To see what other outcomes you could get, you could repeat the trial many more times (you may not get all the possible outcomes, though):

```
trials=matrix(rep(0,16),ncol=4)
for(i in 1:100){ # This is doing 100 trials
trials[i,]=sample(1:4, 4, replace=F)
}
trials
```

- **Question 1.** List 10 possible different outcomes of the sample space based on what you get in the simulation. How many possible outcomes do you think there are in total, even if you did not get all of them in your simulation?
- **Question 2.** Think of possible subsets that you could get. For example, A = {“A2 gets the correct date”}. List all the possible outcomes in A. Then find out how many times in your simulation you got that event happening. To find out the latter, type in R

```
A2match=matrix(trials[trials[,2]==2],ncol=4)#extract the rows where
A2 gets correct date
nrow(A2match) # how many rows are such that 2 gets the right date
```

- **Question 3.** How many elements are there in the event B = {“A1 and A1 get the right date”}. How many times in your 100 trials did this event happen? To find the latter, type in R

```
A1A2match=matrix(trials[trials[,2]==2&
trials[,1]==1],ncol=4)
nrow(A1A2match)
```

2.10 Chapter Exercises

Exercise 1. A company is allowed to interview candidates until two qualified candidates are found. But budget constraints dictate that no more than 10 candidates can be interviewed. List the outcomes in sample space.

Exercise 2. There are 25 students in a classroom, 10 are electrical engineer majors, 5 are statistics majors, and 13 have other majors. Using W to denote those that are EE majors and M to denote those that are statistics majors, symbolically denote the following events, and identify the number of students in each set:

a. the set of all students that have are double majors in EE and statistics
b. the set of all students who are in only one of those two majors
c. the set of all students that are not in any of those majors

Exercise 3. Consider the Venn diagrams and events A, B, associated with Figure 2.7. List the elements in the following events:

a. $A^c \cap B^c$
b. $(A \cap B^c) \cup (B \cap A^c)$
c. $A^c \cup B^c$
d. $B \cup A^c$

Exercise 4. Sketch the region corresponding to the following event $(A \cup B)^c \cap C$ in Figure 2.10

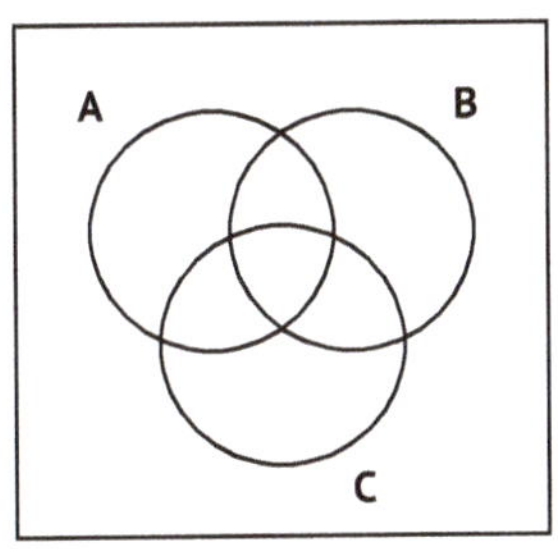

Figure 2.10

Exercise 5. The web site http://www.csun.edu/~ac53971/pump/20090310_dice.pdf contains activities similar to some of the ones seen in this chapter, but conducted with a TI-84 calculator. At the end of the activity, there is a section on the game Shooting Craps. Answer question 6 in that activity. You may conduct the simulation, but you must provide also the theoretical probability.

Exercise 6. (This exercise is an adaptation of a problem by Ross 2010, page 108, problem 3.69.) A certain organism possesses a pair of each of 5 different genes (which we will designate by the first 5 letters of the English alphabet). Each gene appears in 2 forms (which we designate by lowercase and capital letters). The capital letter will be assumed to be the dominant gene in the sense that if an organism possesses the gene pair xX, then it will outwardly have the appearance of the X gene. For instance, if X stands for brown eyes and x for blue eyes, then an individual having either gene pair XX or Xx will have brown eyes, whereas one having gene pair xx will have blue eyes. The characteristic appearance of an organism is called its phenotype, whereas its genetic constitution is called its genotype. (Thus 2 organisms with respective genotypes aA, bB, cc, dD, ee and AA, BB, cc, DD, ee would have different genotypes but the same phenotype.) In a mating between two organisms, each one contributes, at random, one of its gene pairs of each type. Consider the experiment consisting of the mating between organisms having genotypes aA, bB, cC, dD, eE and aa, bB, cc, Dd, ee, what are the possible outcomes for genotype of the progeny? List this sample space. Separately, list the sample space if the interest is in the possible phenotypes. (Ross 2010)

Exercise 7. The image in Exercise 4, Figure 2.10, shows a Venn diagram of three events (A, B, and C), in a sample space. Each of the cells delimited by the solid curves represents an event. All the cells shown comprise a partition of the sample space. Use the notation and operations on sets learned in this chapter to list all the sets in the partition.

Exercise 8. It is possible to derive formulas for the number of elements in a set which is the union of more than two sets, but usually it is easier to work with Venn diagrams. For example, suppose that the data science club reports the following information about 30 of its members: 19 work part time, 17 take stats, 11 volunteer on Volunteer day, 12 work part time and take stats, 7 volunteer and work part time, 5 take stats and volunteer and 2 volunteer, take stats, and work part time. Using Figure 2.10, fill in the number of elements in each subset working from the bottom of the list given in this problem to the top.

Exercise 9. (Based on Khilyuk, Chilingar, and Rieke 2005, page 37) A protect-the-bay program is trying to prevent eutrophication (excessive nutrient enrichment that produces an increasing biomass of phytoplankton and causes significant impact on water quality and marine life). To measure biologic water quality the protect-the-bay program uses mean chlorophyll concentration on the surface, mean chlorophyll concentration on the photic layer, and mean chlorophyll concentration of the water column. If each of these are ranked as high or normal, what are the possible outcomes in the sample space of biological water quality?

Exercise 10. A psychologist has some mice that come from lab A and some that come from lab B. The psychologist ran 50 mice through a maze experiment and reported the following: 25 mice were from lab A, 25 were previously trained, 20 turned left (at the first choice point), 10 were previously trained lab A mice, 4 lab A mice turned left, 15 previously trained mice turned left, and 3 previously trained lab A mice turned left. Draw an appropriate Venn diagram and determine the number of lab B mice who were not previously trained and who did not turn left. Put how many mice are in each piece of your Venn diagram and label your events clearly. Make your plot very large, so we can clearly see the numbers that you write. (Goldberg 1960, 25, problem 3.8)

Exercise 11. Persons are classified according to blood type and Rh quality by testing a blood sample for the presence of three antigens: A, B, and Rh. Blood is of type AB if it contains both antigens A and B, of type A if it contains A but not B, of type B if it contains B but not A, and of type O if it contains neither A nor B. In addition, blood is classified as Rh+ if the Rh antigen is present, and Rh- otherwise. If we let A, B, and Rh denote the sets of people whose blood contains the A, B, and Rh antigens respectively, then all people can be classified into one of the eight categories indicated using a Venn diagram with three events that intersect inside a box representing *S*. (i) Draw A Venn diagram as indicated. Make it big enough so that the different classes mentioned above are clearly put into one part of the diagram. For example, the area corresponding to A- should be clear in your Venn diagram. (ii) A laboratory technician reports the following probabilities for blood samples of people:

50% contain antigen A
52% contain antigen B
40% contain antigen Rh
20% contain both A and B
13% contain both A and Rh
15% contain both B and Rh
5% contain all three antigens

Find: (i) the proportion of type A^- persons; (ii) the proportion of O^- persons; (iii) the proportion of B^+ persons. (Based on Goldbert 1960, 22–23)

Exercise 12. A tract of land in the Alabama Piedmont contains a number of dead shortleaf pine trees, some of which had been killed by the littleleaf disease, some by the southern pine beetle, and some by fire. Suppose that out of 500 trees,

- 70 have littleleaf disease alone
- 50 have southern pine beetle alone
- 10 were killed by fire alone
- 100 were killed littleleaf disease and southern pine beetle
- 160 were killed by littleleaf disease and fire
- 90 were killed by pine beetle and fire
- 20 were killed by all three factors

What proportion of trees were killed by littleleaf disease? (Johnson 2000, chapter 7)

2.11 Chapter References

Cĭrca, Simon. 2016. *Probability for Physicists*. Springer Verlag.
Denny, Mark, and Steven Gaines. 2000. *Chance in Biology*. Princeton University Press.
Goldberg, Samuel. 1960. *Probability: An Introduction*. New York: Dover Publications, Inc.
Johnson, Evert W. 2000. *Forest Sampling Desk Reference*. CRC Press
Khilyuk, Leonid F., George V. Chilingar, and Herman H. Rieke. 2005. *Probability in Petroleum and Environmental Engineering*. Houston: Gulf Publishing Company.
Mosteller, Frederick, Robert E. K. Rourke, and George B. Thomas. 1961. *Probability and Statistics*. Addison Wesley Publishing Company.

Parzen, Emanuel. 1960. *Modern Probability Theory and Its applications*. New York: John Wiley and Sons, Inc.

Pfeiffer, Paul E. *Concepts of Probability Theory*. Second revised edition. New York: Dover Publications Inc.

Ramachandran, Kandethody, and Chris P. Tsokos. 2015. *Mathematical Statistics with Applications* in R. Elsevier.

Ross, Sheldon. 2010. *A First Course in Probability*. 8th Edition. Pearson Prentice Hall.

Siegrist, Kyle. 1997. *The Random Project*. http://www.randomservices.org/random/

Weber, Wendy, Carrie Even, and Susannah Weaver. 2015. "Odd or Even? The Addition and Complement Principles of Probability." American Statistical Association. https://www.amstat.org/ASA/Education/STEW/home.aspx (accessed December 2018).

Chapter 3

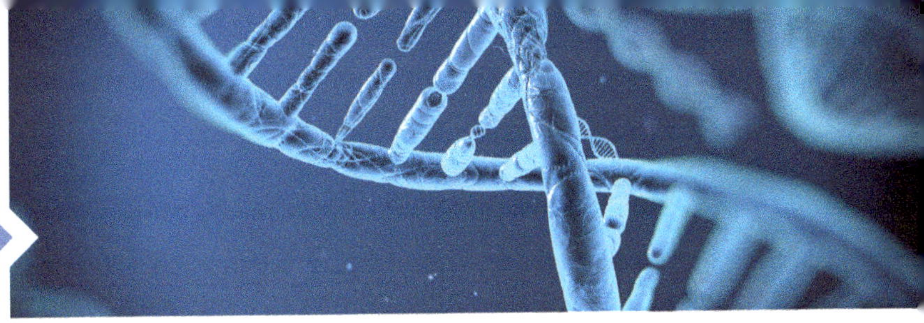

Rational Use of Probability in Data Science

> Buildings should be constructed to withstand any force of nature, but that would be clearly cost prohibitive, so a statistical approach to reasonableness in design has to be employed. Engineering is a probabilistic enterprise.
> **(Palmer 2011, 148)**

- The taxicab problem was made famous by Tversky and Kahneman (1982). These two psychologists studied the judgment of probability in people. Before you research these two scholars and their taxicab problem, think about it yourself and propose a solution plan. Revisit it again after you have studied the chapter. Then you may research other interesting puzzles posed by these authors. Here is the taxicab problem.

A cab was involved in a hit and run accident at night. Two cab companies, the Green and the Blue, operate the city. You are given the following information:

85% of the cabs in the city are Green and 15% are Blue

A witness identified the cab as blue. The court tested the reliability of the witness under the same circumstances that existed on the night of the accident and concluded that the witness correctly identified each one of the two colors 80% of the time and failed 20% of the time.

What is the probability that the cab involved in the accident was Blue rather than Green? How would you answer this question?

(Tversky and Kahneman 1982)

3.1 Modern mathematical approach to probability theory

Chapter 2 described the mathematical notions with which we may state the postulates of a mathematical model of a random phenomenon or experiment, namely an experiment, a sample space, and a collection of events. To complete a probability model we need a probability measure that we will denote by P. This probability measure must assign, to each event *A* in *S*, regardless of whether it is elementary or complex, a probability *P*(*A*). In this chapter, we talk about this probability measure and the properties that it must have to allow us to compute the probability of complex events from elementary events. The results found in this chapter are powerful and big aids in decision-making under uncertainty.

As indicated in Chapters 1 and 2, Kolmogorov gave probability an axiomatic foundation, thus making it mathematical and general enough to handle almost any problem that involves uncertainty. The axioms found in Definition 3.1.1 are attributed to him.

Definition 3.1.1

Probability is a function P defined on sets of the larger set containing all logically possible outcomes, S, such that this function satisfies Kolmogorov's axioms, which are:

1. If *A* is an event in the sample space S, $P(A) \geq 0$ *for all events A*
2. $P(S)=1$ *for the certain event S*
3. Axiom of countable additivity: If $A_1, A_2, \ldots$ .. is a collection of pairwise disjoint or mutually exclusive events, all defined on a sample space S, then

$$P(\cup_{i=1}^{\infty} A_i) = \sum_{i=1}^{\infty} P(A_i).$$

Axiom 3 is saying that the probability of the union of mutually exclusive events is the sum of their probabilities.

This axiomatic approach made it easier for people from many different backgrounds and levels of training to talk about probability. Regardless of how they obtained their probabilities (experimentation, subjective, model-based), and how they defined probability (classical definition, frequentist definition, subjective definition) the probability is a function defined on the events in the sample space that must satisfy the axioms.

The assumption of the existence of a set function *P*, defined on the events of a sample space *S*, and satisfying Axioms 1, 2, and 3, constitutes the modern mathematical approach to probability theory. For any sample space S, many different probability functions can be defined that satisfy the axioms.

Example 3.1.1 (Ross 2010)

Consider the experiment that consists of tossing a coin. Let's find a couple of probability functions and see how we do it. Let $S = \{H, T\}$. A reasonable probability function is $P(\{H\}) = P(\{T\})$. What would $P(\{H\}), P(\{T\})$ have to be for *P* to be a probability function? Since $S = \{H \cup T\}$, and $\{H\}$ and $\{T\}$ are disjoint, we have by axiom 3 and then by axiom 2, $P\{H \cup T\} = P(H) + P(T) = 1$, and therefore $P(\{H\}) = P(\{T\}) = 1/2$ is a probability function that satisfied the Axioms. Another good function is

$$P(\{H\}) = \frac{1}{3};\ P(\{T\}) = 2/3.$$

Because there can be many probability functions defined on a sample space, the task of the data scientist is to determine from incomplete data what is the probability function behind a particular experiment, i.e., generating the observed data. That is the task of the mathematical statistician and applied probability modelers. The probability function is a model. For example, when faced with a coin toss, we do not know if the coin is fair or not. By observing data we would be able to ascertain what model is the coin following: the fair coin one, or another model like the 1/3, 2/3 one.

We assume in this book that if E_i, $i = 1,2, \ldots$ is an event in S, $P(E_i)$ is defined for all the events E_i, $i = 1,2, \ldots$ of the sample space. In more advanced courses, the reader will find out that when the sample space is an uncountably infinite set, not always are the sets measurable. The complications arising from them are beyond the level of mathematics assumed in this book and need not concern us now. Books that talk about probability using the measure theoretic approach are numerous. Examples are Billingsley (1979), Roussas (2014), and Chung (1974).

It should be emphasized that one can speak of the probability of an event only if the event is a subset of a definite sample space S.

Example 3.1.2

Consider a game of dart played by throwing a dart at a board and receiving a score corresponding to the number assigned to the region in which the dart lands. The probability of the dart hitting a particular region is proportional to the area of the region. Thus, a bigger region has a higher probability of being hit.

If we make the assumption that the board is always hit, then we have

$$P(scoring\ i\ points) = \frac{area\ of\ region\ i}{area\ of\ dart\ board}.$$

The sum of the areas of the disjoint regions equals the area of the dart board. Thus the probabilities assigned to the 5 outcomes sum up to 1 and satisfies the axioms of probability. Dart games exercises appear in numerous probability textbooks, including Ross(2010). The idea of finding theoretical probabilities by calculating how much of an area is covered by random throwing of a dart is behind modern computational methods based on Markov Chain Montecarlo Simulations. The Metropolis-Hastings algorithm to compute posterior distributions in a Bayesian context is an example.

The Metropolis-Hastings algorithm and other MCMC methods allow estimation of probabilities when the probability models are very complex and can not be handled by mathematics alone.

3.1.1 Properties of a probability function

We will now show how one can derive from axioms 1 to 3 some of the important properties of probability. In particular, we will show how axiom 3 suffices to compute the probabilities of events constructed by means of complementations and unions of other events in terms of the probabilities of these other events.

If P satisfies the axioms then it will satisfy all the properties seen next. We will actually use the axioms and the very important concept of partition defined in Chapter 2 to prove some of these properties. Others will be left as exercises, but will require use of the same methodology.

Theorem 3.1.1

Let P be a probability function and let A be any event in a sample space S. Then

$$(1)\ P(\phi) = 0$$

$$(2)\ P(A^C) = 1 - P(A)$$

Proof (The proof is from Ross (2010).)

If we consider a sequence of events $E_1, E_2 \ldots\ldots$ *where* $E_1 = S, E_i = \phi, i > 1$, *then as the events are mutually exclusive and as* $S = \cup_{i=1}^{\infty} E_i$, *we have from Axiom 3 that* $P(S) = \Sigma_{i=1}^{\infty} P(E_i) = P(S) + \Sigma_{i=1}^{\infty} P(\phi) = 1$, *implying that the null event* ϕ *has probability 0 of occurring.*

It also follows, using the same argument, that for any finite sequence of mutually exclusive events $E_1, E_2, \ldots., E_n$, *the probability of the union of n events is the sum of the probabilities of each event.*

$$P\left(\cup_{i=1}^{n} E_i\right) = \sum_{i=1}^{n} P(E_i).$$

This follows from Axiom 3 by defining E_i *to be the null event for all values of i greater than n. Axiom 3 is equivalent to the equation above when the sample space is finite. However, the added generality of Axiom 3 is necessary when the sample space consists of an infinite number of points. Although not the subject of the first part of the book, this result will be meaningful in the second part.*

To prove that

$$P(A^C) = 1 - P(A).$$

Realize that since A and A^C are disjoint, then, by Axiom 3,

$P(A \cup A^C) = P(A) + P(A^C)$ but $P(A) + P(A^C) = P(S) = 1$ by Axiom 2, from where the statement given in the theorem follows.

Theorem 3.1.2

If P is a probability function and A and B are any sets, then

1. $P(B \cap A^C) = P(B) - P(A \cap B)$. The probability that "only B" happens.
2. $P(A \cup B) = P(A) + P(B) - P(A \cap B)$. The probability that A or B happens.
3. If A is contained in B, then $P(A) \leq P(B)$.

Proof

1. Make B the union of two mutually exclusive events: $B = (A \cap B) \cup (B \cap A^C)$. Then by Axiom 3,
   $P(B) = P(A \cap B) + P(B \cap A^C)$. It follows then that $P(B \cap A^C) = P(B) - P(A \cap B)$.

2. $(A \cup B) = (A \cap B^c) \cup (A \cap B) \cup (B \cap A^c)$, which are mutually exclusive events. So by Axiom 3, $P(A \cup B) = P(A \cap B^c) + P(A \cap B) + P(B \cap A^c) = P(A) - P(A \cap B) + P(A \cap B) + P(B) - P(A \cap B) = P(A) + P(B) - P(A \cap B)$. You will have noticed that we used result (1) to obtain this final result.
3. The proof of statement (3) is left as an exercise.

Corollary 1

If A and B are disjoint, $P(A \cup B) = P(A) + P(B)$.

If $E_1, E_2 \ldots \ldots E_m$ are mutually exclusive, the probability of the event consisting of their union is the sum of their probabilities.

Theorem 3.1.3

The probability of the union of n events equals the sum of the probabilities of these events taken one at a time, minus the sum of the probabilities of these events taken two at a time, plus the sum of the probabilities of these events taken three at a time, and so on.

$$P(E_1 \cup E_2 \cup \ldots\ldots \cup E_n) = P(E_1) + P(E_2) + \ldots.. + P(E_n) - \sum_{i<j} P(E_i \cap E_j)$$

$$+(-1)^{r+1} \sum_{i_1<i_2<\ldots.<i_r} P(E_1 \cap E_2 \cap \ldots.. \cap E_r) + \ldots\ldots + (-1)^{n+1} P(E_1 \cap E_2 \cap \ldots.. \cap E_n).$$

The summation $\sum_{i_1<i_2<\ldots<i_r} P(E_1 \cap E_2 \cap \ldots.. \cap E_r)$ is taken over all of the $\binom{n}{r}$ possible subsets of size r of the set {1,2,...,n}.

Example 3.1.3

A data festival has received three fancy markers from one of the sponsors. The fancy markers will be given to three students chosen at random from the five winners of the Festival: Ana (a), Betsy (b), Charles (c), Dai (d), and Ezra (e). What is the probability that both Ana and Betsy are chosen, or both Charles and Ezra are chosen, or Betsy, Charles, and Dai are chosen?

The sample space consists of 60 possible selections of the 3 winners, but each of the 10 possible sets of three individuals listed below appears 6 times in different orders, so we can think of the outcomes possible as

$$\{abc, abd, abe, acd, ace, ade, bcd, bce, bde, cde\},$$

each with probability 6/60 or 1/10.

The event "Ana and Betsy are chosen" is $A = \{abc, abd, abe\}$.
The event "Charles and Ezra are chosen" is $B = \{ace, bce, cde\}$.
The event "Betsy, Charles, and Dai are chosen" is $C = \{bcd\}$.
Because the events A, B, C are mutually exclusive,

$$P(A \cup B \cup C) = P(A) + P(B) + P(C) = 0.3 + 0.3 + 0.1 = 0.7.$$

Example 3.1.4

Suppose that on a random week day, after dinner, Rabindranath watches CNN 2/3 of the time, watches BBC 1/2 of the time, and watches both CNN and BBC 1/3 of the time. On a randomly selected weekday evening (i) What is the probability that Rabindranath watches only CNN? (ii) What is the probability that Rabindranath watches neither station?

Let A be the event that Rabindranath watches CNN. $P(A) = 2/3$.
Let B be the event that Rabindranath watches BBC. $P(B) = 1/2$.

$$P(A \cap B) = 1/3.$$

To answer question (i) we make use of Theorem 3.1.2.

$$(i)\ P(A \cap B^c) = P(A) - P(A \cap B) = 1/3.$$

To answer question (ii) we make use of Theorem 3.1.1 and 3.1.2. First we compute

$$P(A \cup B) = P(A) + P(B) - P(A \cap B) = \frac{2}{3} + \frac{1}{2} - \frac{1}{3} = \frac{5}{6} \text{ by Theorem 3.1.2.}$$

Then, using Theorem 3.1.1, we compute

$$(ii)\ P(A \cup B)^c = 1 - P(A \cup B) = \frac{1}{6}.$$

Theorem 3.1.4

Let $E_1, E_2, \ldots, E_n$, be events that form a partition of the sample space S as indicated in Figure 3.1. Let B be any event, also shown in Figure 3.1. Then

$$P(B) = P(E_1 \cap B) + P(E_2 \cap B) + \ldots + P(E_n \cap B).$$

Proof

The events $E_1 \cap B, E_2 \cap B, \ldots, E_n \cap B$ are mutually exclusive and their union is B, thus forming a partition of the event B. By axiom 3, the probability of their union is the sum of their probabilities.

3.1.2 Exercises

Exercise 1. Prove that $P(A \cup B \cup C) = P(A) + P(B) + P(C) - P(A \cap B) - P(A \cap C) - P(C \cap B) + P(A \cap B \cap C)$, but you are allowed to use only the results already seen in Section 3.1.1, as needed.

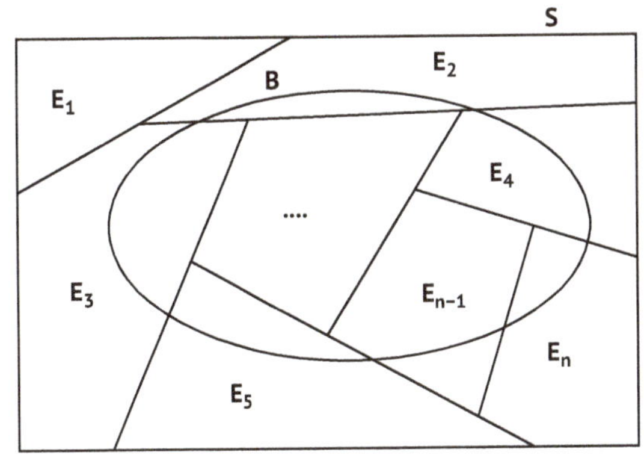

Figure 3.1 A partition of event B.

Exercise 2. Prove statement (3) of Theorem 3.1.2 in Section 3.1.1, namely that If A is contained in B, then

$$P(A) \leq P(B).$$

Exercise 3. According to the American Time Use survey, in 2016 a total of 83 percent of employed persons did some or all of their work at their workplace, and 22 percent did some or all of their work at home. Assuming that those probabilities represent the population of all workers, what percentage of employed persons used at least one of those alternatives?

Exercise 4. Prove that

$$P(E \cup F \cup G) = P(E) + P(F) + P(G) - P(E^c \cap F \cap G) - P(F^c \cap E \cap G) - P(G^c \cap F \cap E) - 2P(E \cap F \cap G).$$

Exercise 5. In a remote organization, 28.1% among the total adult population are current smokers. Moreover, 37.7% of the adult population works in a place indoors where there is no rule against smoking at work. The proportion of all adults that are current smokers and work in an indoor place with no rule against smoking at work is 5%. What is the probability that a randomly selected adult in this remote organization is a current smoker or works in a place with no ban on smoking at work?

Exercise 6. (This exercise is based on a similar problem from Ross (2010).) *The Talented Mr. Ripley* is a mystery novel by Patricia Highsmith (1955). Mr. Ripley is a very lucky killer, who gets away with two murders but sometimes gets anxious at the thought of being discovered. There is no probability mentioned in the book, but it is lurking in the thought process. Suppose Mr. Ripley thinks of these possibilities: A (the father will call me today), B (the police are on the way to arrest me), $A \cap B$ (the father calls and the police are on the way to arrest me), $A \cup B$ (at least one of A and B will happen). This person assesses: $P(A) = 0.30$, $P(B) = 0.40$, $P(A \cap B) = 0.20$, and $P(A \cup B) = 0.60$ as answers. Are Mr. Ripley's imaginary answers consistent with the axioms of probability? Why or why not? If not, which axiom is violated?

Exercise 7. (This problem is from Khilyuk, Chillingar, and Rieke (2005, page 58).) Consider an urban water-supply system. It can fail because of either the lack of water or damage of supplying pipes. On any given day, the supply system can be in one of the following two states: proper functioning (event A) or failure (event B). Reliability of the system can be defined as the probability of proper functioning on any given day $P(A)$. Based on the information presented below, one needs to evaluate the reliability of the water supply system,

$$P(A) = \text{reliability}.$$

Let $W =$ event lack of water. $P(W) = 0.014$. Let D be the event line damage $P(D) = 0.030$. IT is known that $P(D \cap W) = 0.011$. Calculate $P(A)$.

Exercise 8. A big department store offers the clients two types of payment style options. One option, A, is paying in cash, and 40% of its customers choose that; the other option, B, is paying with a credit card, which 60% of customers choose. It is possible to pay both with credit card and cash, an option employed by 10% of customers. There are many other options such as gift cards, debit cards, checks, and EBT cards. (i) What percentage of this department store's customers will use only one of the A, B options? (ii) What percentage of its customers do not use either A or B?

3.2 Calculating the probability of events when the probability of the outcomes in the sample space is known

Suppose we know, by some luck, the probabilities of each of the logically possible elementary outcomes of the experiment given in a discrete sample space, or we assume what they are as, for example, when we assume that outcomes are equally likely. If s_i is a simple outcome in the sample space S, and A is an event defined in S, then if any outcome in A occurs, the event A occurs. As a consequence, the probability of A can be found by adding the probabilities of all the simple outcomes of S that are in A, i.e.,

$$P(A) = \sum_i P(s_i \in S \mid s_i \in A) = \sum P(\textit{outcomes in } A).$$

Box 3.1

Probability of an event

The probability of any event (no matter how complicated it may be) can be computed if:

a. you can identify the individual outcomes of the sample space that are in the event.
b. you know the probability of each of the outcomes in the sample space.

If those two conditions are satisfied, then the probability of the event is the sum of the probabilities of the outcomes of S that are in the event. Sometimes we know the probabilities of the outcomes in S, sometimes we have to compute them ourselves. A book on the theory of probability like this helps you do the latter.

Example 3.2.1

After shipping five computers to users, a computer manufacturer realized that two out of the five computers were not configured properly (i.e., were defective), without knowing specifically which ones. The manufacturer allocated resources to recall only two randomly chosen computers in succession out of the five for examination.

Let s represent a defective computer and f a nondefective one. Then the sample space of this recalling experiment is

$$S = \{ss, sf, fs, ff\},$$

where ss reflects that the first computer was defective and the second computer recalled was also defective.

The event A that the second recalled computer is a defective computer contains the following outcomes:

$$A = \{ss, fs\}.$$

We were told by someone that went through the trouble of finding them out that the probabilities of each outcome in the sample space are:

$$P(ss) = 1/10;\ p(sf) = 3/10;\ P(fs) = 3/10;\ p(ff) = 3/10.$$

Then the probability that the second recalled computer is a defective computer is

$$P(A) = 1/10 + 3/10 = 4/10.$$

Example 3.2.2

A job requires that prospective employees go through a security clearance check to make sure that they qualify for a job where they must sign a nondisclosure agreement. Let s represent "passing the security clearance check" and f denote "not passing the test." In a company that narrows down the pool of qualified applicants to the three most qualified ones, the possible outcomes of the security check are

$$S = \{sss, ssf, sfs, sff, fss, fsf, ffs, fff\},$$

where sss means that all three candidates pass the security clearance check, and sff means that the first passes but the other two don't. Based on past records, the company knows that

$$P(\{sss\}) = \frac{8}{27};\ P(ssf) = P(sfs) = P(fss) = \frac{4}{27};\ P(ffs) = P(fsf) = Psff) = \frac{2}{27};\ P(fff) = \frac{1}{27}.$$

Let D be the event that two of the qualified candidates pass the security clearance check. Then

$$P(D) = \frac{4}{27} + \frac{4}{27} + \frac{4}{27} = \frac{12}{27}.$$

Example 3.2.3

A person can be born in any of the 12 months of the year. Consider four persons. Each of them can be born in any of the 12 months. In total, there are 12^4 4-tuples, each indicating the birth months of four people. If we assume each of those is equally likely to happen, the probability of each is $\frac{1}{12^4}$. What is the probability of the event that the four people were born in different months?

We first figure out how many elements are in the event. For the first person, there could be 12 months, for the second the remaining 11, for the third the remaining 10, and for the fourth the remaining 10. So in total there are (12) (11) (10) (9) possible outcomes. If we multiply 11,880 times $\frac{1}{12^4}$, we get 0.5729.

Box 3.2

Probability and odds

When you roll a fair six-sided die, the probability that a specific face will show up is 1/6, which means one out of six equally likely outcomes. We may also say that the chance is 1 in 6, or the odds are 1/5 to 1. In other words, the odds of an event A is the probability that the event will occur divided by the probability that the event will not occur:

$$\textit{Odds of A happening to not happening} = \frac{P(A)}{P(A^c)} \textit{ to } 1$$

Informal and non-probability based definition of odds

Odds are not defined properly sometimes (http://senseaboutscienceusa.org/know-the-difference-between-odds-and-probability/). In some places, you may have seen this argument: "Higher odds correspond to smaller probabilities," or "if the probability of an event is p, the chance of the event is 1 in 1/p, and the odds of the event are 1/p to 1." For example, if you play a lottery in which you choose 6 numbers from among 49 numbers, "the probability of winning the next draw's jackpot is $\frac{1}{\binom{49}{6}}$, that is, the chance is 1 in 13,983,816 and the jackpot odds are 13,983,816 to 1." (Henze and Riedwyl 1998, 14–16) One could argue that it is easier for people to understand the dimensionality of their luck using this informal approach to describe odds, but the reader should be aware that such definitions are not the probabilistic definition of odds.

3.2.1 Exercises

Exercise 1. When each outcome in the sample space is equally likely to happen we may use the classical definition of probability to calculate the probability of events. Consider an individual who seeks advice regarding one of two possible courses of action from three consultants, who are each equally likely to be wrong or right. This individual follows the recommendation of the majority. What is the probability that this individual makes the wrong decision?

Exercise 2. People these days use apps to measure the distance they run, the number of calories burned in the exercise, and other metrics. People also post this kind of information on websites where they can receive kudos for great performance. In a particular demographic group, the probabilities of doing these activities (measure or not measure, post or not post) are given by the following table, where for example, 0.07 means that the individual measures effort and posts in Runners World United. All the probabilities add to one, as these outcomes define the whole sample space for the experiment: observing posts and observing these types of measurements.

	Measure distance	Measure calories	Measure pulse	Measure effort
Posts on Facebook	0.02	0.06	0.02	0.03
Posts on Runners World united	0.2	0.2	0.01	0.07
Posts on personal web site	0.1	0.02	0.01	0.01
Posts on Your body your mind website	0.05	0.1	0.05	0.05

(i) What is the probability that an individual measures calories? (ii) What is the probability that an individual Posts in personal web site?

Exercise 3. In a population of 485 smokers, 238 of the smokers divorced and 247 of the non-smokers did not divorce. Calculate the odds of a smoker being divorced to not being divorced.

3.3 Independence of events. Product rule for joint occurrence of independent events

The notion of independent and dependent events plays a central role in probability theory and in statistics. See Definition 3.3.1.

Example 3.3.1

(This example is based on Example 2.17, page 39 in Scheaffer (1995).) A frequent flyer with lots of frequent flyer miles is considering four different countries for the next vacation: Malaysia, China, Ethiopia, and South Africa. The available frequent flyer miles allow for travel to any of these countries. The frequent flyer cannot decide, so a random draw is done by numbering the country and rolling a fair four-sided die. Let A denote the event that Malaysia or China is chosen; let B denote the event that Malaysia or Ethiopia is chosen, and let C denote the event that Malaysia is selected. Are A and B independent? Are A and C independent?

Definition 3.3.1

Let A and B be two events defined on the same sample space. The events A and B are independent if

$$P(A \cap B) = P(A)P(B)$$

A collection of events $A_1, A_2, \ldots A_n$, are independent if

$$P(A_1 \cap A_2 \cap \ldots \cap A_n) = P(A_1)P(A_2)\ldots P(A_n)$$

and any group of those sets are also independent, i.e.,

$$P(A_1 \cap A_2) = P(A_1)P(A_2)$$

$$P(A_1 \cap A_2 \cap A_3) = P(A_1)P(A_2)P(A_3)$$

and so on; that is, if n events are independent, any subset of these events are also independent.

Randomly selected means that the probability of selecting each of the countries is the same, 1/4 . Answer:

$$P(A) = P(\{1\} \cup \{2\}) = \frac{1}{4} + \frac{1}{4} = \frac{1}{2}$$

$$P(B) = P(\{1\} \cup \{3\}) = \frac{1}{4} + \frac{1}{4} = \frac{1}{2}$$

$$P(C) = P(\{1\}) = \frac{1}{4}$$

$$P(A \cap B) = P(\{1\}) = \frac{1}{4}$$

Thus, A and B are independent, because

$$P(A \cap B) = P(A)P(B)$$

But A and C are not independent because

$$P(A \cap B) \neq P(A)P(B)$$

Example 3.3.2

(This problem is from Cĭrca (2016, 15).) The spin in a quantum system can have two projections: $+1/2$ (spin "up") or $-1/2$ (spin "down"). The orientation of the spin is measured twice in a row. We make the following event assignments:

$$A = \{up\ in\ the\ first\ measurement\};\ B = \{up\ in\ the\ second\ measurement\};$$

$$C = \{both\ spins\ in\ same\ direction\}$$

The sample space for the measured pairs of orientation is:

$$S = \{uu,\ ud,\ du,\ dd\},$$

while the chosen three events correspond to its subsets

$$A = \{uu,\ ud\};\ B = \{uu, du\};\ C = \{uu, dd\}$$

If each of these pairs is equally likely, i.e., each has probability 1/4,

$$P(A) = P(B) = P(C) = \frac{2}{4} = \frac{1}{2}$$

$P(A \cap B) = P(A \cap C) = P(B \cap C) = P(\{uu\}) = 1/4$ and $P(A \cap B \cap C) = 1/4$. Since $P(A \cap B) = P(A)P(B) = P(A \cap C) = P(A)P(C) = P(B \cap C) = P(B)P(C)$, events A, B and C are pair-wise independent. But

$$P(A \cap B \cap C) = \frac{1}{4} \neq \frac{1}{8} = P(A)P(B)P(C),$$

So the events are not mutually independent.

Example 3.3.3

(This problem is from Horgan (2009, 88, example 6.3).) A computer student keeps a backup copy of the programs developed, on the hard disk or a memory stick. There is a 10% chance that the hard disk will crash but a 50% chance that the memory stick will be lost. What is the probability that the student`s work will get lost?

Let $E = \{$event that the work gets lost$\}$; $D = \{$hard disk will crash$\}$;

$$M = \{\textit{memory stick will be lost.}\}$$

The hard disk and the memory stick are independent units, hence

$$P(E) = P(D \cap M) = P(D)P(M) = (0.1)(0.5) = 0.05$$

There is a 5% chance that the student`s work will be lost. A better backup system should be considered.

Example 3.3.4

Reliability is the probability that a system will work, given probabilities about the components of this system. It applies to industry, or to standardized exams with more than one part, and many other things. There are various types of configurations of the components in a system. In a series system all the components are in series and they all have to work for the system to work. If one component fails, the system fails. A robotic tool is given a one-year guarantee after purchase. The tool contains 4 components in series, each of which has a probability 0.01 of failing during the warranty period. The tool fails if any one of the components fails. Assuming that the components fail independently, what is the probability that the system works. That is, what is the reliability of this system? What is the probability that the system fails?

Let E_i, $i = 1,...,4$ be the event that component i works. Then $P(E_i) = 0.99$, $i = 1,\ldots,4$.

Let E be the event that the system works. Then $P(E) = P(E_1 \cap E_2 \cap E_3 \cap E_4) = P(E_i)^4 = 0.960596$.

The reliability of the system is then 0.960596.

The complement event (E^C) that the system fails has then probability 0.039404.

Example 3.3.5

There are many cards problems in numerous probability books, because there are many games that use cards, and it is interesting to use probability to predict possible winnings and losses or perhaps just the outcomes. If the reader is not familiar with cards, the following image may help see what the contents of a deck of cards is (See Figure 3.2). Assuming that the deck has not been tampered with, each card has equal probability of being selected. One could define many events around this deck. For example, let A be the event "selecting a Queen" and B be the event "selecting a heart." In the context of this section, we are interested in whether A and B are independent. We use the definition.

$$P(A \cap B) = \frac{1}{52}; \quad P(A)P(B) = \left(\frac{4}{52}\right)\left(\frac{13}{52}\right) = \frac{1}{52}.$$

Because $P(A \cap B) = P(A)P(B)$ we can conclude that the events A and B are independent.

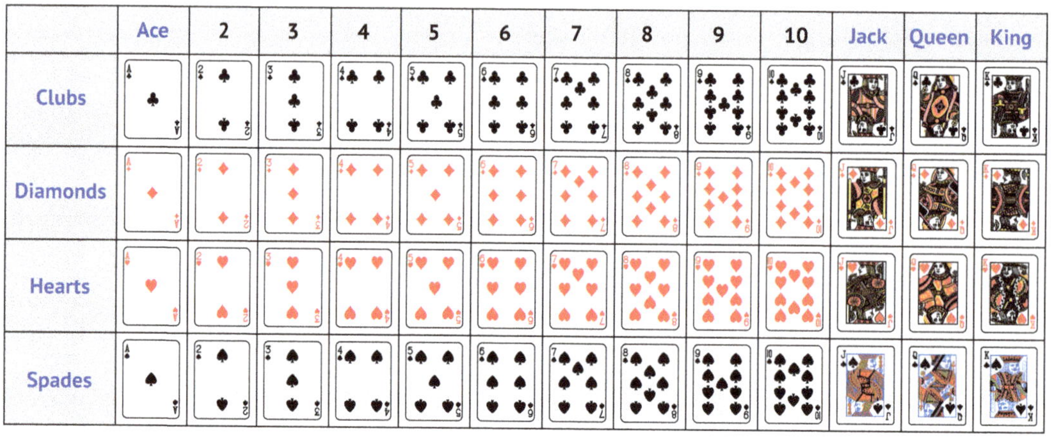

Figure 3.2 A deck of cards.

3.3.1 Exercises

Exercise 1. (This is from Parzen (1960, 90).) Consider an automobile accident on a city street in which car I stops suddenly and is hit from behind by car II. Suppose the three persons, whom we call *A'*,*B'*, and *C'*, witness the accident. Suppose the probability that each witness has correctly observed that car I stopped suddenly is estimated by having the witnesses observe a number of contrived incidents about which each is then questioned. Assume that it is found that *A'* has probability 0.9 of stating that car I stopped suddenly, *B'* has probability 0.8 of stating that car I stopped suddenly, and *C'* has probability 0.7 of stating that car I stopped suddenly. Let *A*, *B*, and *C* denote, respectively, the events that persons *A'*,*B'*, and *C'* will state that car I stopped suddenly. Assuming that *A*, *B*, and *C* are independent events, what is the probability that (i) *A'*, *B'*, and *C'* will state that car I stopped suddenly, (ii) exactly two of them will state that car I stopped suddenly?

Exercise 2. Actuarial science is the science that estimates risks of insurance and other financial endeavors. Their first preliminary probability exam is close to the level of this book. You may consider trying an online exam when you are done studying the book. These current sample exams are at https://www.soa.org/Education/Exam-Req/Syllabus-Study-Materials/edu-exam-p-online-sample.aspx One of the actuarial applications of probability is to mortality. The probability of an individual's death is important in the actuarial fields of life insurance and pensions. Life insurance pays out when a death occurs and a pension pays until death occurs. Therefore, calculating the expected present value of the liability to the insurer or pension provider requires understanding of the probabilities of death. For more information, see chapter 9 of Introduction to Actuarial and Financial Methods (Garret 2015).

To keep this problem simple, suppose an individual is alive at age 35. What is the probability that the person will die between the ages of 40 and 45? Assume that the probability of dying in a given year is 0.1 and the probability of being alive is 0.9. (Usually, in actuarial science, the probabilities vary each year or there are probabilities calculated from empirical data that tell the actuary what is the probability, for example, that a person aged 25 will live up to age 40. But we do not have those in this problem.)

Exercise 3. (This is Example 3.1.3 (applying Axiom 3) from Keeler and Steinhorst (2001).) Given recent flooding between Town *A* and Town *B*, the local telephone company is assessing the value of an independent trunk line between the two towns. The second line will fail independently of the first because it will depend on different equipment and routing (assume a regional disaster is unlikely). Under current conditions, the present line works 98 out of 100 times someone wishes to make a call. If the second line performs as well, what is the chance that a caller will be able to get through?

Exercise 4. Consider again the deck of cards of Figure 3.2. Assume it is well shuffled. A card will be dealt off the top of the deck. Let event A be "the card is the number 7" and let B be the event " the card is a club." Are these two events independent?

Exercise 5. A six-sided fair die is rolled once. Hence the probability that the number 6 turns up on the dice is $p = 1/6$. Is the probability of seeing a 6 equal to 2/6 if the die is rolled twice?

Exercise 6. A flight system consists of five radio communication devices in series. The system will work if the five components work. Let R_i represent the event that component i works, $i = 1,...,5$. The $P(R_i) = 0.87$ for all i. What is the probability that the system will work (also known as the reliability of the system?).

Exercise 7. According to an article published by the Asian Journal of Transfusion Science (Agrawal and Tiwari 2014) the four major blood ABO types are present in the following proportions in India

Type	O	B	A	AB
Proportion	0.3712	0.3226	0.2288	0.0774

Note that type AB is a separate type.

After defining what mutually exclusive blood types means for a randomly chosen person from India, find the probability that two randomly chosen individuals from India share the same blood type.

3.4 Conditional Probability

An event G could have $P(G)$, but then after observing that event B occurs that probability could change. We say that we update the probability of G after information on event B reaches us. The former $P(G)$ is the prior probability, also called total probability of G. To mark the

Definition 3.4.1

Let G and B be two events. The conditional probability of event G given that B has occurred is denoted by $P(G \mid B)$ and is defined as

$$P(G|B) = \frac{P(G \cap B)}{P(B)}$$

The symbol | means "given that"

Similarly,

$$P(B|G) = \frac{P(G \cap B)}{P(G)}$$

Visit the applet http://www.randomservices.org/random/apps/ConditionalProbabilityExperiment.html to see what these probabilities represent in the context of two events of the sample space. The grey area is the equivalent of event G. The purple area is the part of B that is in G, that is the $(B \cap G)$.

distinction between this probability and the one we will have after observing B, we have definition 3.4.1 of conditional probability of an event.

It follows from the definition of independence in section 3.3 that if G and B are independent events,

$$P(G|B) = \frac{P(G)P(B)}{P(B)} = P(G)$$

and

$$P(B|G) = \frac{P(G)P(B)}{P(G)} = P(B).$$

That is, if two events G and B are independent, the probability of G is not affected by the information contained in B and the probability of B is not affected by the information contained in G.

However, if the events are not independent, we have an alternative way of computing the joint probability of two events, namely:

$$P(G \cap B) = P(G \mid B)P(B)$$

and

$$P(G \cap B) = P(B \mid G)P(G).$$

This is the general product rule.

Example 3.4.1

In 2016–2017, approximately 32% of enrolled undergraduate students were recipients of Pell Grants (https://trends.collegeboard.org/student-aid/figures-tables/undergraduate-enrollment-and-percentage-receiving-pell-grants-over-time#Key%20Points). According to Facts and Figures at the author's institution, 34% of undergraduates at UCLA receive Pell Grants. The probability that a randomly chosen student from the US receives a Pell Grant can be approximated by 32%. But if we learn that the student is from UCLA, we should update that probability to 34%.

Let A be the event that a randomly chosen undergraduate in the US receives a Pell Grant. Let B be the event that an undergraduate student is at UCLA. Then

$$P(A) = 0.32,$$
$$P(A|B) = 0.34.$$

3.4.1 An aid: Using two-way tables of counts or proportions to visualize conditional probability

A two-way table enables us to understand in a simple way how many conditional probabilities encountered in real life are computed.

Example 3.4.2

Rossman and Short (1995) used the following example of Table 3.1 that classifies members of the 1994 US Senate according to their political party and sex. The following table summarizes their findings.

Table 3.1

	Men	Women	Row Total
Republicans	42	2	44
Democrats	51	5	56
Column Total	93	7	100

Rossman and Short asked: Is it legitimate to say that "most Democratic senators are women" and "most women senators are Democrats"? What conditional probability statements are these?

The first statement is saying $P(Women|Democrat) > P(Men \mid Democrat)$

The second statement is saying $P(Democrat|Women) > P(Republican \mid Women)$

Having translated the statements, we can use our tools to analyze the legitimacy of these statements:

$$\left[P(Women|Democrat) = \frac{5}{56}\right] < \left[P(Men|Democrat) = \frac{51}{56}\right].$$

Thus the first statement is not legitimate. It is not supported by the table.

Let us look at the second statement. We find that:

$$\left[P(democrat|Women) = \frac{5}{7}\right] > \left[P(Republican|Women) = \frac{2}{7}\right].$$

Thus, the second statement is legitimate.

The author agrees with Rossman and Short that the use of two-way tables is more conducive to the organization and interactive calculation of the appropriate conditional probabilities than trees, but nevertheless, trees are introduced in the next section.

Example 3.4.3

(This example is from Keeler and Steinhorst (2001).) Many couples take advantage of ultrasound exams to determine the sex of their baby before it is born. Some couples prefer not to know beforehand. In any case, ultrasound examination is not always accurate. About 1 in 5 predictions are wrong. In one medical group, the proportion of girls correctly identified is 9 out of 10 and the number of boys correctly identified is 3 out of 4. The proportion of girls born is 48 out of 100. What is the probability that a baby predicted to be a girl is actually a girl?

To answer this question, think about the next 1,000 births handled by this medical group. By answering the following questions, we complete Table 3.2: (i) How many of the 1,000 should be girls? How many should be boys? (ii) Of the girls, how many will the test indicate are girls? Of the boys, how many will the test indicate are girls?

Table 3.2

	Girls	Boys	Row Total
Ultrasound says girl	432	130	562
Ultrasound says boy	48	390	438
Column Total	480	520	1000

From the numbers in the table you can compute the probability requested:

$$P(girl|ultrasound\ says\ girl) = \frac{432}{562}$$

3.4.2 An aid: Tree diagrams to visualize a sequence of events

Tree diagrams are another tool for computing conditional and total probabilities. They are often useful in clarifying our thinking about sequences of events.

We will quote the detailed information about a probability tree found in Wild and Seber (2000). Look at Figure 3.3 at the same time.

> The probability written beside each line segment in the tree is the probability that the right-hand event on the line segment occurs given the occurrence of all events that have appeared along the path so far (reading from left to right). Each time a branching occurs in the tree, we want to cover all eventualities, so the probabilities besides any "fan" of line segments should add to unity.
>
> Because the probability information on a line segment is conditional on what has gone before, the order in which the tree branches should reflect the type of information that is available. Unconditional probability information should go in the first set of branches. The readily available probability information in the second branch depends (i.e., is conditional on) what happened in the first branch and so on....
>
> Rules for use are: (a) Multiply along a path to obtain the joint probability that all the events in that path occur (using the multiplication rule for conditional probabilities).
>
> Add the probabilities of all whole paths in which an event occurs to obtain the probability of that event occurring (using the addition rule for mutually exclusive events). This gives the total probability of the event of interest.

Wild and Seber (2000) use the following example to illustrate the use of trees.

Example 3.4.4

In 1992, 14% of the population of Israel was Arabic, and of those, 52% were described as living below the poverty line. On the other hand, 86% of the population of Israel was Jewish that year, and of those 11% were described as living below the poverty level. Let B represent the

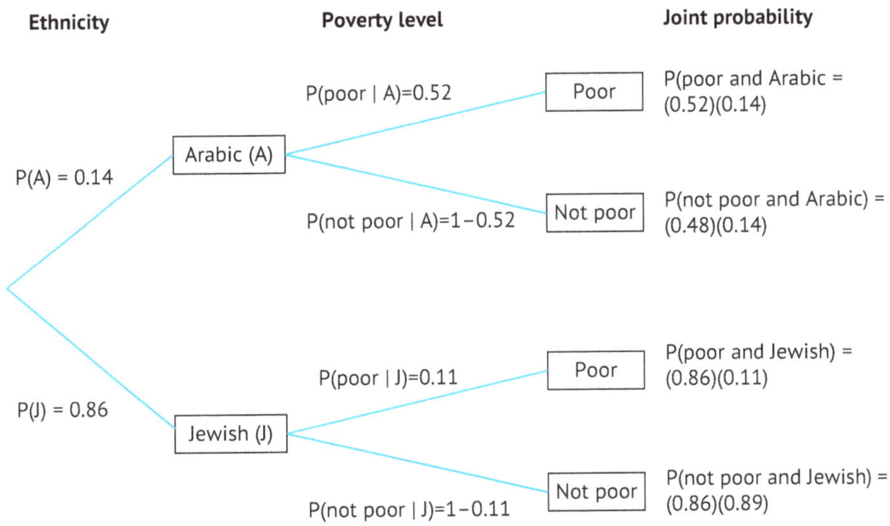

Figure 3.3 Tree diagram to represent total, conditional and joint probabilities (Based on an example from Wild and Seber, 2000).

event "poor", i.e., living below the poverty line, A denote the event "Arabic" and J denote the event "Jewish." A tree representation first splits the population of Israel first on ethnicity, because we have unconditional (or total) probabilities for this, and then on poverty, because the information on this event is conditional on ethnic group.

In the tree, the total probability of poor is obtained by adding the joint probabilities of all the branches in which the event "poor" appears. For example,

$$P(poor) = P(poor\ \ and\ \ arabic) + P(poor\ \ and\ \ Jewish) = (0.14)(0.52) + (0.86)(0.11) = 0.1674.$$

Similarly, we can extract conditional probabilities, for example

$$P(Arabic \mid poor) = \frac{P(Arabic\ \ and\ \ poor)}{P(poor)} = \frac{(0.14)(0.52)}{0.1674} = 0.4348865.$$

3.4.3 Constructing a two way table of joint probabilities from a tree

We may obtain a two-way table of probabilities from a tree or from a count table. We will illustrate this by extracting the joint probabilities from the tree in Figure 3.3 into Table 3.2.

Table 3.3

	Arabic (*A*)	**Jewish (*J*)**	**Total Probability**
Poor (*B*)	$P(A \cap B) = (0.14)\ (0.52)$	$P(J \cap B) = (0.86)\ (0.11)$	$P(B) = P(A \cap B) + P(J \cap B)$
Not poor B^C	$P(A \cap B^C) = (0.14)\ (0.48)$	$P(J \cap B^C) = (0.86)\ (0.89)$	$P(B^C) = P(A \cap B^C) + P(J \cap B^C)$
Total probability	$P(A) = P(A \cap B) + P(A \cap B^C)$	$P(J) = P(J \cap B) + P(J \cap B^C)$	1

We can see that we can calculate the same probabilities as with the tree. One thing should be clear. In a two-by-two table, the conditional probabilities do not appear directly.

Regarding the independence of the event ethnicity and poor poverty level, we can determine whether these two characteristics of the population of Israel are independent by using the definition of independence. If ethnicity and poverty level were independent, then all the joint probabilities reflected in the table should be equal to the products of the respective total probabilities. If this is true for one of the joint probabilities, then we must show that all the cells satisfy the condition. If this is not true for one cell, then we are done proving that they are not independent. For example:

$$P(\text{poor and Arabic}) = (0.14)\,(0.52) = 0.0728.$$

But

$$P(\text{poor})\,P(\text{Arabic}) = (0.14 * 0.52 + 0.86 * 0.11)(0.14 * 0.52 + 0.14 * 0.48) = 0.023436.$$

Thus, because

$$P(\text{poor and Arabic}) \neq P(\text{poor})\,P(\text{Arabic}),$$

ethnicity and poverty status in Israel in 1992 were not independent.

3.4.4 Conditional probabilities satisfy axioms of probability and have the same properties as unconditional probabilities

Since $\{A \cap B\} \subset B$, then $P(A \cap B) < P(B)$. Also, $P(A \cap B) > 0$ and $P(B) > 0$ so

$$0 \leq P(A|B) = \frac{P(A \cap B)}{P(B)} \leq 1$$

Second,

$$P(S|B) = \frac{P(S \cap B)}{P(B)} = \frac{P(B)}{P(B)} = 1$$

Third, if $A_1, A_2, \ldots.$, are mutually exclusive events, then so are $A_1|B, A_2|B, \ldots.$, and

$$P(\cup_{i=1}^{\infty} A_i|B) = \frac{P((\cup_{i=1}^{\infty} A_i) \cap B)}{P(B)} = \sum_{i=1}^{\infty} \frac{P(A_i \cap B)}{P(B)} = \sum_{i=1}^{\infty} P(A_i \mid B)$$

The above axioms imply that all the properties of probability apply to conditional probabilities.

1. $P(A|B) = 1 - P(A^c|B)$.
2. $P(A \cup C|B) = P(A|B) + P(C|B) - P((A|B) \cap (C|B))$.
3. $P(A \cap C^c|B) = P(A|B) - P(A \cap C|B)$.

3.4.5 Conditional probabilities extended to more than two events

What are we to mean by the conditional probability of the event C, given that the events A and B have occurred, denoted by $P(C \mid A, B)$?

$$P(C \mid A \cap B) = \frac{P(C \cap A \cap B)}{P(A \cap B)} = \frac{P(A \cap B \mid C)P(C)}{P(A \cap B \mid C)P(C) + P(A \cap B \mid C^c)P(C^c)}$$

if $P(A \cap B) > 0$.

Example 3.4.2

Cells carry the 23 pairs of chromosomes present in the human body. Each chromosome has a simple DNA molecule in it. Each DNA molecule carries several hundreds, or thousands, of genes. Chromosomes go in pairs in human beings. A pair of genes (one in each chromosome) is necessary for a trait. This makes humans "diploid." For example, the trait for hair color or breast cancer status is determined by two genes, one in each pair of chromosomes.

There are many types of the same gene. Each type is called "allele." Two alleles are needed for the trait. All the possible pairs of alleles for a trait constitute the set of genotypes for the trait. The observable characteristics of the trait that result from the genotype is the set of phenotypes.

In the 90s, when the study of genetic counseling for breast and ovarian cancer acquired importance, it was believed that the trait "breast-ovarian cancer susceptibility" is determined by a gene with two alleles in chromosome 13 or a gene with two alleles in chromosome 17. At each of the loci where these genes reside we can assume that the Mendelian model of segregation operates. And we can assume that the two genes are independent. The Mendelian model that applies at each locus is illustrated in Table 3.4.

Table 3.4

Genotype i	**(a1a1)**	**a1a2**	**a2a2**
Population probability	p^2	$2p(1-p)$	$(1-p)^2$
Penetrance:P(affected|i)	1	1	0
P(normal|i)	0	0	1
P(offspring=1|parents=			
(a1a1, a1a1)	1	0	0
(a1a1, a1a2)	1/2	1/2	0
(a1a2, a1a2)	1/4	1/2	1/4
(a1a1, a2a2)	0	1	0
(a1a2, a2a2)	0	1/2	1/2
(a2a2,a2a2)	0	0	1

A typical problem in epidemiology is to estimate the probability that a person is affected (has breast-ovarian cancer), based on genealogies of families and the Mendelian model. Notice how there are many conditional probabilities involved. In population genetics conditional probability plays a very important role.

3.4.6 Exercises

Exercise 1. At the University of California Los Angeles in 2017 there were 102,242 applicants to the freshman class. Of these, 16,456 were admitted. Of these, 6,037 enrolled. http://www.admission.ucla.edu/Prospect/Adm_fr/Frosh_Prof17.htm

If the pattern repeats for 2020, what is the probability that in 2020 an admitted student enrolls?

Exercise 2. Of the graduate students at a university, 70% are engineering students and 30% are students of other sciences. Suppose that 20% and 25% of the engineering and the "other" population, respectively, smoke cigarettes. What is the probability that a randomly selected graduate student is

a. an engineer who smokes?
b. an "other" student who smokes?
c. a smoker?

Exercise 3. (This problem is from the Society of Actuaries (2007).) A public health researcher examines the medical records of a group of 937 men who died in 1999 and discovers that 210 of the men died from causes related to heart disease. Moreover, 312 of the 937 men had at least one parent who suffered from heart disease, and, of these 312 men, 102 died from causes related to heart disease. Determine the probability that a man randomly selected from this group died of causes related to heart disease, given that neither of his parents suffered from heart disease.

Exercise 4. (This problem is inspired by Redelmeier and Yarnell (2013).) Do car crashes increase in days close to Tax Day, which is April 15th in the United States? Consider 30 million and 300 days in which the time of the year and the number of crashes were observed. In 200 of those days, there were crashes and it was close to tax day; in 10 million of those days there were no crashes and it was around tax day. In 100 of those days there were crashes and the days were not around tax day, but were days of similar characteristics to tax day otherwise. And finally, in 20 million of those days there were no crashes and no tax days. What would be the estimate of the condition probability of a crash given it is a day close to tax day?

Exercise 5. (From Khilyuk, Chilingar, and Rieke (2005, 66).) Oysters are grown on three marine farms for the purpose of pearl production. The first farm yields 20% of total production of oysters, the second yields 30%, and the third yields 50%. The share of the oyster shells containing pearls in the first farm is 5%, the second farm is 2%, and the third farm is 1%. (i) What is the probability of event A that a randomly chosen shell contains a pearl? (ii) Under the conditions presented above, a randomly chosen shell contains the pearl. What is the probability of the event that this shell is grown in the first farm?

Exercise 6. The approximately 100 million adult Americans (age 25 and over in 1985) were roughly classified by education and age as follows (Statistical Abstract of the United States 1987, 122) (The numbers in the middle are proportions of adult Americans).

		Age		
		25–35 years old	**35–55 years old**	**55–100 years old**
	None	0.01	0.02	0.05
EDUCATION	Primary	0.03	0.06	0.1
	Secondary	0.18	0.21	0.15
	College	0.07	0.08	0.04

(i) If an adult American is chosen at random, what is the probability of getting a 25–35-year-old college graduate? (ii) What is the probability that a 35–55-year-old has completed secondary education?

Exercise 7. By using the definition of conditional probability, show that

$$P(ABC) = P(A)\, P(B|C)\, P(C|AB).$$

Exercise 8. About 52% of the population of China lived in urban areas in 2012. In 2012, the upper-middle class accounted for just 14% of urban households, while the middle-middle class accounted for almost 50%. About 56% of the urban upper-middle class bought electronics and household appliances, as compared to 36% of the middle-middle class. If this continued like this in the near future, what would be the probability that a randomly chosen household in China is an upper-middle class urban person that purchases appliances and electronics? This information was obtained from https://www.mckinsey.com/industries/retail/our-insights/mapping-chinas-middle-class.

3.5 Law of total probability

There are many circumstances in which you would like to know the probability of an event, but you cannot calculate it directly. You may be able to find it if you know its probability under some conditions. The desired probability is a weighted average of the various conditional probabilities. To see how we can achieve this, consider two events B and G defined in the sample space.

$$B = (B \cap G) \cup (B \cap G^c).$$

The two events, $(B \cap G)$, $(B \cap G^c)$ are mutually exclusive. Therefore, by the third axiom,

$$P(B) = P(B \cap G) + P(B \cap G^c).$$

By results in section 3.4, we can express the joint probabilities $P(B \cap G)$, $P(B \cap G^c)$ in terms of the conditional probabilities:

$$P(B) = P(B|G)\ P(G) + P(B|G^c)\ P(G^c).$$

This is the law of total probability.

More generally, if we have a partition of the sample space into n events G_i, $i = 1, 2, \ldots, n$ then

$$P(B) = \sum_{i=1}^{n} P(B \mid G_i)P(G_i).$$

Example 3.5.1

(Horgan (2009, Example 6.5).) Enquiries to an online computer system arrive on five communication lines. The percentage of messages received through each line are:

Line	1	2	3	4	5
% received	20	30	10	15	25

From past experience, it is known that the percentage of messages exceeding 100 characters on the different lines are:

Line	1	2	3	4	5
% exceeding 100 characters	40	60	20	80	90

What is the overall proportion of messages exceeding 100 characters?

Let A be the event that the percentage of messages exceeds 100 lines. Then we want

$$P(A) = P(A \mid L1)P(L1) + P(A \mid L2)P(L2) + P(A \mid L3)P(L3) + P(A \mid L4)P(L4) + P(A \mid L5)P(L5) = 0.625.$$

3.5.1 Exercises

Exercise 1. Automobile recalls by car manufacturers are associated mostly with three defects: engine (E), brakes (B) and seats (Seats). A database of all recalls indicates that the probability of each of these defects is:

$$P(E) = 0.05,\ P(B) = 0.50 \text{ and } P(Seats) = 0.3.$$

Let R be the event that a car is recalled. It is known that $P(R|E) = 0.9$, $P(R|B) = 0.8$, $P(R \mid Seats) = 0.4$ and $P(R|\text{other defect}) = 0.4$. What is the probability that a randomly chosen automobile is recalled?

Exercise 2. (This is based on a problem by Pfeiffer (1978, 51).) A noisy communication channel transmits a signal which consists of a binary coded message; i.e., the signal is a sequence of 0's and 1's. Noise acts upon one transmitted symbol at a time. Let A be the event that a 1 is sent and B the event that a 1 is received at a certain time. The following is known.

$$P(A) = p, \quad P(B^c|A) = p_1, \quad P(B|A^c) = p_2$$

What is the probability that there is an error?

Exercise 3. (Chowdhury, Flentje, and Bhattacharya 2010, page 132). An earth dam may fail due to one of three causes, namely: (a) overtopping; (b) slope failure; (c) piping and subsurface erosion. The probabilities of failure due to these causes are respectively 0.7, 0.1, and 0.2. The probability that overtopping will occur within the life of the dam is 10^{-5}. The probability that slope failure will take place is 10^{-4} and the probability that piping and subsurface erosion will occur is 10^{-3}. What is the probability of failure of the dam, assuming that there are no other phenomena which can cause failure?

3.6 Bayes theorem

▶ You are asked to identify the source of a defective part. You know that the part came from one of three factories. Factory A produces 60% of the parts, factory B 30% and factory C 10% of the parts. It is known that 10% of the parts produced by factory A are defective, 30% of the parts produced by factory B are defective, and 40% of the parts produced by machine C are defective. Where did the defective part come from? What do you think? Write your conclusion down on a piece of paper.

To evaluate your conclusion, do the following exercise: imagine that there are 100 parts. Consider Table 3.5.

Table 3.5 To be completed by the reader.

	Defective	Not Defective	Row Total
Factory A			
Factory B			
Factory C			
Column total			

To guide your work completing the table, answer the following questions: (i) Of every 100 parts produced, how many were made by Factory A, B and C? Fill these in as the row totals of the table. (ii) Of those parts produced by factory A, how many would you expect to be defective? Repeat for Factories B and C, recording your results in the "Defective" column. (iii) How many of the total of 100 parts in your table are defective? Enter the result as the column total for the "Defective Column" (iv) Of the number of parts expected to be defective, what proportion were produced by Factory A, by Factory B, by Factory C?

3.6.1 Bayes Theorem

Let G and B be two events in S. By the definition of conditional probability in section 3.4, it follows that

$$P(G|B)\,P(B) = P(B|G)\,P(G).$$

And, from this result, it follows that

$$P(G|B) = \frac{P(B|G)P(G)}{P(B)},$$

where P(G|B) is called the posterior probability of G, P(G) is the prior probability of G when written in a Bayes rule formula, P(B|G) is the known probability of B after seeing G. P(B) is the total probability of B, calculated as indicated in Section 3.5.

It also follows that

$$P(B|G) = \frac{P(G|B)P(B)}{P(G)},$$

where now P(B|G) is the posterior probability of B given G, and P(B) is the prior probability.

These last two results are called Bayes Theorem and reflect the relation between conditional probabilities. If we know P(B|G) we can obtain the probability of P(G|B) with Bayes theorem. and vice versa. Bayes rule indicates how probabilities change in light of evidence. The author of Bayes Theorem was Thomas Bayes (1702–1761), a mathematician and minister who published little in mathematics, but what he wrote has been very significant in numerous decision making problems. One area where conditional probability and Bayes Theorem play a very important role is criminology. Bayesian filtering, or recalculating the probability of something given new information, plays a very important role in DNA processing and solving crimes.

The conditional probability applet found at http://www.randomservices.org/random/apps/ConditionalProbabilityExperiment.html illustrates Bayes theorem using Venn diagrams. The purple area divided by the area of the event marked in color grey is the conditional probability.

Example 3.6.1

As an example of application of Bayes theorem and conditional probability in Criminology, the reader is encouraged to complete the activity at

https://education.ti.com/en/activity/detail?id=510617F0CEE24132860AC2E01779C503

In this activity, suspects are narrowed down using Bayes theorem to update the probability of their guilt given new information.

Example 3.6.2

Bayes theorem is the basis for many filtering programs, most notably those that filter spam from our e-mail inboxes. Spam is unsolicited commercial e-mail. Significance, a statistical magazine published at the time by the Royal Statistical Society of the United Kingdom, reported in an article written by Joshua Goodman and David Heckerman (Goodman and Heckerman, 2004) that about 50% of the mail on the internet at the time was spam. Given that it costs very little to send spam, even a tiny response rate makes spam economically viable. It pays to familiarize yourself with the problems and solutions associated with spam mail, because you may end up paying a high cost. One of the most popular techniques for stopping spam is Bayesian spam filtering. Many mail clients incorporate a Bayesian spam filter today. The mentioned article explains how it works, at a basic level.

We present in this example a simplified version of a Bayesian spam filter.

A department in a major company keeps all emails received by employees. The first month that the company did this, there were 10000 emails. The IT person concluded that during that month:

- 90% of the emails that are spam contained the word sex in the subject field
- 7% of the emails that were not spam contained the word sex in the subject field
- 20% of the emails received were spam

This information is called in machine learning a "training set." It can be used to create a filter that would automatically decide which emails should not be allowed to enter the server the next month. The filter would operate according to the following rule: if the probability that an email that contains the word sex in the subject field is spam is larger than the probability that this email is not spam, reject the email. That seems like a good rule. However, we do not know those probabilities. They must be calculated somehow. Bayesian spam filtering offers a methodology for that.

$$P(spam|word\ sex) = \frac{P(word\ sex|spam\ message)P(spam)}{P(word\ sex)} = 0.7627,$$

where, using the law of total probability,

$$P(word\ sex) = P(word\ sex|spam)P(spam) + P(word\ sex|no\ spam)P(no\ spam).$$

Most messages contain many words. Actual spam filters, compute many conditional probabilities:

$$P(sex \cap click \cap \ldots \cap other\ words|spam\ message)$$
$$= P(word\ sex|spam)P(word\ click|spam)\ldots..P(other\ word\,|\,spam).$$

This computation is an expression of what is called conditional independence, an assumption that is much criticized in spam filtering, but is widely used. It would be more realistic to assume that the words are not conditionally independent, but that would complicate the analysis.

As we said, these probabilities are learned from what is called a "training set" of messages, where the number of times the word appears in all spam messages is counted and divided by the total number of spam messages). This is a common machine learning technique for creating models that help us predict whether a new message is spam or not.

You should read the article mentioned above, for more details. Naïve Bayes filtering is very widely used.

Example 3.6.3

Pittman (1993) starts discussion of conditional probability with the following example:

If you bet that 2 or more heads will appear in 3 tosses of a fair coin, you are more likely to win the bet given the first toss lands heads than given the first toss lands tails. Why?

If there are at least two heads, then the following event A happens:

$$A = \{2 \text{ or more heads in 3 tosses}\} = \{HHH, HHT, HTH, THH\}$$

and $P(A) = 4/8 = 1/2$, if we assume that all outcomes are equally likely. But, given that the first toss lands heads (say W), which happens 4 out of 8 times, event A occurs if there is at least one head in the next two tosses, with a chance of 3/4. So it is said that the conditional probability of A given W is 3/4. The mathematical notation for the conditional probability of A given W is $P(A|W)$, read "P of A given W." In the present example,

$$P(A|W) = 3/4.$$

because W = {HHH, HHT, HTH, HTT} can occur in 4 way, and just 3 of these outcomes make A occur. These three outcomes define the event {HHH, HHT, HTH} which is the intersection of A and W, denoted A and W, $A \cap W$ or simply AW Similarly, if the event W^C = "First toss lands tails" occurs, event A happens only if the next two tosses land heads, with probability 1/4. So

$$P(A|W^C) = 1/4.$$

Conditional probabilities can be defined as follows in any setting with equally likely outcomes:

$$\text{Counting formula for } P(A|B) = \frac{\text{number of outcomes in B that are also in A}}{\text{total number of outcomes in B}}.$$

For a finite set S of equally likely outcomes, and events A and B represented by subsets of S, the conditional probability of A given B is the number of outcomes in both A and B divided by the number of outcomes in B. But this is true only in that setting.

Example 3.6.4

Bayes theorem is widely used by statisticians in forensic science to identify drug traces, fiber matching from clothes and guilt of suspects in light of evidence. In statistics, the given conditional probabilities are obtained from data, and are called the likelihood function. The following example is a forensic application.

Murder convictions in a country are equally likely to have been committed wearing Nikeia shoes or not. Murders periodically occur. It is known that 10% of the alleged murders committed by alleged suspects wearing Nikeia shoes end up in a conviction and that 15% of the murders committed by those not wearing Nikeia shoes end up in conviction. If a randomly chosen murder suspect is convicted, what is the probability that this person was wearing Nikeia shoes?

Let I denote the suspects that wear Nike shoes and II those who do not wear Nike shoes and let C denote the event that a suspect is convicted of murder.

$$P(I) = 0.5 = P(II),$$
$$P(C|I) = 0.1,$$
$$P(C|II) = 0.15.$$

By the law of total probability,

$$P(C) = P(CI) + P(CII) = (0.1)\ (0.5) + (0.15)\ (0.5) = 0.125.$$

By Bayes Theorem,

$$P(I|C) = \frac{P(C|I)P(I)}{P(C)} = \frac{(0.1)(0.5)}{0.125} = 0.4\,.$$

The probability that a convicted suspect was wearing Nikeia shoes is 0.4.

Example 3.6.5

(This example is from Berry 1996, 150).) Legal cases of disputed paternity in many countries are resolved using blood tests. Laboratories make genetic determinations concerning the mother, father, and alleged child. Most labs apply Bayes' rule in communicating the testing results. They calculate the probability that the alleged father is in fact the child's father given the genetic evidence.

Suppose you are on a jury considering a paternity suit brought by Suzy Smith's mother against Al Edged. The following is part of the background information. Suzy's mother has blood type *O* and Al Edged is type *AB*. All probability calculations are done conditional on this information.

You have other information as well. You hear testimony concerning whether Al Edged and Suzy's mother had sexual intercourse during the time that conception could have occurred, about the timing and frequency of such intercourse, about Al Edged's fertility, about the possibility that someone else is the father, and so on. You put all this information together in assessing the probability that Al is Suzy's father.

The evidence of interest is Suzy's blood type. If it is *O*, then Al Edged is excluded from paternity-he is not the father, unless there has been a gene mutation or a laboratory error. Suzy's blood type turns out to be *B*; call this event *B*. According to Bayes' rule, if *F* is the event that Al Edged is the father,

$$\mathrm{P(F|B)} = \frac{\mathrm{P(B|F)P(F)}}{\mathrm{P(B|F)P(F)} + \mathrm{P(B|F^c)(1 - P(F))}}$$

According to Mendelian genetics, $P(B|F) = 1/2$. The blood bank's estimate of $P(B|F^C) = 0.09$ as the proportion of B genes to the total number of ABO genes in their previous cases. A typical value among Caucasians is 9%. So

$$P(F|B) = \frac{0.5P(F)}{0.5P(F) + 0.09(1 - P(F))}$$

The P(F) comes from all the other, nonblood, evidence. Here are the possible values of the posterior probability P(F|B) under different assumptions for P(F).

P(F)	0	0.100	0.250	0.500	0.750	0.900	1.000
P(F|B)	0	0.382	0.649	0.847	0.943	0.980	1.000

The reason such a large increase is possible is that Suzy's paternal gene B is relatively rare.

Blood banks and other laboratories that analyze genetic factors in paternity cases have a name for the Bayes factor in favor of F: Paternity Index.

Example 3.6.6 Bayesian methods in Astronomy.

Astronomy is a data-driven science with an unprecedented amount of data. It is believed that Bayesian classifiers on large astronomy data sets will be a driving force for astronomy in the 21st century.

Bayesian classifiers are able to predict class membership probabilities, such as the probability that a given galaxy belongs to a particular morphological type. Such classification schemes, based on posterior probabilities are a better fit with the reality of the objects we encounter in astronomy since we expect a continuous and fuzzy transition between galaxy types rather than a closed definition.

Moreover, additional prior information is important in determining class divisions –as in star formation activity and galaxy colour, for example. Conversely, as more data become available, previous classification schemes can be updated, like the recent case of Pluto, which was downgraded to the category of dwarf planet.

(De Souza and Ishida 2014, 60)

Box 3.3

Big data, rare events

The idea reflected in Bayes theorem, that prior probabilities may not be ignored is very important in the current research on the human genome. As Dimitry A. Kondrashov puts it:

> *Very often, modern biomedical research involves digging through a large amount of information, like an entire human genome, in search for associations between different genes and a phenotype, like a disease. It is a priori unlikely that any specific gene is linked to a given phenotype, because most genes have very specific functions, and are expressed quite selectively, only at specific times or in specific types of cells. However, publishing such studies results in splashy headlines ("Scientists find a gene linked to autism!", and so a lot of false positive results are reported, only to be publicized later, in much less publicized studies.*
> (Kondrashov 2016, 164)

Example 3.6.7

(The following example is based on a study by Hilbe and Riggs (2014), somewhat simplified.)

A near-earth object (NEO) is a comet and asteroid in orbits that allows it to enter the Earth's neighborhood. The level of hazard (H) to the earth is a function of NEO size, as indicated by crater diameters (D) data from Mars and the Moon. We have some idea of P(D|H) and of P(H) given historical records. The question of interest is: what is the probability of a specific level of Earth impact hazard H, given a detected NEO of specified diameter D. Hilbe and Rigg (2014) say that the following Bayes formula connects historical hazard level to current NEO orbit characteristics.

$$P(H|D) = \frac{P(D|H)P(H)}{P(D)} = \frac{P(D|H)P(X)P(I)}{P(D)}$$

Where P(X) is the probability that the Earth presents a viable collision cross-section to a NEO, and P(I) the probability that the orbits of the Earth and a given NEO intersect. Since I and X are independent, we multiply their probabilities to get P(H). X and I are conditions needed to have hazard.

3.6.2 Exercises

Exercise 1. (This exercise is from Bennet (1998, 3).) If a test to detect a disease whose prevalence is one in a thousand has a false positive rate of 5 percent, what is the chance that a person found to have a positive test result actually has the disease, assuming you know nothing about the person's symptoms or signs? (Bennet 1998)

Exercise 2. In Example 3.5.1, compute the probability that a message that has more than 100 characters came from line 2.

Exercise 3. A bin contains 25 light bulbs. Let G be the set of light bulbs that are in good condition and will function for at least 30 days. Let T be the set of light bulbs that are totally defective and will not light up. And let D be the set of light bulbs that are partially defective and will fail in the second day of use. If a randomly chosen bulb initially lights, what is the probability that it will be working still after a week?

Exercise 4. Two methods, A and B, are available for teaching a certain industrial skill. The failure rate is 30% for method A, and 10% for method B. Method B is more expensive, however, and hence is used only 20% of the time. Method A is used the other 80% of the time. A worker is taught the skill by one of the two methods, but he fails to learn it correctly. What is the probability that he was taught by using method A?

Exercise 5. In a US campus, 20% of data analysts use R, 30% use SPSS and 50% use SAS. 20% of the R programs run successfully as soon as they are typed, 70% of the SPSS programs run successfully as soon as they are typed and 80% of the SAS programs run successfully as soon as they are typed. (i) What is the probability that a program runs successfully as soon as it is typed? (ii) If a randomly selected program runs successfully as soon as it is typed, what is the probability that it has been written in SAS?

3.7 Mini quiz

Question 1. We are given that $P(A) = 0.3$, $P(B) = 0.7$ and $P(A \cap B) = 0.1$. Thus

a. A and B are Independent
b. A and B are mutually exclusive
c. $P(A|B) = 0.1428571$
d. $P(A|B) = P(A)$

Question 2. Which of the following sequences is most likely to result from flipping a fair coin five times?

a. HHHTT
b. THHTH
c. THTTT
d. HTHTH
e. All four sequences are equally likely

Question 3. Which of the following sequences is least likely to result from flipping a fair coin five times?

a. HHHTT
b. THHTH
c. THTTT
d. HTHTH
e. All four sequences are equally likely

Question 4. A high end neighborhood has two types of residents. It is known that 30% of residents are only architects, 50% are only runners, and 10% are both architects and runners. Let A denote the event that a resident, randomly chosen, is an architect, and let R denote the event that the resident is a runner. The $P(A^c \cap B^c)$ is

a. 0.6
b. 0.5
c. 0.2
d. 0.1

Question 5. You are given $P(A) = 0.4$ and $P(B) = 0.3$. Which of the following cannot be a possible value for $P(A \cup B)$?

a. 0.8
b. 0.3
c. 0.6
d. 0.5
e. 0.4

Question 6. The price of the stock of a very large company on each day goes up with probability p or down with probability $(1 - p)$. The changes on different days are assumed to be independent. Consider an experiment where we observe the price of the stock for three days. And consider event A that the stock price goes up the first day. What is the probability of event A?

a. p^3
b. $3(1-p)^2 + (1-p)^3$
c. $p^3 + 2p(1-p)^2 + (1-p)p^2$
d. $p^3 + 2p^2(1-p) + p(1-p)^2$

Question 7. In the Campos de Dalias in Almeria, Spain, 59.7% of the hectares in the agricultural region is covered by invernaderos (greenhouses under which most of agricultural production takes place under a very controlled and highly technological method) and the rest is based on traditional agricultural methods. 30% of the hectares in the invernadero area are dedicated to producing tomatoes. What is the probability that a randomly chosen hectare in the agriculture region of Campos de Dalias is of the invernadero type and produces tomatoes? (Junta de Andalucia 2016)

a. 0.1791
b. 0.08
c. 0.2
d. 0.12

Question 8. An incoming lot of silicon wafers is to be inspected for defectives by an engineer in a microchip manufacturing plant. Suppose that, in a tray containing twenty wafers, four are defective. Two wafers are to be selected randomly for inspection. The probability that at least one of the two is defective is

a. 0.5
b. 0.1011
c. 0.6316
d. 0.3684

Question 9. (This problem is from Castañeda et al. (2012).)
Mr. Rodrigues knows that there is a chance of 40% that the company he works with will open a branch office in Montevideo (Uruguay). If that happens, the probability that he will be appointed as the manager in that branch office is 80%. If not, the probability that Mr. Rodriguez will be promoted as a manager to another office is only 10%. Find the probability that Mr. Rodriguez will be appointed as the manager of a branch office from his company.

a. 0.62
b. 0.38
c. 0.84211
d. 0.1118

Question 10. (This problem is from Johnson (2000).)
There are three suppliers of loblolly pine seedlings. All three obtain their seed from areas in which longleaf pine is present, and, consequently, some cross-fertilization occurs forming Sonderegger pines. Let B1 represent the first supplier, B2 the second, and B3 the third. B1 supplies 20% of the needed seedlings, of which 1% are Sonderegger pines. B2 supplies 30% of which 2% is Sonderegger pines, and B3 supplies 50% of which 3% is Sonderegger pines. In this situation, what is the probability that a blindly chosen seedling will be Sonderegger pine?

a. 0.03
b. 0.023
c. 0.3
d. 0.091

3.8 R code

Some theoretical probabilities are hard to get analytically. But doing a simulation may provide some insight and very accurate answers. In this chapter's R session you will be doing a simulation to calculate probabilities of matching. In this section, we will do the probability version of the R activity in Chapter 2, but with a different application.

3.8.1 Finding probabilities of matching

Four students were talking, preparing a presentation for their capstone project. They all had the same MacBook, with no distinguishing features on the outside of it, and the four MacBooks were standing on the table, closed. Suddenly class time arrived, and they all run with a computer, randomly chosen among the four on the table. What is the probability that all students picked up their actual computer?

We will do 1,000 trials of an experiment.

- Probability model: insert four numbers 1 to 4 in a box. The numbers will be drawn without replacement.
- Trial: Select four numbers without replacement.
- What to record: whether all the numbers match the students' computers (1 = yes, 0 = no).
- What to compute: number of yes/total number of trials.

```
trials=matrix(rep(0,400000),ncol=4,nrow=100000)
matching=c(rep(0,100000))
for(i in 1:100000){
trials[i,]=sample(1:4,4, replace=F)
if(sum( trials[i,]==sort(trials[i,]) ) ==4 ) { matching[i] =1}
else {matching[i]=0}
}
table(matching) # see how many full matches
head(trials) # double check your first numbers
head(matching) # double check your first numbers
1/24 # theoretical solution
sum(matching)/100000 # simulation answer
```

3.8.2 Exercises

Exercise 1. Using code used in Lesson 2, find the probability that student 1 gets the right computer. Compare that with the theoretical probability.

Exercise 2. Modify the code given in section 3.8.1 to find the empirical probability that there are no matches. What probability does your simulation produce? Compare it with the theoretical probability.

Exercise 3. Modify the code given in section 3.8.1 to find the empirical probability that student 1 or 3 get their computer. Compare with the theoretical probability.

3.9 Chapter Exercises

Exercise 1. Prove the following theorem: If E and F are independent, then so are the following pairs of events: (a) E and F^c ; (b) E^c and F; (c) E^c and F^c.

Exercise 2. True or False, and explain

a. If A and B are independent, they must also be mutually exclusive.
b. If A and B are mutually exclusive, they cannot be independent

Exercise 3. (This exercise is based on Horgan (2009, Section 6.3).) A microprocessor chip contains is a very important component of every computer. Once in a while tech news magazines report a defect in a chip and producers have to respond by giving some indication of the damage that we should expect due to the defect. In 1994 a flaw was discovered in the Intel Pentium chip. The chip would give an incorrect result when dividing two numbers. But Intel initially announced that such an error would occur in 1 in 9 billion divides. Consequently, it did not immediately offer to replace the chip. Horgan (2010) demonstrates what a bad decision that was. She shows, using the product rule for independent events, that the probability of at least one error can be as large as 0.28 in just 3 billion divides, which is not uncommon in many computer operations.

$$P(\textit{at least } 1 \textit{ error in } 3 \textit{ billion divides}) = 1 - \left(1 - \frac{1}{9 \textit{ billion}}\right)^{3 \textit{ billion}} = 0.2834687.$$

Calculate the probability of at least one error in 5 billion divides.

Exercise 4. Application to 2008 Elections. The following are results from Election Day.

	Obama	McCain	Other candidates
White	43%	55%	2%
Black	95%	4%	1%
Hispanic	66%	31%	3%
Asian	62%	35%	3%
Other	r	s	t

[Note: We are assuming that the groups are mutually exclusive here and in what follows.] From national results, we know that 52% of the votes went to Obama, 46% to McCain, and 2%

It is also known that

- Whites made up 74% of the voters So P(white voters) = 0.74
- Blacks made up 13% of the voters P(Black voters) = 0.13
- Hispanics made up 8% of the voters P(Hispanic voters) = 0.08
- Asians made up 2% of the voters P(Asian voters) = 0.02

Find the values of r, s, and t in the table given above.

Exercise 5. (This problem is based on Society of Actuaries (2007, Problem 2).) The probability that a visit to the dentist ends up in neither X-rays nor tooth pulling out is 35%. Typically, 30% of visits with this prognosis result in the tooth being pulled out and 40% in just X-rays. Determine the probability that a visit to a dentist clinic results in both a tooth pulled out and X rays.

Exercise 6. (Society of Actuaries (2007).) You are given $P(A \cup B) = 0.7$ and $P(A \cup B^c) = 0.9$ Determine $P(A)$.

Exercise 7. (This problem is from Feller (1950, 9).) A deck of bridge cards consists of 52 cards arranged in 4 suits of 13 cards each. There are 13 face values (2, 3, ..., 10), jack, queen, king, and ace in each suit. The 4 suits are called spades, clubs, hearts and diamonds. The last two are the red, the first two are the black. Cards of the same face value are called of the same kind. Playing bridge means distributing the cards to four players, to be called North, South, East and West (or N,S, E, W, for short) so that each receives 13 cards. Playing poker, by definition, means selecting 5 cards out of the pack. What is the probability that all cards in a poker hand are different?

Exercise 8. (This problem is based on Winter and Carlton (2000, page 60).) A lie detector test is accurate 70% of the time, meaning that for 30% of the times a suspect is telling the truth, the test will conclude that the suspect is lying, and for 30% of the times a suspect is lying, the test will conclude that he or she is telling the truth. A police detective arrested a suspect who is one of only four possible perpetrators of a jewel theft. If the test result is positive, what is the probability that the arrested suspect is actually guilty? Show work.

Exercise 9. An incoming lot of cell phones is to be inspected for defects by an engineer in a cell phone manufacturing plant. Suppose that, in a tray containing twenty cell phones, four are defective (D) and sixteen are working properly (W) so that the $P(W) = 16/20$ and $P(D) = 4/20$. Two cell phones are to be selected with replacement. After listing the sample space, find the probabilities of the following events: (a) neither is defective; (b) at least one of the two are defective.

Exercise 10. An actuary studying the insurance preferences of automobile owners makes the following conclusions: (i) an automobile owner is twice as likely to purchase collision coverage as disability coverage, (ii) the event that an automobile owner purchases collision coverage is independent of the event that he or she purchases disability coverage, and (iii) the probability that an automobile owner purchases both collision and disability coverage is 0.15.

So the actuary asks: What is the probability that an automobile owner purchases neither collision nor disability coverage?

Exercise 11. A collection of 100 computer programs was examined for various types of errors (bugs). It was found that 20 of them had syntax errors, 10 had input/output (I/O) errors that were not syntactical, five had other types of errors, six programs had both syntax errors and I/O errors, three had both syntax errors and other errors, two had both I/O and other errors, while one had all three types of errors. A program is selected at random from the collection—that is, selected in such a way that each program was equally likely to be chosen. Let Y be the event that the selected program has errors in syntax, I be the event that it has I/O errors, and O the event that it has other errors. What is the probability that the randomly chosen program has some type of error?

Exercise 12. If $P(A) = 0.41$, $P(B) = 0.35$ and $P(A \cap B) = 0.1$ what is:
(i) $P(A \cup B)$; (ii) $P(A^c \cup B)$; (iii) $P(A^c \cup B^c)$; (iv) $P(A \cap B^c)$; (v) $P(A^c \cap B^c)$?

Exercise 13. Not all taxpayers want to finance new infrastructure projects. For that reason, public opinion is constantly measured by local governments in order to determine the chance of new projects implementation. Public opinion in a city regarding the opening of a car pool lane in its most congested highway is reflected in the following table.

	Yes	**No**
Center of the city	0.150	0.250
Suburbs	0.250	0.150
Rural areas	0.050	0.150

The table reflects the opinion of adults eligible to vote and is saying, for example, that 15% of the town adults eligible to vote live in the center of the city and are in favor of the car pool lane.

With this information, answer the following questions: (i) What is the probability that a randomly chosen eligible voter disapproves of the car pool lane? (ii) What is the probability that a randomly chosen eligible voter does not live in the center of the city and disapproves of the car pool lane? (iii) What is the probability that a voter from the suburbs disapproves of the car pool lane?

Exercise 14. (This problem is from Rossman and Short (1995).) Consider the case of Joseph Jamieson, who was tried in a 1987 criminal trial in Pittsburgh's Common Pleas Court on charges of raping seven women in the Shadyside district of the city over a period from April 18, 1985, to January 30, 1986. Fienberg (1990) reports that by analyzing body secretion evidence taken from the scenes of the crimes, a forensic expert concluded that the assailant had the blood characteristics and genetic markers of type B, secretor, PGM $2+1-$. She further testified that only .32% of the male population of Allegheny County had these blood characteristics and that Jamieson himself was a type B, secretor, PGM $2+1-$. The natural question to ask is how a juror should update the probability of Jamieson's guilt in light of this quantitative forensic evidence (event E). Note that, in this case, $Pr(E|G) = 1$ and $Pr(E|\text{not } G) = .0032$, since if Jamieson did not commit the crimes, then some other male in Allegheny County presumably did. Plugging these into Bayes' Theorem as presented above and simplifying leads to the expression

$$P(G \mid E) = \frac{P(E \mid G)P(G)}{P(E|G)P(G) + P(E|G^c)P(G^c)}$$

where Pr(G) represents the juror's subjective assessment of Jamieson's guilt prior to hearing the forensic evidence. Calculate the posterior or updated probability of guilt for the following values of the prior probability. The first value is given.

Prior Probability	0.5	0.2	0.1	0.01	0.001
Updated Probability	0.9968				

Exercise 15. The US Bureau of Labor Statistics provided the data below as a summary of employment in the United States for 1989.

Classification	Total (in millions)
Civilian noninstitutionalized population	154
Civilian Labor force	102
Employed	98
Unemployed	4
Not in the labor force	52

Note: an employed and unemployed person are by definition in the labor force. See the US Bureau of Labor Statistics glossary. It is important to know the glossary to answer this problem.

Suppose that an arbitrarily selected person from the civilian noninstitutionalized population was asked, in 1989, to fill out a questionnaire on employment. Find the probability of the following events. (i) The person was in the labor force. (ii) The person was employed. (iii) The person was employed and in the labor force. (iv) The person was not in the labor for or unemployed.

Separately, answer the following question: what is the probability that a person in the labor force was employed?

Exercise 16. Here is some information about the first-year class at University of Washington in a given year of the past. (a) 50% are mostly masculine; (b) 25% have a car; (c) 60 of those with a car drive to school; (d) 40% are blonde; (e) 80% are from the state of Washington; (f) 10% are from Oregon; and (g) 5% are from California. If I pick a student at random, what is the chance that this student is from outside the state of Washington?

Exercise 17. (This problem is based on Berry and Chastain (2004).) Testosterone (T) is the naturally occurring male hormone produced primarily in the testes. Epitestosterone (E) is an inactive form of testosterone that may serve as a storage substance or precursor that gets converted to active T. The normal urinary range for the T/E ratio for any person has been set by scientists to be 6:1 (meaning that 99% of normal men will have that or lower). (i) If 1% of nonusers of testosterone as a doping agent have a urinary T/E ratio above the established normal range, what would be the probability that the test for testosterone doping is a false positive? How many of the 90,000 athletes tested annually would be accused of testosterone doping even though they did not dope?

Anti-doping screening is done to detect whether an athlete has used testosterone as a doping agent. In the context of a disease like, for example, AIDS, and a test to screen for AIDS, we define sensitivity of a test as the value of the following probability: $P(+ \mid D)$, i.e., the probability of a true positive result in the test for someone that has AIDS. We define the specificity of the test as the $P(- \mid \text{no } D)$, i.e., the probability that the person without the disease gets a negative test result. (ii) Define sensitivity and specificity in the context of the anti-doping screening.

(iii) If doping suddenly became very prevalent in the population of athletes, how would that affect the probability that an athlete that tests positive is a user? (assume the same quality of the test before and after the increase in prevalence). Note that prevalence is the proportion of the population of athletes that dopes.

Exercise 18. In 2002, a group of medical researchers reported that on average, 30 out of every 10,000 people have colorectal cancer. Of these 30 people with colorectal cancer, 18 will have a positive hemoccult test. Of the remaining 9,970 people without colorectal cancer, 400 will have a positive test. (i) If a randomly chosen person has a negative test result, what is the probability that the person is free of colorectal cancer? (ii) If it is learned that there were 2,400 patients tested, about how many should we expect to be free of colorectal cancer?

Exercise 19. Of the new homes on the market in a neighborhood of California, 21% have pools, 64% have garages, and 17% have both. (a) If you pull up a house with a garage, what is the probability that it has a pool? (ii) Are having a garage and a pool disjoint events? (iii) Are having a garage and a pool independent events?

Exercise 20. A box contains three black tickets numbered 1, 2, 3, and three white tickets numbered 1,2,3. One ticket will be drawn at random. You have to guess the number on the ticket. You catch a glimpse of the ticket as tit is drawn out of the box. You cannot make out the number but see that the ticket is black. (i) What is the chance that the number on it will be 2? (ii) The same but the ticket is white. (iii) Are color and number independent?

Exercise 21. Someone is going to toss a coin twice. If the coin lands heads on the second toss, you win a dollar. (i) If the first toss is heads, what is your chance of winning the dollar? (ii) If the first toss is tails, what is your chance of winning the dollar? (iii) Are the tosses independent?

Exercise 22. (This exercise is from Ross (2010).) A certain organism possesses a pair of each of 5 different genes (which we will denote by the first five letters of the English alphabet). Each gene appears in 2 forms (which we designate by lowercase and capital letters). The capital letter will be assumed to be the dominant gene in the sense that if an organism possesses the gene pair xX, then it will outwardly have the appearance of the X gene. For instance, if X stands for brown eyes and x for blue eyes, then an individual having either gene pair XX or Xx will have brown eyes, whereas one having gene pair xx will have blue eyes. The characteristic appearance of an organism is called its phenotype, whereas its genetic constitution is called its genotype. (Thus 2 organisms with respective genotypes aA, bB, cc, dD, ee and AA, BB, cc, DD, ee would have different genotypes but the same phenotype.) In a mating between two organisms, each one contributes, at random, one of its gene pairs of each type. The 5 contributions of an organism (one of each of the 5 types) are assumed to be independent and are also independent of the contributions of its mate. In a mating between organisms having genotypes aA, bB, cC, dD, eE and aa, bB, cc, Dd, ee what is the probability that the progeny will (i) phenotypically and (ii) genotypically resemble

- The first parent
- The second parent
- Either parent
- Neither parent.

To guide your decision, refer back to Figure 2.1 in Chapter 2.

Exercise 23.
A study of the relationship between smoking and lung cancer found that 238 individuals smoked and had lung cancer, 247 individuals smoked and had no lung cancer, 374 individuals did not smoke and had lung cancer, and 810 individuals did not smoke and did not have lung cancer. There were a total of 1,669 people randomly chosen to participate in the study. Since smoking is a risk factor for lung cancer, the epidemiology literature refers to probability as risk when associated to a risk factor. For example, the risk of lung cancer among smokers is the probability that a smoker has lung cancer. The risk of lung cancer among nonsmokers is the probability that a nonsmoker has lung cancer. The relative risk is the ratio of those probabilities. Are those (i) conditional, (ii) total, (iii) joint probabilities? Select one and calculate the risks mentioned using the information given.

Exercise 24.
A blood test for hepatitis is 90% accurate. If a patient has hepatitis, the probability that the test will be positive is 0.9 and if the patient does not have hepatitis the probability that the test is negative is 90%. The rate of hepatitis in the general population is 1 in 10,000. Jaundice is a a medical condition with yellowing of the skin or whites of the eyes, arising from excess of the pigment bilirubin and typically caused by obstruction of the bile duct, by liver disease, or by excessive breakdown of red blood cells. The physician knows that this type of patient has a probability of ½ of having hepatitis.

(i) What is the probability that a person who receives a positive blood test result actually has hepatitis? (ii) A patient is sent for a blood test because he has lost his appetite and has jaundice. If this person receives a positive test result, what is the probability that the patient has hepatitis?

Exercise 25. A concert in Vienna is paid for by Visa (25% of customers), Mastercard (10% of customers), American Express (15% of customers), Apple Pay (35%) and PayPal (15%). If we choose two persons that will attend the concert and already bought tickets, what is the probability that the two persons will have paid by PayPal?

Exercise 26. At the time of writing this book, the Brexit deal in the United Kingdom was being debated. It turns out that the United Kingdom has tried before to leave the European Union. The British National Referendum of 1975 asked whether the United Kingdom should remain part of the European Union. At that date, which was shortly after an election which the Labor Party had won, the proportion of the electorate supporting Labor (L) stood at 52%, while the proportion supporting the conservatives (C) stood at 48%. There were many opinion polls taken at the time, so we can take it as known that 55% of Labor supporters and 85% of Conservative voters intended to vote "Yes" (Y) and the remainder intended to vote "No" (N). Suppose that, knowing all this, you met someone at the time who said that she intended to vote "Yes", and you were interested in knowing which political party she supported. If the information above is all you had available, how would you determine which party this person is more likely to support?

Exercise 27. The prosecutor's fallacy is $P(A|B) = P(B|A)$. Under what conditions would that equality be true?

Exercise 28. (This exercise is based on Skorupski and Wainer (2015).) In 1992, the population of women in the United States was approximately 125 million. That year, 4,936 women were murdered. Approximately 3.5 million women are battered every year. In 1992, 1,432 women were murdered by their previous batterers. Let B be the event "woman battered by her husband, boyfriend or lover," M the event "woman murdered." What is the probability that a murdered woman was murdered by her batterer?

Exercise 29. (This problem is based on Schneiter (2012).) The following website contains an activity for K–12 students to illustrate Buffon's needle problem, as an example of geometric probabilities.

Geometric probabilities is a field in which probabilities are concerned with proportions of areas (lengths or volumes) of geometric objects under specified conditions.

Go to http://www.amstat.org/education/stew and find the activity "Exploring Geometric Probabilities with Buffon's Coin Problem," by Schneiter. Complete the student's activity pages. In addition to that, discuss the definition the author gives of "theoretical probability." Is that the only definition of probability?

Exercise 30. To do blind grading, a professor asks students to write a code on the front page of their exam and the second page. The first page will be torn off. The code must have five characters, each being one of the 26 letters of the alphabet (a–z) or any of the ten integers (0–9). The code must start with a letter. If we select a student's code at random, what is the probability that the code starts with a vowel or ends with an odd number?

3.10 Chapter References

Agrawal, A., and A. K. Tiwari, N. Mehta, P. Bhattacharya, R. Wankhede, S. Tulsiani, and S. Kamath. 2014. "ABO and Rh(D) group distribution and gene frequency; the first multicentric study in India." *Asian J. Transfus Sci* 8, no. 2 (Jul-Dec): 121–125.

Bennet, Deborah J. 1998. *Randomness*. Harvard University Press.

Berry, Donald A. 1996. *Statistics*. Duxbury Press.

Berry, Donald A. and Chastain, LeeAnn. 2004. "Inference About Testosterone Abuse Among Athletes." *Chance* 17, no. 2.

Billingsley, Patrick. 1979. *Probability and Measure*. Anniversary Edition. Wiley.

Castañeda, Liliana Blanco, Arunachalam Viswanathan, and Delvamuthu Dharmaraja. 2012. Introduction to Probability and Stochastic Processes with Applications.Wiley.

Chowdhury, Robin, Phil Flentje, and Gautam Bhattacharya. 2009. *Geotechnical Slope Analysis*. CRC Press.

Cĭrca, Simon. 2016. *Probability for Physicists*. Springer Verlag.

Chung, Kai Lai. 1974. *A Course in Probability Theory*. Second Edition. Academic Press.

De Souza, Rafael, and Emille Ishida. 2014. "Making sense of massive unknowns." *Significance* 11, no. 5. https://rss.onlinelibrary.wiley.com/doi/full/10.1111/j.1740-9713.2014.00785.x

Feller, William. 1950. *An Introduction to Probability Theory and its Applications*. Wiley and Sons.

Garret, S. J. 2015. *Introduction to Actuarial and Financial Mathematical Methods*. Elsevier.

Goodman, Joshua, and David Heckerman. 2004. "Fighting Spam with Statistics." *Significance*. https://rss.onlinelibrary.wiley.com/doi/10.1111/j.1740-9713.2004.021.x

Henze, Norbert, and Hans Riedwyl. 1998. *How to Win More: Strategies for Increasing a Lottery Win*. A.K. Peters, Massachusetts

Hilbe, Joseph M., and Jamie Riggs. 2014. "Will this century see a devastating meteor strike?" *Significance* 11, no. 5. https://rss.onlinelibrary.wiley.com/doi/full/10.1111/j.1740-9713.2014.00785.x

Horgan, Jane M. 2009. *Probability with R*. Wiley.

Johnson, Evert W. 2000. *Forest Sampling Desk Reference*. CRC Press

Junta de Andalucia. 2016. Estrategia de Gestion de Restos Vegetales en la horticultura de Andalucia. Hacia una Economia Circular.

Keeler, Carolyn, and Kirk Stenhorst. 2001. "A New Approach to Learning Probability in the First Statistics Course." *Journal of Statistics Education* 9, no. 3.

Khilyuk, Leonid F., George V. Chillingar, and Herman H. Rieke. 2005. *Probability in Petroleum and Environmental Engineering*. Elsevier.

Kondrashov, Dimitry A. 2016. *Quantifying Life: A Symbiosis of Computation, Mathematics, and Biology*. University of Chicago Press.

Palmer, David. 2011. "Will the big machine arrive? Engineering the Miami Tunnel." *Significance* 8, no. 4 (December).

Parzen, Emanuel. 1960. *Modern Probability Theory and Its Applications*. New York: John Wiley and Sons, Inc.

Pfeiffer, Paul E. *Concepts of Probability Theory*. Second revised edition. New York: Dover Publications Inc.

Pittman, Jim. 1993. *Probability*. Springer Texts in Statistics.

Redelmeier, Donald A., and Christopher J. Yarnell. 2013. "Can Tax Deadlines Cause Fatal Mistakes?" *Chance* 26, no. 2 (April): 8–14.

Ross, Sheldon. 2010. *A First Course in Probability*. 8th Edition. Pearson Prentice Hall.

Rossman, Allan J., and Thomas H. Short. 1995. "Conditional Probability and Education Reform:. Are They Compatible?" *Journal of Statistics Education* 3, no. 2.

Roussas, George G. 2014. *Measure-Theoretic Probability.* Second Edition. Academic Press.

Scheaffer, Richard L. 1995. *Introduction to Probability and Its Applications*, Second edition. Duxbury Press.

Schneiter, Kady. 2012. Exploring Geometric Probabilities with Buffon's Coin Problem. American Statistical Association. http://www.amstat.org/education/stew/

Skorupski, William P., and Howard Wainer. 2015. "The Bayesian flip: Correcting the prosecutor's fallacy." *Significance* 12, no. 4 (August): 16–20.

Society of Actuaries/Casualty Actuarial Society. 2007. Exam P Probability. Exam P Sample Questions. P-09-07. https://www.soa.org/Files/Edu/edu-exam-p-sample-sol.pdf

Tversky, Amos, and Daniel Kahneman. 1982. "Evidential Impact of Base Rates"" In *Judgment Under Uncertainty: Heuristics and Biases by Daniel* Kahneman, Paul Slovic, and Amos Tversky. Cambridge University Press.

Wild, Christopher J., and George A. Seber 2000. *Chance Encounters. A First Course in Data Analysis and Inference.* John Wiley & Sons.

Winter, Mary Jean, and Ronald J. Carlson. 2000. *Probability Simulations.* Key Curriculum Press.

Chapter 4

Sampling and Repeated Trials

▶ Sarah, a public health researcher, wants to conduct a study to determine whether a breast cancer prevention intervention that she has designed is effective or not. Her intervention consists of providing educational programs via periodic telephone calls to women to encourage them to follow up on their breast abnormality with a doctor. In order to conduct her study, Sarah has recruited many women from medical centers located in the area targeted. All the women have a suspected breast abnormality detected either by themselves at home or by their primary care physician, and are referred to an oncologist by their primary care physician. Oncologists are medical doctors specialized in cancer diagnostic and treatment. Sarah has recruited 200 women that have a suspected breast abnormality and are in their first visit to an oncologist, which makes them eligible for her study. How should Sarah design her study in order to obtain reliable conclusions about the effectiveness of her treatment?

Figure 4.1 Breast cancer awareness symbol.

4.1 Sampling

Sampling is a way of collecting observations from a larger population to learn about that population from the information contained in the sample. It is a kind of experiment. Sampling can be done in many ways. However, it is widely accepted nowadays that the only kind of sampling from which reliable conclusions can be obtained is probability sampling. Conclusions not based on probabilistic sampling should not be considered reliable. Most of the information we have nowadays via official statistics produced by governments or polls and surveys produced by private organizations are based on probabilistic sampling. It is not coincidentally then that a basic random phenomenon with whose analysis we are concerned in probability theory is that of *finite sampling*.

In the physical and life sciences and engineering it is repeated experimentation what helps trust conclusions from experiments. It is not coincidentally either that another basic phenomenon with whose analysis we are concerned in probability theory is that of *repeated experimentation*.

Chapter 4 is intended to give you a methodology to approach applied probability problems concerned with random samples and with repeated experimentation in a formal way. This chapter presents a few models that are often used to solve a very wide array of applied probability problems in the context of sampling from populations and repeated sampling. The methods learned in this chapter may be applied to other probability problems as well.

The chapter uses some notations from combinatorial analysis as the mathematical counterpart of counting samples and repeated experimentation.

4.1.1 n-tuples

A basic tool for the construction of sample spaces in the context of sampling is the notion of an n-tuple.

Definition 4.1.1

An n-tuple $(o_1,o_2,\ldots,o_n)$ is an array of n symbols, $o_1,o_2,\ldots,o_n$ which are called, respectively, the first component, the second component, and so on, up to the nth component, of the n-tuple. The order in which the components of an n-tuple are written is of importance (and consequently one sometimes speaks of ordered n-tuples). Two n-tuples are identical if and only if they consist of the same components written in the same order. The usefulness of n-tuples derives from the fact that they are convenient devices for reporting the results of drawing of a sample of size n.

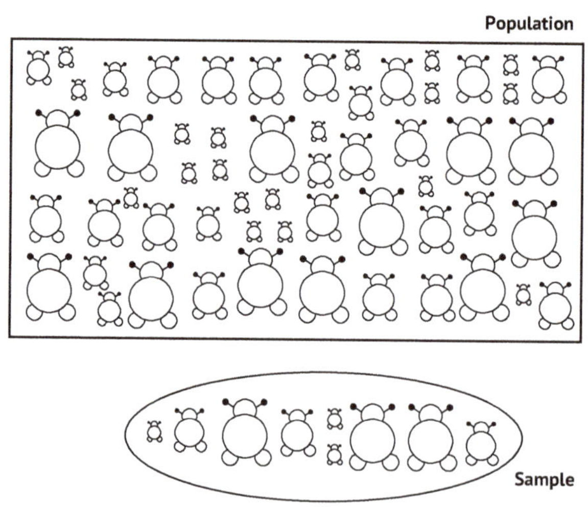

Figure 4.2 Sampling from populations results in an n-tuple. Depending on the sample drawn, the n-tuple will be different.

Example 4.1.1

Rolling once a red, a green, and a blue six-sided fair dice results in a 3-tuple (o_1, o_2, o_3). One possible outcome is (4, 2, 3), i.e., a 4 for the red die, a 2 for the green die, and a 3 for the blue die. Another logical outcome could be (6,4,5), i.e., 6 for the red, 4 for the green, and a 5 for the blue.

Example 4.1.2

Drawing the five winning numbers from a lottery consisting of 52 possible number results in a 5-tuple $(o_1, o_2, o_3\, o_4, o_5)$. One possible outcome is (2,16,45,13,9).

4.1.2 A prototype model for sampling from a finite population

Statisticians talk about populations. In probability books, the equivalent concept is an *urn with numbered balls* as a prototype for a population. In fact, when sampling from populations, it is customary to number the population and pretend the population is an urn from which we are drawing the sample.

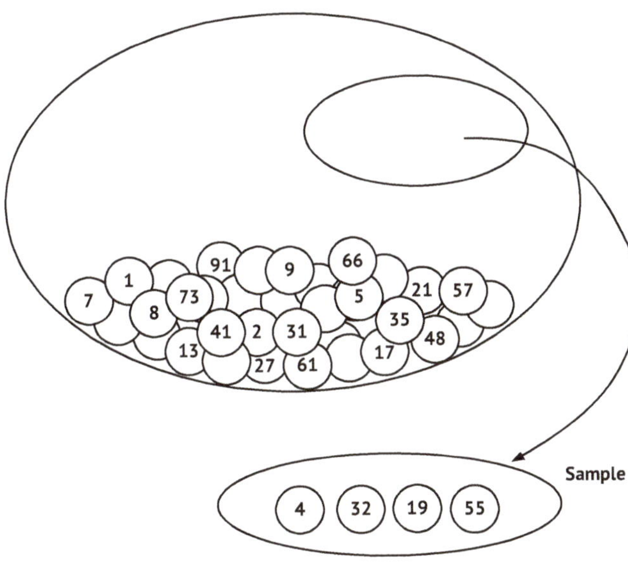

Figure 4.3 An urn model helps visualize the process of drawing random samples from populations.

Definition 4.1.2

Suppose we have an urn containing M balls, which are numbered 1 to M. Suppose we draw a ball from the urn one at a time, until n balls have been drawn. For brevity, we say we have drawn a *sample (or an ordered sample) of size n*. Of course, we must also specify whether the sample has been drawn *with replacement or without replacement.*

To report the result of drawing a sample of size n, an *n-tuple* $(o_1, o_1, \dots ..o_n)$ is used, in which o_1 represents the number of the ball drawn on the first draw, and so on, up to o_n, which represents the number of the ball drawn in the *nth* draw.

Definition 4.1.3

The drawing is said to be done *with replacement*, and the sample is said to be drawn with replacement, if after each draw the number of the ball drawn is recorded, but the ball itself is returned to the urn. If the drawing is done with replacement, the number of samples (of n-tuples) that one can draw equals

$$M^n.$$

This is because for the first draw there are M numbers, for the second, there are also M numbers, and so on. If the drawing is done with replacement, the size of the sample, n, can be any number.

Sampling with replacement is equivalent to not changing the conditions of the experiment.

Statisticians do surveys and they sample from large populations in order to learn about the population from the sample. Surveys use sampling without replacement. A simple random sample is a sample drawn without replacement. However, if we make the assumption that the population

Box 4.1

Sampling with or without replacement

When drawing n balls (objects) from an urn (a population) containing M balls (objects of study) with replacement, there are

$$M^n$$

possible ordered samples and, when doing the same operation without replacement, there are

$$(M)_n = M(M-1)\dots(M-n+1) = \frac{M!}{(M-n)!}$$

Definition 4.1.4

The drawing is said to be done *without replacement*, and the sample is said to be drawn without replacement, if the ball drawn is not returned to the urn after each draw, so that the number of balls available in the urn for the kth draw is $M - K + 1$. In this case, the number of samples (of n-tuples) that one can draw equals $M(M - 1) \dots . M - n + 1$. Various notations have been used to denote this product. Suffice to notice three of them:

$$(M)_n = M(M-1)\dots(M-n+1) = \frac{M!}{(M-n)!}$$

is infinite, then this process is equivalent to drawing with replacement. Statisticians have come up with "finite population corrections" when this assumption is not valid, however. There is a whole area called Sampling Theory that deals with these issues.

Box 4.2

Math tidbit

If M = 7 and n = 3,

$$M! = 7\times6\times5\times4\times3\times2\times1$$

Think of this as the number of ways in which a race with seven horses could end.

$$\frac{M!}{(M-n)!} = \frac{7\times6\times5\times4\times3\times2\times1}{4\times3\times2\times1} = M(M-1)\dots(M-n+1).$$

Think of this as the number of ways in which we could select a first-prize winner ($100,000), second-prize winner ($50,000), and third-prize winner ($25,000) from seven contestants. A convenient notation is (M_n) but there could be other notations for the same thing.

Finally,

$$\frac{M!}{(M-n)!n!} = \frac{7\times6\times5\times4\times3\times2\times1}{(4\times3\times2\times1)\times(3\times2\times1)} = \frac{M(M-1)\dots(M-n+1)}{n!}$$

A convenient notation for that is $\begin{pmatrix} M \\ n \end{pmatrix}$.

is the number of ways in which we could select three people at random to get a free movie pass to the latest summer blockbuster. The three are getting the same prize. (Albert, Aakifah, Sidharta) is the same as (Aakifah, Albert, Sidharta) and (Sidharta, Albert, Aakifah) are three of the 6 ways we can order these three names.

Example 4.1.3 Sampling without replacement

All 24 samples of size n = 3 from an urn containing M = 4 balls, numbered 1 to 4, if we draw *without replacement*, can be found.

An *n-tuple* $(o_1, o_2, \ldots, o_n)$ in this case, is a 3-tuple (o_1, o_2, o_3). The sample space S would be:

$$S = \{(1,2,3), (1,2,4), (1,3,2), (1,3,4), (1,4,2), (1,4,3), (2,3,1), (2,3,4), (2,4,1), (2,4,3), (2,1,3), (2,1,4),$$

$$(3,4,1), (3,4,2), (3,1,2), (3,1,4), (3,2,1), (3,2,4), (4,1,2), (4,1,3), (4,2,3), (4,2,1), (4,3,1), (4,3,2)\}$$

And we can see that the number of ordered 3-tuples (the number of random samples) in this sample space can be calculated, using the various notations introduced earlier,

$$(4)_3 = 4 \times 3 \times 2 = \frac{4!}{1!} = 24.$$

If we assume that each of the balls is equally likely to be chosen, which would be a reasonable assumption if the balls are well mixed and the drawing is done at random, the probability of obtaining the first number is 1/4. This leaves three equally likely to be drawn balls, given as probability for the second ball of 1/3, and by the same token, the third ball has probability 1/2. The probability of each of the 24 3-tuples is $(1/4)(1/3)(1/2) = 1/24 = 0.0416667$, and we can see that each of the samples in the sample space would have the same probability. When we multiply 24 by (1/24) we get 1, as we should. Recall that P(S) = 1, by axiom.

With this information, we can put the concepts learned so far at work to find probabilities of events.

We said in Chapter 3, Section 3.2, that the probability of an event is the sum of the probabilities of the outcomes in the event. Let *A* be the event that the numbers 1,2,3 are in the sample. What is the probability of *A*? We observe that there are six samples with the number (1,2,3), each sample with probability 1/24. So the probability is 6(1/24) = 1/4 = 0.25.

We also said in Chapter 1, Section 1.3, that when the outcomes are equally likely, we can just calculate the probability alternatively by counting the number of favorable outcomes and dividing by the total number of outcomes in the sample space. This result can also be seen as the number of ways the numbers 1,2,3 can be ordered (3!) divided by the total number of samples in the sample space, or

$$P(A) = \frac{3!}{24} = 6(1/24) = 1/4 = 0.25.$$

Example 4.1.4

If the sampling is done with replacement, then the 64 samples of size 3 from an urn containing 4 balls, numbered 1 to 4, can also be found.

$$S = \{(1,1,1), (2,2,2), (3,3,3), (4,4,4), (1,1,2), (2,1,1), (1,2,1), (1,1,3), (1,3,1)(3,1,1), (1,1,4),$$

$$(1,4,1), (4,1,1), (2,2,1)\ (1,2,2), (2,1,2), (2,2,3), (2,3,2), (3,2,2), (2,2,4), (2,4,2), (4,2,2),$$

$$(3,3,1), (3,1,3), (1,3,3), (3,3,2), (3,2,3), (2,3,3), (3,3,4), (3,4,3), (4,3,3), (4,4,1), (4,1,4), (1,4,4),$$
$$(4,4,2), (4,2,4), (2,4,4), (4,4,3), (4,3,4), (3,4,4),$$

$$(1,2,3), (1,3,2), (2,1,3), (2,3,1), (3,1,2), (3,2,1),$$

$$(1,2,4), (1,4,2), (2,1,4), (2,4,1), (4,1,2),(4,2,1)$$

$$(1,3,4), (1,4,3),(3,1,4), (3,4,1), (4,1,3),(4,3,1)$$

$$(2,3,4),(2,4,3), (3,2,4),(3,4,2),(4,2,3),(4,3,2) \}.$$

We can see that the number of ordered n-tuples in this sample space can be calculated, using the notations seen earlier, as $4 \times 4 \times 4 = 4^3 = 64$ n-tuples. If we assume that each of the balls is equally likely to be chosen, which would be a reasonable assumption if the balls are well mixed and the drawing is done at random, the probability of obtaining the first number is 1/4. Because the ball is put back in the urn, the probability of the second number is 1/4, and, by the same token, the number in the fourth ball has probability 1/4. The probability of each of the 64 3-tuples is $(1/4)(1/4)(1/4) = 1/64 = 0.015625$, and we can see that each of the samples in the sample space would have the same probability. Again, we check that $64 \times 0.015625 = 1$, because $P(S)$ must be 1, by axiom.

With this information, we can put the concepts learned so far to work to find probabilities of events, like we did in Example 4.1.3.

We said in Chapter 3 that the probability of an event is the sum of the probabilities of the outcomes in the event.

Let A be the event that the numbers 1,2,3 are in the sample. What is the probability of A now that we are sampling with replacement? We observe that there are six samples with the numbers (1,2,3), each sample with probability 1/64. So the probability is 6(1/64) = 0.09375.

$$P(A) = \frac{3!}{64} = 6(1/64) = 0.09375.$$

4.1.3 Sets or samples?

When we introduced sets in Chapter 2, we said that the order of elements in a set does not matter. In other words, {1,2,3} is the same set as the set {1,3,2} or equal to any of the sets {3,1,2}, {3,2,1}, {2,1,3}, {2,3,1}. Thus we must distinguish between the sets of three numbers and the sample of three numbers that result in drawing 3 balls from the urn containing 4 balls with the numbers 1,2,3,4. Consider first *sampling without replacement*.

Table 4.1

Sets	Samples
{(1,2,3)}	(1,2,3),(1,3,2),(2,1,3),(2,3,1),(3,1,2),(3,2,1)
{(1.2.4)}	(1,2,4),(1,4,2),(2,1,4),(2,4,1),(4,1,2),(4,2,1)
{(2,3,4)}	(2,3,4),(2,4,3),(3,2,4),(3,4,2),(4,3,2),(4,2,3)
{(1,3,4)}	(1,3,4),(1,4,3),(3,1,4),(3,4,1),(4,1,3),(4,3,1)

We can see that the number of sets of 3 that can be obtained from an urn with 4 balls numbered 1 to 4 is, when drawing without replacement,

$$\binom{4}{3} = \frac{4\times3\times2}{3!} = \frac{4!}{3!1!} = 4.$$

Which contains more information? The set notation or the listing of the corresponding samples? To compute probabilities, the sample listing is the most informative. In practice, without concern about the probability, it depends what the sampling is done for. If the numbers in the balls have been assigned to individuals (for example, Jean Claude got number 1, Ching Ti got number 2, Francisca got number 3 and Rakiyah got number 4), and the drawing is done to decide who will be the first, second, and third to be called when needed for combat, then the samples are each representing distinct things. (1,2,3) means Jean Claude will go first if there is need for someone in combat. Next time there is need, Ching Ti will go, and so on. On the other hand, if the sample is (3,2,1), things look different for Francisca now because she will go first. In other words, the information in sample (1,2,3) is not the same as that in (3,2,1). If we just used the set notation we would be losing a lot of information.

If the drawing is done to select three people for a committee representing the school, with no particular title for any of the members in the sample, then they might as well be represented by the set, without loss of information about the content.

Returning to the combat situation. When sampling with replacement, it is possible that the same person is called to combat repeatedly. For example, (1,1,1) means that Jean Claude gets called to combat first, then the next time someone is needed he could be called again, and the next time he would be the one going as well. Sample (4,1,4) means that the first time it is Rakiyah who goes to combat, the second time is Jean Claude, and the third time is Rakiyah, again. Thus, what model to use, with or without replacement, depends on what the context is for your problem.

Regardless of the specification used, when computing probability, one must keep in mind the larger, sample specifications to compute the probabilities correctly.

Example 4.1.5

Given the collection of $4 \times 3 \times 2$ possible samples in Example 4.1.3, the number of sets can be found by dividing $4 \times 3 \times 2$ by $3 \times 2 \times 1$ or 3!, where 3! represents the number of ways in

> Box 4.3
>
> **Vietnam War draft**
>
> The Vietnam War draft lottery https://www.usatoday.com/vietnam-war/draft-picker) was sharply criticized by statisticians for not using a true probability sampling method. The birth month and day were placed in the bowl in such a way that 18-year-old men born in the last months of the year had lower draft numbers, and therefore a greater chance of being drafted than those born earlier in the year. Most of the drafted soldiers would end up fighting in the jungles of Vietnam. Starr Norton (Norton (2017)) gives a survey of some statistical analyses done of the resulting samples.
>
> https://www.tandfonline.com/doi/full/10.1080/10691898.1997.11910534
>
> The lottery has been mentioned in numerous textbooks and articles on some aspect of probability. For example, Wild and Seber (2000, 145).
>
> Knowing probability theory well helps understand the shortcomings that may transpire in some surveys and polls conducted nowadays.

Figure 4.4 Vietnam War draft lottery.

Source: https://commons.wikimedia.org/wiki/File:1969_draft_lottery_photo.jpg..

which the numbers 1,2,3 can be ordered. The notation that is usually adopted to represent this operation is the binomial coefficient

$$\frac{4\times3\times2}{3!}=\frac{(4)_3}{3!}=\frac{4!}{1!3!}=\binom{4}{3}=4\cdot$$

where $\binom{4}{3}$ is read "4 choose 3." The probability of the event A can be written in terms of this notation. That is, the probability of obtaining the set {(1,2, 3)} can be calculated as

$$P(A)=6\frac{1}{24}=\frac{1}{\binom{4}{3}}=1/4.$$

The number of sets of S of size k, multiplied by the number of samples of size k that can be drawn without replacement from a subset of size k, is equal to the number of samples of size k that can be drawn without replacement from an urn containing balls numbered 1, 2, 3, 4.

There are $\binom{4}{3}=4$ subsets of size 3 that can be formed, namely, {1,2,3},{1,2,4}, {1,3,4},{2,3,4}.

From each of these subsets one may draw, without replacement, 6 samples so that there are twenty four possible samples of size 3 to be drawn without replacement from an urn containing 4 balls.

Example 4.1.6

(This exercise is based on an exercise in Mosteller, Rourke and Thomas (1967, 69).) For a chronic disease, there are 5 standard ameliorative treatments: a, b, c, d, and e. A doctor has resources for conducting a comparative study of three of these treatments. If he chooses the three treatments for study at random from the five, what is the probability that (i) treatment a will be chosen, (ii) treatments a and b will be chosen, (iii) at least one of a and b will be chosen?

Using the urn prototype, where the urn contains the M = 5 treatments, the number of samples is

$$(5)_3 = 5 \times 4 \times 3 = 60 \textit{ samples.}$$

We may assume that each of these samples is equally likely to occur. On the other hand, there are

$$\binom{5}{3} = 10 \textit{ sets of treatments} .$$

Notice that $10 \times 6 = 60$.

(i) Let A be the event "treatment *a*" is chosen. Treatment *a* appears in $\binom{4}{2} = 6$ sets.

You may want to check that by either writing the whole collection of possible samples, or by realizing that forcing treatment *a* to be in the sample, the other two treatments can only be formed in $4 \times 3 = 12$ ways, resulting in $12/2 = 6$ different sets. So the probability of selecting treatment a is given by the number

$$P(A) = \frac{\binom{4}{2}}{\binom{5}{3}} = \frac{6}{10}.$$

(ii) Let B be the event "treatments *a* and *b* are chosen." The probability that treatments *a* and *b* are chosen, using similar reasoning as in (i), is

$$P(B) = \frac{\binom{3}{1}}{\binom{5}{3}} = \frac{3}{10}.$$

(iii) Let C be the event "*b* is chosen." The probability that at least one of *a* and *b* will be chosen has as a complement that none of the two are chosen. If *a* and *b* are not chosen, there is only one set with the other three. The probability is

$$P(A \cup C) = 1 - \frac{\binom{3}{3}}{\binom{5}{3}} = 1 - \frac{1}{10} = \frac{9}{10}.$$

Example 4.1.7

All possible samples of size $n = 3$ from a population of 30 people if we draw without replacement can be approached using the urn models. The urn contains $M = 30$ balls numbered 1 to 30. The sample space of this experiment would be too long to be enumerated. It would contain:

$$(30)_3 = 30 \times 29 \times 28 = \frac{30!}{27!} = 24360 \ samples.$$

There are $3 \times 2 \times 1 = 3! = 6$ samples with the same numbers in them. Thus there are

$$\binom{30}{3} = \frac{30!}{27!3!} = \frac{24360}{3!} = 4060 \ sets \ of \ 3 \ numbers.$$

Let A denote the event "individuals 1, 2, 3 are in the sample," The probability of obtaining individuals 1, 2, 3 is

$$P(A) = \frac{6}{24360} = \frac{1}{\binom{30}{3}} = \frac{1}{4060}.$$

Box 4.4

Sampling in statistics

Statisticians use sampling for different purposes. When sampling to conduct a survey and gather data to learn about a population, the sampling is, in fact, done without replacement. However, a common assumption is that the population is so large, that extracting a random sample from a very large population is equivalent to drawing with replacement. The methods of Mathematical Statistics define a simple random sample as one drawn without replacement.

On the other hand, statisticians use a sampling method to determine the properties of the methods used by mathematical statistics. This sampling method, the bootstrap, relies on samples actually conducted with replacement, samples taken from the sample.

Nowadays, with the large computing power in our hands, sampling is also the source of modern machine learning methods such as Bayesian estimation with Markov chain Monte Carlo methods.

Statistics and tools of data science are now used in almost any type of enquiry regarding data. Ask yourself: what is the role of sampling in my area of interest? and do some research to find out. For example, in Section 4.1.5 we illustrate how the prototype urn model of sampling has led to competing theories about the equilibrium state of a physical system in physics. Similar competing models can be found, for example, in linguistics.

4.1.4 An application of an urn model in computer science

Suppose five terminals are connected to an online computer system by attachment to one communication line. When the line is polled to find a terminal ready to transmit, there may be

0, 1,2, 3, 4, 5 terminals in the ready state. One possible sample space to describe the system state consists of n-tuples $(o_1, o_2, \ldots\ldots, o_5)$, where each o_i is either 0 (terminal i is not ready) or 1 (terminal i is ready). The sample point (0,1,1,0,0) corresponds to terminals 2, 3 ready to transmit but terminals 1,4,5 not ready. There are 32 possible outcomes in this sample space. But if it is known that exactly three of the terminals are ready, then the number of sample points in S that apply are just 10.

If the terminals are polled sequentially until a ready terminal is found, the number of polls required can be 1, 2, or 3. One poll will be required if terminal 1 is ready (no need to continue polling) and the remaining two ready terminals occur in the remaining 4 positions. Two polls will be needed if terminal one is not ready, terminal 2 is and the remaining two ready terminals are in the remaining three positions.

$$P(1\ poll\ needed) = \frac{\binom{4}{2}}{10} = \frac{6}{10}.$$

$$P(2\ polls\ needed) = \frac{\binom{3}{2}}{10} = \frac{3}{10}.$$

$$P(3\ polls\ needed) = \frac{\binom{2}{2}}{10} = \frac{1}{10}.$$

4.1.5 Exercises

1. An urn contains 20 balls, numbered 1 to 20. Suppose we draw a ball from the urn one at a time, until 6 balls have been drawn. How many samples (or n-tuples), can we draw and what is the probability of a single sample when the drawing is done: (i) with replacement, (ii) without replacement? Explain your answer.

2. An urn contains 20 balls, numbered 1 to 20. Suppose we draw a ball from the urn one at a time, until 6 balls have been drawn. How many sets of 6 balls are there and what is the probability of a set when the drawing is done: (i) with replacement, (ii) without replacement? Explain your answer.

3. (Inspired by Roxy Peck (2008, page 285).) The instructor of a probability class which has 40 students enrolled comes to class daily with a little box containing balls numbered 1 to 40. The instructor also brings a roster that has the students' names sorted by last name. The first student in the list is number 1, the last one number 40. For example,

1 Ayala, Maria
2 Coelho, Brenda
3 Chen, Cynthia
....
....

......

....

40 Vidal, Arturo.

To make the class more interactive and engage students, the teacher asks three questions to three randomly chosen students in the class. To select the students the teacher draws three balls at random from the box containing the 40 balls, and then looks at the roster to see the names of the students. One possible outcome could be (2,3,40)- the first question is asked to Coelho, the second question to Chen, and the third to Vidal. Which of the sampling methods defined in this section 4.1 would you prefer the teacher to use? Why? Explain your reasoning.

4.1.6 An application of urn sampling models in physics

Social sciences, physical sciences, and the humanities draw random samples from populations to learn about those populations. In physics, for example, a problem of interest is to determine the *equilibrium state of a physical system* composed of a very large number *n* of "particles" of the same nature: electrons, protons, photons, mesons, neutrons, etc. For simplicity, assume that there are M *microscopic states* in which each of the particles can be (for example, there are *M* energy levels that a particle can occupy). To describe the *macroscopic state of the system*, suppose it suffices to state the number of particles in each of the microscopic states. The equilibrium state of the system of particles is defined as that macroscopic state with the highest probability of occurring. To compute the probability of any macroscopic state, a model is needed. The models that prevail in physics are the ones seen in this chapter, but adapted to the context in which this physical problem occurs.

> *Let's see how the model is used in this context. There are many particles,* n. *Think of the numbered balls in the urn now as indicating the state* j *that a particle will occupy,* $j = 1,2,3,\ldots,M$ *Drawing a random sample of size* n *with replacement from the urn, for example, if* $n = 20$ *and* $M = 10$, *the sample could be* (3,1,4,8,1,1,10,9,9,1,4,5,7,1,3,6,8,10,10,9), *is indicating that particle 1 is going to state 3, particle 2 is going to state 1, particle 3 goes to state 4, particle 4 goes to state 8, particle 5 goes to state 1 and so on. It is also indicating a macroscopic state, i.e., the number of particles that go into each state (5 particles are in microscopic state 1, 0 are in microscopic state 2, and 2 particles are in microscopic state 3, and so on). We can rewrite the macrostate for the given example as* $(n_1 = 5,\ n_2 = 0, n_3 = 2, n_4 = 2, n_5 = 1, n_6 = 1, n_7 = 1, n_8 = 2, n_9 = 3, n_{10} = 3)$. *In total, there are* 10^{20} *allocations of 20 particles to 10 states. In Physics jargon, there are* 10^{20} *macrostates. Since we are drawing with replacement, for the first particle we have M states, for the second one we have M states, ..etc. Then obviously, there could be more than one particle in a given state. If we consider the particles distinguishable, i.e., as if they were arriving in order and the order of arrival matters, then this result is known as Maxwell-Boltzmann's model. If nobody is keeping track of the order of arrival (imagine they all arrive at once), the particles are not distinguishable, and then we have Bose-Einstein model. If the sampling had been done with replacement, and indistinguishable balls, we would have had the Fermi-Dirac model. In Physics, it is considered that Maxwell-Boltzmann's model is a good approximation to reality.*

The specific application of the urn model to the macroscopic state of particles in Physics is known as "the occupancy problem." You will find it described in books as "distributing balls among urns," or "occupancy problem."

(Parzen 1960)

Parzen uses the physics example just described to illustrate the occupancy problem, and contains more details about it.

4.2 Inquiring about diversity

We may use the n-tuple sampling approach to solve probability problems where we inquire about the diversity of the elements in the population.

Consider the following problem. Two balls are drawn without replacement from an urn containing six balls, of which four are white and two are red. Find the probability that (i) both balls will be white, (ii) both balls will be the red, (iii) at least one of the balls will be white.

To set up a mathematical model for the experiment described using what we have already learned in section 4.1 of this chapter, assume that the balls in the urn are indistinguishable; in particular, assume that they are numbered 1 to 6. Let the white balls bear numbers 1 to 4, and let the red balls be numbered 5 and 6.

The sample space of the experiment, S, is the set of $6 \times 5 = 30$ 2-tuples (o_1, o_2) whose components are any numbers, 1 to 6,subject to the restriction that no two components of a 2-tuple are equal.

$$S = \{(1,2),(1,3),(1,4),(1,5),(1,6),(2,1),(2,3),(2,4),(2,5),$$

$$(2,6),(3,1),(3,2),(3,4),(3,5),(3,6),(4,1),(4,2),(4,3),(4,5),$$

$$(4,6),(5,1),(5,2),(5,3),(5,4),(5,6),(6,1),(6,2),(6,3),(6,4),(6,5)\}.$$

We may assume that all the samples are equally likely, i.e., each of them has probability 1/30.

Now let A be the event that both balls drawn are white, let B be the event that both balls drawn are red, and let C be the event that at least one of the balls drawn is white. The problem at hand can then be stated as one of finding (i) $P(A)$, (ii) $P(A \cup B)$, and (iii) $P(C)$. It should be noted that $C = B^c$ so that $P(C) = 1 - P(B)$. Further, A and B are mutually exclusive, so that $P(A \cup B) = P(A) + P(B)$.

Now, because the white balls bear numbers 1 to 4, the event A is

$$A = \{(1,2),(1,3),(1,4),(2,1),(2,3),(2,4),(3,1),(3,2),(3,4),(4,1),(4,2),(4,3)\},$$

whereas, because the red balls bear the numbers 5 and 6,

$$B = \{(5,6),(6,5)\}$$

Using the definition of probability of an event as equal to the sum of the probability of the outcomes in the event (i.e., assuming all the samples are different),

$$(i)\ P(A) = \frac{4 \times 3}{6 \times 5} = \frac{12}{30} = \frac{2}{5}.$$

If we instead consider the number of subsets

$$(i) P(A) = \frac{\binom{4}{2}}{\binom{6}{2}} = \frac{2}{5} = 0.4.$$

Now event B.

$$P(B) = \frac{2 \times 1}{6 \times 5} = \frac{2}{30},$$

or if we do not consider them distinguishable,

$$P(B) = \frac{\binom{2}{2}}{\binom{6}{2}} = \frac{1}{15} = 0.0666667.$$

So

$$(ii) P(A \cup B) = P(A) + P(B) = 0.4 + 0.0666667 = 0.4666667,$$

and

$$P(C) = 1 - P(B) = 1 - 0.0666667 = 0.933333.$$

We could solve the problem using the product rule, for example,

$$P(A) = (4/6)(3/5) = 2/5 = 0.4,$$

$$P(B) = (2/6)(1/5) = 1/15.$$

4.2.1 The number of successes in a sample. General approach

A basic question when sampling is the following. An urn contains M balls, of which M_w are white and M_R are red. A sample of size n is drawn either without replacement (in which case $n \leq M$), or with replacement. Let k be an integer between 0 and n (that is, $k = 0,1,2,\ldots,n$). What is the probability that the sample will contain exactly k white balls?

This problem is a prototype of many problems, which, as stated, do not involve the drawing of balls from an urn. For that reason, we can omit the reference to the color of balls and speak instead of scoring k successes. The question asked can then be rephrased as what is the probability of the event A_k that one will score exactly k successes when one draws a sample of size n from an urn containing M balls, of which M_w are white. In the case of sampling without replacement, the following equivalent expressions will give the $P(A_k)$:

$$P(A_k) = \begin{pmatrix} n \\ k \end{pmatrix} \frac{M_w(M_w-1)\ldots(M_w-k+1)(M-M_w)(M-M_w-1)\ldots(M-M_w-(n-k)+1)+1)}{M(M-1)\ldots..(M-n+1)}$$

$$= \frac{\begin{pmatrix} M_w \\ k \end{pmatrix}\begin{pmatrix} M-M_w \\ n-k \end{pmatrix}}{\begin{pmatrix} M \\ n \end{pmatrix}},$$

whereas in the case of sampling with replacement,

$$P(A_k) = \begin{pmatrix} n \\ k \end{pmatrix} \frac{(M_w)^k(M-M_w)^{n-k}}{(M^n)}.$$

For many purposes, it is useful to write these expressions in terms of

$$p = \frac{M_w}{M},$$

the proportion of white balls in the urn. The formula for $P(A_k)$ can then be compactly written, in the case of sampling with replacement, as

$$P(A_k) = \begin{pmatrix} n \\ k \end{pmatrix} (p)^k(1-p)^{n-k}.$$

This formula is only approximately correct in the case of sampling without replacement, but the approximation gets better as M increases.

Example 4.2.1

If we were to draw three cards from a box containing 52 cards, 26 of which are black, and we draw without replacement, what is the probability of obtaining three black cards?

The urn size is $M = 52$ cards. There are $M_b = 26$ black cards and $M - M_b = 26$ other cards. Let A_3 be the event that consists of obtaining 3 black bards.

$$P(A_3) = \begin{pmatrix} 3 \\ 3 \end{pmatrix} \frac{26(25)(24)}{52(52-1)(52-3)} = \frac{\begin{pmatrix} 26 \\ 3 \end{pmatrix}\begin{pmatrix} 26 \\ 0 \end{pmatrix}}{\begin{pmatrix} 52 \\ 3 \end{pmatrix}} = 0.1176471.$$

Example 4.2.2

Acceptance sampling of a manufactured product. Suppose we are to inspect a lot of size M of manufactured articles of some kind, such as light bulbs, screws, resistors, or anything else that is manufactured to meet certain standards. An article that is below standards is said to be defective. Let a sample of size n be drawn without replacement from the lot. A basic role in the theory of statistical quality control is played by the following problem. Let k and M_D be integers such that $k \le n$, $M_D \le M$. What is the probability that the sample will contain k defective articles if the lot contains M_D defective articles? This is the same problem as Example 4.2.1 with defective articles playing the role of cards.

Example 4.2.3

A box of chocolates is to be inspected for defects by a chocolatier in a chocolate factory. Suppose that, in a box containing twenty chocolates, four are defective and sixteen are not defective (note: a defective chocolate will be one that has some scratch or discoloration due to mixing of chocolate types, etc.). A sample of two chocolates is to be selected randomly for inspection without replacement. We will compute the probability that: (i) neither is defective, (ii) at least one is defective, (iii) neither is defective given that at least one is nondefective.

The urn size is $M = 20$ chocolates. There are $M_d = 4$ defective and $M - M_d = 16$ nondefective. Let A_2 be the event that consists of obtaining 2 nondefective chocolates. The sample size is n = 2.

$$(i)\ P(A_2) = \binom{2}{2}\frac{16(15)}{20(19)} = \frac{\binom{16}{2}\binom{4}{0}}{\binom{20}{2}} = 0.6315789.$$

Let B be the event "at least one of the two is defective," which is the same event as "one or two of them are defective."

$$(ii)\ P(B) = \binom{2}{1}\frac{16(4)}{20(19)} + \binom{2}{2}\frac{4(3)}{20(19)} = \frac{\binom{16}{1}\binom{4}{1}}{\binom{20}{2}} + \frac{\binom{16}{0}\binom{4}{2}}{\binom{20}{2}} = 0.36842.$$

Notice that B is the complement of A_2.

Consider the event C to be "at least one is nondefective" The probability of the event "neither is defective given that at least one is nondefective" is

$$P(A_2|C) = \frac{P(A_2 \cap C)}{P(C)} = \frac{P(A_2)}{P(C)} = \frac{0.6315789}{0.9684211} = 0.6521738,$$

since

$$P(C) = (16/20)(15/19) + 2(16/20)(4/19) = 0.9684211.$$

Example 4.2.3

Simple-minded game warder. Consider a fisherman who has caught 10 fish, 2 of which were smaller than the law permits to be caught. A game warden inspects the catch by examining two that he selects randomly from among the fish. What is the probability that he will not select either of the undersized fish? This problem is an example of those previously stated, involving sampling without replacement, with undersized fish playing the role of white balls, and M $= 10, M_w = 2$, n $= 2$, k $= 0$. There are $8 \times 7/2 = 28$ sets of good fish. There are $10 \times 9/2 = 45$ sets of size n $= 2$. The probability that the game warder selects two lawful fish is 28/45 = 0.62222. Using the product rule, we can also calculate this probability as (8/10)(7/9).

Example 4.2.4

A simple-minded die. Another problem, which may be viewed in the same context but which involves sampling with replacement, is the following. Let a fair six-sided die be tossed four times. What is the probability that one will obtain the number 3 exactly twice in the four tosses? This problem can be stated as one involving the drawing (with replacement) of balls from an urn containing balls numbered 1 to 6, among which ball number 3 is white and the other balls red (or, more strictly, nonwhite). In the notation of the problem introduced at the beginning of the section this problem corresponds to the case $M = 6, M_w = 1, n = 4, k = 2$. Sampling with replacement, there are $6^4 = 1296$ samples. The number of sets of 4 with two whites is $4 \times 3/2 = 6$.

Example 4.2.5

Five employees of a firm are ranked from 1 to 5 in their abilities to program a computer. Three of these employees are selected to fill equivalent programming jobs. If all possible choices of three (out of the five) are equally likely, find the probabilities of the following events: (i) A = the employee ranked number 1 is selected, (ii) B = employees ranked 4 and 5 are selected

(i) Let an urn contain numbered balls, one for each of the ranked employees. The question can be rephrased as "what is the probability of obtaining k = 1 special employee if we sample n = 3 employees without replacement?

$$P(A) = \frac{\binom{1}{1}\binom{4}{2}}{\binom{5}{3}} = 0.6.$$

(ii) Let the urn now contain M = 5 employees, with $M_d = 2$ (*special employees* 4,5), $M - M_d = 3$(*employees* 1,2,3)The question can be rephrased as "what is the probability of obtaining k = 2 special employees if we sample n = 3 employees without replacement?

$$P(A) = \frac{\binom{2}{2}\binom{3}{1}}{\binom{5}{3}} = 0.3.$$

4.2.2 The difference between k successes and successes in k specified draws

Let a sample of size 3 be drawn without replacement from an urn containing six balls, of which four are white. The probability that the first and second balls drawn will be white and the third ball red is equal to

$$\frac{4 \times 3 \times 2}{6 \times 5 \times 4}.$$

However, the probability that the sample will contain exactly two white balls is equal to

$$\binom{3}{2}\frac{4\times3\times2}{6\times5\times4},$$

to account for the $\binom{3}{2}$ possible locations of the two balls, i.e., to account for all the outcomes in S that contain two white balls.

4.3 Independent trials of an experiment

Scientists conduct experiments repeatedly under identical conditions to learn about the effects of medicines, the resistance to stress of materials, the way the brain responds to stimuli, to name a few examples. The following formulation is one of the most clear formulations found in the probability theory literature for beginners, and as such, we present it here verbatim.

> *Suppose an experiment is under consideration. As we know, we think instead of its mathematical counterpart, the sample space S, where*
>
> $$S=\{o_1,o_2,\ldots,o_N\}.$$
>
> *where* o_j, $j=1, 2,\ldots,N$ *is the outcome of the experiment.*
>
> *We assume that an acceptable assignment of probabilities has been made to the simple events of S; i.e., to each* $\{o_j\}$ *there is assigned a nonnegative number* $P(\{o_j\})$ *in such a way that*
>
> $$\sum_{j=1}^{N} P(o_j)=1.$$
>
> *The outcomes do not need to be equally likely. Now let us think of performing this experiment and then performing it again. The succession of two experiments is a new experiment that we want to describe mathematically. In order to avoid confusing reference to original experiments and this new experiment, it is convenient to refer to the original experiments as trials and to describe the new experiment as made up of two trials, each represented by (or corresponding to) the sample space S. This new experiment is mathematically defined, as are all experiments, by a sample space. The elements (outcomes) of this new sample space are all the ordered pairs* (o_j,o_k) *denoting the occurrence of outcome* o_j *at the first trial and* o_k *at the second trial. Thus the sample space for the experiment is the Cartesian product of* $S\times S$. *Since the sample space* S *for each of the two trials making up the experiment has* N *elements, there are* N^2 *ordered pairs in* $S\times S$.
>
> *Before probability questions can be answered for the experiment, we must make some acceptable assignment of probabilities to the* N^2 *simple events of* $S\times S$; *i.e., we must assign a nonnegative number to* $\{(o_j,o_k)\}$ *for each* j *and* k *in such a way that the sum of all* N^2

numbers is 1. As we know, there are infinitely many ways of doing this. But if we say that the two trials are independent, then by definition there is one and only one way that we must use; the assignment must be made so that

$$P(\{(o_j,o_k)\}) = P(\{(o_j)\})P(\{(o_k)\}) \text{ for } j = 1, 2,..,N \text{ and } k = 1, 2,\ldots,N$$

This formula expresses the probability of the simple event $\{(o_j,o_k)\}$ *of* $S \times S$ *as the product of the probabilities of the simple events* $\{o_j\}$ *and* $\{o_k\}$ *of* S.

The result we obtained for two trials can be generalized to any number of trials. That is, suppose n is a positive integer and let S_j (for $j = 1, 2,\ldots,n$) *be a sample space with outcomes* $o_1^{(j)}, o_2^{(j)},\ldots,o_N^{(j)}$. *By the experiment consisting of the succession of n trials, the first corresponding to* S_1, *the second to* S_2, *etc. we mean the sample space* $S_1 \times S_2 \times \ldots \times S_n$ *whose elements are all the* $N_1N_2\ldots N_n$ *ordered n-tuples (see Definition, 4.1.1).*

(Parzen 1960)

We will use Parzen's formulation in the following examples of this section, with the understanding that in the rest of the book, after this section, the sample space made of several trials will also be denoted by a simple S.

Example 4.3.1

The toss of a coin twice or the toss of two coins is an example of independent trials. Each toss is a trial represented by the sample space $S = \{H,T\}$. Suppose the two simple events have been assigned the probabilities P({H}) = 2/3 and P({T}) = 1/3; i.e., the coin is not fair. The outcomes of the two trials are given by $S \times S = \{HH,HT,TH,TT\}$. If the two tosses are independent, then the probability of each outcome is

$$P(\{(HH)\}) = (2/3)^2 \; ; P(\{(TT)\}) = (1/3)^2 \; ; P(\{HT)\}) = P(\{(TH)\}) = (1/3)(2/3).$$

Tossing two coins is equivalent to drawing a sample of two numbers with replacement from an urn containing three numbers 1 to 3, where 1 = H, 2 = H and 3 = T.

Example 4.3.2

If we expand Example 4.3.1 and instead of 2 we do three trials,

$$S \times S \times S = \{HHH, HHT, HTH, HTT, THH, THT, TTH, TTT\},$$

where

$$\begin{aligned} P(\{(HHH)\}) &= (2/3)^3; P(\{(TTT)\}) = (1/3)^3; P(\{HHT)\}) = P(\{(THH)\}) = P(\{(HTH)\}) \\ &= (1/3)(2/3)^2; P(\{THT)\}) = P(\{(TTH)\}) = P(\{(HTT)\}) = (2/3)(1/3)^2. \end{aligned}$$

Notice that three trials of this experiment "tossing the coin" is equivalent to drawing a sample of three numbers with replacement from an urn containing three numbers 1 to 3, where 1 = H, 2 = H and 3 = T.

Example 4.3.3

A quiz has four questions of multiple-choice type. There are three possible answers for each question, but only one answer is right. Assuming a student guesses at random for his answer to each question and that his successive guesses are independent, what is the probability that he gets more right than wrong answers?

The sample space for each trial (answering a question) is $S = \{R, W\}$, where R denotes a right answer, W a wrong answer. We are given that

$$P(\{R\}) = \frac{1}{3}, \quad P(\{W\}) = \frac{2}{3}$$

For the four-question test, the sample space is

$$S \times S \times S \times S = \left\{ \begin{array}{c} RRRR,\ RRRW,\ RRWR,\ RRWW,\ RWRR,\ RWRW,\ RWWR,\ RWWW,\ WRRR,\ WRRW, \\ WRWR,\ WRWW,\ WWRR,\ WWRW,\ WWWR,\ WWWW \end{array} \right\}.$$

Let A denote the event "3 or 4 right", which is the subset:

$$A = \{RRRR,\ RRRW,\ RRWR,\ RWRR,\ WRRR\}.$$

The probability of this event is

$$P(A) = (1/3)^4 + (1/3)^3(2/3) + (1/3)^3(2/3) + (1/3)^3(2/3).$$

Example 4.3.4

From a population of n people, one person is selected at random. Another person is then selected at random from the full group; i.e., we allow the same person to be selected at both trials. Each selection (trial) is defined by the sample space $S = \{1, 2, \ldots, n\}$, where each person is identified by a positive integer. Each of the n simple events of S is assigned probability $1/n$; i.e., $P(\{j\}) = 1/n$ for $j = 1, 2, \ldots, n$. The experiment made up of these two trials is called selecting a sample of two with replacement from the population and is represented by the Cartesian product set

$$S \times S = \{(j,k) \mid j \in S, k \in S\}.$$

To say that the two trials (i.e., the selection of the first person and the selection of the second person) are independent is to require that assignment of probabilities to the simple events of $S \times S$ as follows:

$$P(\{j,k\}) = P(\{j\})P(\{k\}) = \left(\frac{1}{n}\right)\left(\frac{1}{n}\right) = \frac{1}{n^2}.$$

The independence of the trials means that each simple event of $S \times S$ is assigned the same probability $1/n^2$. Thus we have the formal mathematical counterpart of our intuitive feeling that selecting a random sample of size two with replacement can be considered as a succession of two independent trials.

Example 4.3.5

Consider a plan manufacturing chips of which 10% are expected to be defective. A sample of 3 chips is randomly selected. What is the probability of obtaining at most 2 (2 or less) defectives?

The sample space for each trial (selecting a chip) is:

$$S = \{D, W\}.$$

where D denotes a defective chip, W a non-defective chip. We are given that

$$P(\{D\}) = 0.1;\ P(\{W\}) = 0.9.$$

For the selection of 3 chips, the sample space is

$$S \times S \times S = \{DDD,\ DDW,\ DWD,\ DWW,\ WDD,\ WDW,\ WWD,\ WWW\}.$$

Let A be the event "at most 2 (2 or less) defectives", which is the subset

$$A = \{DDW,\ DWD,\ DWW,\ WDD,\ WDW,\ WWD,\ WWW\}.$$

This is the complement of the event "all three are defective,"

$$A^c = \{DDD\}.$$

Thus

$$P(A) = 1 - P(A^c) = 1 - P(\{DDD\}) = 1 - 0.1^3.$$

4.3.1 Independent Bernoulli Trials

Many problems in probability theory involve independent repeated trials of an experiment whose outcomes have been classified in two categories, called "successes" and "failures," and represented by the letters *s* and *f*, respectively. Such an experiment, which has only two logically possible outcomes, is called a Bernoulli trial. An experiment consisting of a Bernoulli trial can describe a variety of situations: a coin toss (heads or tails), a competition with two outcomes (win or lose), the allele of a gene (normal or mutant). Which of the outcomes is denoted *s* or *f* depends on what in the experiment interests us. For example, if we are interested in observing whether there is a mutation, then a mutation is a success (*s*).

The probability of the outcome *s* is usually denoted by *p*, and the probability of the outcome *f* is usually denoted by *q*, where

$$p \geq 0, \quad q \geq 0, \quad p + q = 1.$$

Consider now *n* independent repeated Bernoulli trials, in which the word "repeated" is meant to indicate that the probabilities of success and failure remain the same throughout the trials. The sample space $S \times S \times \ldots \times S$ contains 2^n n-tuples $(o_1, o_2, \ldots\ldots, o_n)$, in which each o_i is either an *s* or an *f*. The sample space is finite. The probability of every single n-tuple is

$$P\left((o_1, o_2, \ldots\ldots, o_n)\right) = p^k q^{n-k},$$

where *k* is the number of successes s among the components of the n-tuple.

One usually encounters Bernoulli trials by considering a random event E, whose probability of occurrence is p. In each trial one is interested in the occurrence or nonoccurrence of E. A success s corresponds to an occurrence of the event, and a failure f corresponds to a nonoccurrence of E.

Frequently, the only fact about the outcome of a succession of n Bernoulli trials in which we are interested is the number of successes. We now compute the probability that the number of successes will be k, for any integer k from 0, 1, 2,....., n. The event "k successes in n trials" can happen in as many ways as k letters may be distributed among n places—that is, the same as the number of subsets of size k that may be obtained from a set containing n members. Consequently, there are $\binom{n}{k}$ n-tuples containing exactly k successes and $n-k$ failures. Each such description has probability $p^k q^{n-k}$. Thus the probability that n independent Bernoulli trials, with probability p for success, and $q = 1 - p$ for failure, will result in k successes and $n - k$ failures (in which $k = 0, 1, 2, \ldots, n$) is given by

$$P(\text{"}k \text{ successes} \mid n,p) = \binom{n}{k} p^k q^{n-k}.$$

This is called the binomial formula, because by the binomial theorem

$$\sum_{k=0}^{n} \binom{n}{k} p^k q^{n-k} = (p+q)^n = 1.$$

Example 4.3.6

A system has five identical independent components connected in series. A component works with probability $p = 0.9$. The system works if all the components work. What is the probability that the system works?

Observing each component is a Bernoulli trial, where the probability of success is 0.9 if the component works, and the probability of failure is 0.1, if the component does not work. Observing the five components is a repeated experiment consisting of five independent Bernoulli trials.

There are $2^5 = 32$ logically possible configurations of the system, each of them a sequence of five Bernoulli trials.

$$P(\text{"5 working"} \mid n,p) = \binom{5}{5} 0.9^5 0.1^0 = 0.59049.$$

Example 4.3.7

Davy Crocket was a member of the US Congress during the 1830s. There are three books supposedly written by Crocket: *A Narrative of the Life of David Crockett*, *An Account of Col. Crockett's Tour to the North and Down East*, and *Col. Crockett's Exploits and Adventures in Texas*, the latter published after his death. There is serious doubt whether Crockett wrote any of these books. Beyond these books, the only examples of Crockett's supposed writings consist of speeches he made in Congress, and a small number of short letters. The letters were filled with misspellings and poor grammar. In an article in *Chance Magazine* in 1999, David Salsburg and Dena Salsburg ask: Were the three books and the speeches written by the same person? (Salsburg and Salsburg 1999). In order to answer this question, they created a list of contentless words such as "also," "an," "to," etc. Then they went over words in the books and the speeches.

A Bernoulli trial was the observation of a word, with success being whether the word was in the list of contentless words or not. Observing the words in the book was a long sequence of independent Bernoulli trials. They did this for each book. Then, separately, they calculated the proportion of words in the speeches that were contentless. For simplicity, let's assume here that the proportion of the contentless word "also" in the speeches was 0.51 per 1,000 words and that $n = 30$ words were randomly selected in the Texas book, of which $k = 5$ were the word "also." Then the probability of observing five instances of the word "also" in the Texas book under the assumption that Crockett wrote it must be calculated with the probability of that word in the speeches.

$$P(k = 5 \text{ in Texas book} \mid n = 30, p = 0.00051) = \binom{30}{5}(0.00051)^5(0.9949)^{25} \approx 0$$

We would conclude that there is a very small probability of observing the word "also" five times in a random sample of 30 words from the Texas book. Salsburg and Salsburg do statistical analysis using the binomial model and compare the probability that k is 5 in the Texas book under the assumption $p = 0.00051$ and then under the assumption that p is some other value suggested by the word data of the Texas book. The details of their statistical analysis are beyond this book, and concern statistical inference, but let this example illustrate that the models of probability theory play a very important role in disputed authorship.

4.3.2 Exercises

Exercise 1. Demonstrate that the approach used in Example 4.3.4 provides an acceptable assignment of probabilities to the simple events of $S \times S$.

Exercise 2. A pizza restaurant has five ovens. At least four ovens must be working in order to meet customer demand on a given day. The probability of a particular oven working is 0.9. We want to find out the probability of meeting customer demand.

Exercise 3. Factories produce millions of items. Thus, when we sample them to observe quality, we can pretend we are sampling with replacement, although in reality we are sampling at random (without replacement) That is the assumption in industrial reliability and other contexts where populations are very large. An incoming lot of silicon wafers is to be inspected for defective by an engineer in a microchip manufacturing plant. Suppose that, in a tray containing twenty wafers, four are defective and sixteen are working properly. Two wafers are to be selected for inspection with replacement. After listing the sample space, find the probabilities of the following events: (i) neither is defective, (ii) at least one is defective.

Exercise 4. What is the probability of the event A = "having 1 or less defective components" in a system with three components where the probability of defective is 0.3 and the components are independent of each other?

4.4 Mini Quiz

Question 1. A box of 10 computer chips contains 4 defective and 6 nondefective.
We are going to select three at random (without replacement). How many samples are there with two defective and one nondefective?

a. 1440
b. 36
c. 120
d. 10

Question 2. If we roll a die four times, how many samples are there?

a. 360
b. 760
c. 1296
d. 54

Question 3. An audition office has told candidates to prepare excerpts from 10 scripts telling them that the day of the audition they will be asked a random selection of 5 of them. A hopeful actor has practiced only 7 of the scripts. To compute the probability that, the day of the audition, the hopeful actor will be able to do 4 of the scripts using the notation of the model seen in this chapter, what would be the value of M?

a. 5
b. 10
c. 7
d. 3

Question 4. A group of four iPhone fans wait in line for the next version of the iPhone to be released. They slept outside the store because they were told that three of the first four would be chosen randomly to obtain a free iPhone. The same person could not be chosen twice, of course. These fans are numbered. The first one in line is number 1, the second is number 2, the third is number 3, and the fourth is number 4. In how many ways could the three be chosen?

a. 6
b. 4
c. 30
d. 3

Question 5. If there are five states in which a particle can be and there are 10 particles, how many macrostates are there in this physical system?

a. 10,000
b. 100,000
c. 1,000
d. 9765625

Question 6. There are two types of workers in an office: 10 administrative assistants and 5 fund managers. Two workers will be chosen randomly to represent the office on the board of directors and at the town's city hall, one worker to each place. Let A be the event that two fund managers are chosen. What is the probability of A?

a. 0.3156
b. 0.095238
c. 0.6954
d. 0.0910

Question 7. A gardener has been growing tomatoes. This individual takes a box containing 20 beautiful tomatoes to the local market with the hope of selling them. There are 8 tomatoes in the box that look good, but they are from a neighbor's garden. The person buying the tomatoes will not be able to tell just by looking. A customer bought three randomly chosen tomatoes from the box. What is the probability that two of the tomatoes were from the neighbor's garden?

a. 0.2947
b. 0.09824
c. 0.6197
d. 0.1131

Question 8. People arriving to a foreign country must usually pass Customs control. During some periods, the authorities require that a certain number of people be chosen at random and sent to the Customs office for further inspection. They then sit in some room waiting for the authorities to check their records further. Supposedly the authorities are looking for drug dealers that may have stolen someone else's identity. Suppose that a randomly chosen visitor to a major country with millions of visitors has probability 0.005 of being a drug dealer that stole someone's identity. Suppose that three people are randomly chosen. What is the probability that two of the people are found to be drug dealers that have stolen someone's identity?

a. 0.1265
b. 0.0319
c. 0.0051
d. 0.00007462

Question 9. Suppose that the probability of finding the word "furthermore" in the speeches written by Davy Crockett is 0.2. What is the probability of observing four times the word "furthermore" in a random sample of 20 words from the Texas book?

a. 0.8145
b. 0.2181
c. 0.6891
d. 0.0434

Question 10. When going about their business of drawing random samples from populations in order to learn about the population, statisticians prefer to use probability sampling. The equivalent of sampling in practice is, in probability theory, an urn model. How many simple random samples (samples without replacement) of 50 people can be chosen from a population of 1,000,000 to estimate the average number of apps installed in smart phones per person?

a. $\binom{1{,}000{,}000}{50}$

b. $\frac{1{,}000{,}000!}{50!}$

c. 1,298,000,000

d. $\binom{5{,}000{,}000}{50}$

4.5 R corner

R exercise Birthdays.

How would we calculate theoretically (i.e., not by simulation with computers but by the exact mathematical solution) the probability that, in a room of five people, none of them shares the same birthday, assuming 365 days/year? Then how would we find the probability of the complement "at least two people share a birthday"? Use the urn model appropriate for this situation.

Read the article attached at the end of this chapter to find out.

Here, we will use RStudio to do a simulation with a 12-sided fair die to estimate the probability that, in a room of five people, nobody shares the same birth month. Use the following simulation in R. The probability model is a 12-sided fair die. A trial consists of rolling the die five times. We record "0" if some numbers are repeated and "1" if all numbers are different. At the end, we calculate the proportion of trials in which we got TRUE (nobody shared a birth month).

```
# trial 1: get the birth month of 5 people
sample(12, size=5, prob=c(rep(1/12, 12) ), replace=T )
# look at the results. Are there repeated numbers? Yes, then 0
```

```
# Trial 2: get the birth month of 5 people
sample(12, size=5, prob=c(rep(1/12, 12) ), replace=T )
# look at result. Are there repeated numbers? No, record 1.
# We will do many more trials and put the result of each trial in
the row of a matrix, to make the calculations easier on us.
# in the last column of the matrix we will put False
trials=matrix(0,ncol=5,nrow=1000) # 100 trials, 1000 rows
record=matrix(0,nrow=1000,ncol=1)
for(i in 1:1000){
trials[i,] =c(sample(12, size=5, prob=c(rep(1/12, 12) ), replace=T)
record[i]= as.numeric(length(unique(c(trials[i,])))==6)
}

# Calculate proportion of trials in which nobody shared a birth month
Prob=sum(record)/1000
```

4.6 Chapter Exercises

Exercise 1. A first prize of $1000, a second prize of $500, and a third prize of $100 are offered to the best three data mining projects presented at a major data mining competition. There are 10 contestants. How many different outcomes are possible if (i) a contestant can receive any number of awards and (ii) each contestant can receive at most one award?

Exercise 2. You roll two loaded six-sided dice. A single die has probability 0.3 of being a 1, probability 0.3 of being a 5, and the other numbers have probability 0.1. What is more advantageous: betting on a sum of 7 or of 8?

Exercise 3. A bank opens at 9 a.m. on a regular working day with five bank tellers available for assisting entering customers. If three customers are waiting at the door, and each customer chooses a different teller, in how many ways can the tellers be chosen to assist the three customers?

Exercise 4. Let a coin that is weighted so that $P(H) = 2/3$ and $P(T) = 1/3$ be tossed three times. What is the probability of getting a run of two heads?

Exercise 5. If we were to draw three cards from a box containing 52 cards, 26 of which are black, and we draw with replacement, then what is the probability of three black cards? What is the probability if we draw without replacement?

Exercise 6. In the African country of Angola in 2017, 38.2% of the congressional seats were held by women. If we randomly select four seats, what is the probability that we will find one seat held by a woman?

Exercise 7. Suppose there are 40 alumni signed up to travel to Egypt with the university alumni association during the summer. In this group of alumni, 25 received a bachelor of science (BS) degree from the university and 15 received a master of science (MS) degree. We must select a random sample of 7 people. (i) What is the probability that the sample contains 4 BS recipients and 3 MS recipients? (ii) What is the probability that there is at least one BS in the sample?

Exercise 8. If there are 12 strangers in a room, what is the probability that no two of them celebrate their birthdays in the same month?

Exercise 9. A Hollywood producer holds an audition for a movie. The executive office of the studio sends interested parties a set of 15 excerpts from the script of the movie for them to memorize with the information that the audition will consist of a random selection of 5 excerpts. If a candidate has memorized 10 of the excerpts, what is the probability that this candidate will recall (i) all 5 excerpts or (ii) at least 4 of the excerpts?

Exercise 10. *R exercise 1. Rolling a fair six-sided die*

We will first write the command to roll a red and a green fair 6-sided dice. This will be a 2-tuple containing as first number the roll of the red die and as second number the roll of the green die. The sample () function in R does the job of extracting n numbered balls from an urn containing M numbers.

```
my.data=sample(6, size=2, prob=c(rep(1/6, 6) ), replace=T )
```

The 6 = M stands for 1 to 6, size = n = 2 is the number of times we draw the ball, `prob` gives the probabilities assigned to each side of the die; notice that we are saying: repeat 1/6 six times, which means that the number 1 has probability 1/6, the number 2 has probability 1/6 and so on. The argument replace = T is necessary because when rolling a die the model to use is the same as when we were drawing from an urn with six numbered balls with replacement.

It is not very interesting to see just one 2-tuple. As we saw in section 4.1, there are $M^n = 6^2 = 36$ logically possible samples of size 2 that can be drawn with replacement from an urn containing balls numbered 1 to 6. To see more 2-tuples, we can complicate the problem a little bit.

First we create a matrix where we will put the 2-tuples. Each row of the matrix will contain a 2-tuple. We will generate 10 2-tuples so we will create storage for 10 rows, and 2 columns.

```
dice=matrix(0, nrow=10, ncol=2)
```

Then we use a for loop to put in each row a different sample of size 2 (a 2-tuple with the two rolls).

```
for(i in 1:10){
dice[i, ] = sample(6, size=2, prob=c(rep(1/6, 6) ), replace=T )
}
```

The for loop goes row by row, i = 1 to 10, and puts a new 2-tuple in the row. What we want to do in each row is written between { }.The result is the 10 2-tuples (or 10 rolls of the two dice). To view the samples you drew, type
"dice".

Exercise 11. *R exercise 2. Rolling a fair six-sided die*
Now we will write the command to roll an unfair 12-sided die 8 times (the equivalent of drawing an 8-tuple sample from an urn containing six balls numbered 1 to 12.

```
my.data= sample(12, size=8, prob=c(0.5/12, 2/12, 1/12, 3/12, 0.5/12,
0.5/12, 0.3/12, 0.2/12, 2/12,1/12,0.5/12, 0.5/12 ), replace=T )
```

There are 6^8 possible samples (8-tuples). To see a few more, put each sample in the row of a matrix with 10 rows (for 10 samples) and 8 columns.

First we create space to put the samples.

```
unfairdice = matrix(0,nrow=10, ncol=8)
```

Then we use a for loop to put in each row a different sample of size 2 (a 2-tuple with the two rolls).

```
for(i in 1:50){
unfairdice[i,] = sample(12, size=8, prob=c(0.5/12, 2/12, 1/12, 3/12,
0.5/12, 0.5/12, 0.3/12, 0.2/12, 2/12,1/12,0.5/12, 0.5/12 ), replace=T ) }
```

The for loop goes row by row, i = 1 to 10, and puts a new 8-tuple in the row. What we want to do in each row is written between { }.The result is the 10 8-tuples. To view the samples you drew, type

```
unfairdice
```

Notice how we are using the same model to roll two dice of different colors as to roll one die 8 times. In both of R exercise 1 and R exercise 2 we are drawing from an urn containing 6 balls numbered 1 to 6.

Exercise 12. Read the article in Appendix 4.1. Summarize in your own words, and then relate the calculations they do to the formulas seen in this chapter.

Exercise 13. During one year (365 days) 23 earthquakes occurred over an area of interest. What is the probability that two or more earthquakes occurred on the same day of the year? (Note: This is the same problem as the birthday problem.)

4.7 Chapter References

Garrett, S. J. 2015. *Introduction to Actuarial and Financial Mathematical Methods*. Elsevier.

Goldberg, Samuel. 1960. *Probability: An Introduction*. New York: Dover Publications, Inc.

Grinstead, Charles M., and J. Laurie Snell. 1997. *Introduction to Probability*. Second revised edition. American Mathematical Society.

Kemeny, John G., J. Laurie Snell, and Gerald L. Thompson. 1966. *Introduction to Finite Mathematics*. Second Edition. Englewood Cliffs, N.J.: Prentice-Hall, Inc.

Mosteller, Frederic, Robert E. K. Rourke, and George B. Thomas. 1967. *Probability and Statistics*. Addison-Wesley Publishing Company.

Parzen, Emanuel. 1960. *Modern Probability Theory and Its Applications*. New York. John Wiley and Sons, Inc.

Peck, Roxy, Chris Olsen, and Jay Devore. 2008. *Introduction to Statistics and Data Analysis*. Thomson Brooks/Cole.

Pitman, Jim. 1993. *Probability*. New York: Springer–Verlag.

Salsburg, David, and Dena Salsburg. 1999. "Searching for the 'Real' Davy Crockett." *Chance* 12, no. 2: 29–34.

Trumbo, Bruce, Eric Suess, and Clayton Schupp. 2005. "Simulation: Computing the Probabilities of Matching Birthdays." *Stats, The Magazine for Students of Statistics*, Spring 2005, Issue 43, 3–7.

Wild, Christopher J. and George A. Seber. 2000. *Chance Encounters: A First Course in Data Analysis and Inference*. John Wiley & Sons.

APPENDIX 4.1

Simulation: Computing the Probabilities of Matching Birthdays

By Bruce Trumbo, Eric Suess, and Clayton Schupp

The birthday matching problem

Sometimes the answers to questions about probabilities can be surprising. For example, one famous problem about matching birthdays goes like this: Suppose there are 25 people in a room. What is the probability two or more of them have the same birthday? Under fairly reasonable assumptions, the answer is greater than 50:50—about 57%.

This is an intriguing problem because some people find the correct answer to be surprisingly large. Maybe such a person is thinking, "The chance anyone in the room would have my birthday is very small," and leaps to the conclusion that matches are so rare one would hardly expect to get a match with only 25 people. This reasoning ignores that there are $(25 \times 24)/2 = 300$ pairs of people in the room that might yield a match. Alternatively, maybe he or she correctly realizes, "It would take 367 people in the room to be absolutely sure of getting a match," but then incorrectly concludes 25 is so much smaller than 367 that the probability of a match among only 25 people must be very low. Such ways of thinking about the problem are too fuzzy-minded to lead to the right answer.

As with most applied probability problems, we need to start by making some reasonable simplifying assumptions in order to get a useful solution. Let's assume the following:

- *The people in the room are randomly chosen.* Clearly, the answer would be very different if the people were attending a convention of twins or of people born in December.
- *Birthdays are uniformly distributed throughout the year.* For some species of animals, birthdays are mainly in the spring. But, for now at least, it seems reasonable to assume that humans are about as likely to be born on one day of the year as on another.
- *Ignore leap years and pretend there are only 365 possible birthdays.* If someone was born in a leap year on February 29, we simply pretend he or she doesn't exist. Admittedly, this is not very fair to those who were "leap year babies," but we hope it is not likely to change the answer to our problem by much.

Bruce Trumbo (bruce.trumbo@csueastbay.edu) is Professor of Statistics and Mathematics at California State University, East Bay (formerly CSU Hayward). He is a Fellow of ASA and holder of the ASA Founder's Award.

Eric Suess (eric.suess@csueastbay.edu), Associate Professor of Statistics at CSU East Bay, has used simulation methods in applications from geology to animal epidemiology.

Clayton Schupp, (cschupp@walk.ucdavis.edu) an MS student at CSU East Bay when this article was written, is currently a PhD student in statistics at the University of California, Davis.

The solution using basic probability

Based on these assumptions, elementary probability methods can be used to solve the birthday match problem. We can find the probability of no matches by considering the 25 people one at a time. Obviously, the first person chosen cannot produce a match. The probability that the second person is born on a different day of the year than the first is $364/365 = 1 - 1/365$. The probability that the third person avoids the birthdays of the first two is $363/365 = 1 - 2/365$, and so on to the 25th person. Thus the probability of avoiding all possible matches becomes the product of 25 probabilities:

$$P(\text{No Match}) = \prod_{i=0}^{24}\left(1 - \frac{i}{365}\right) = \frac{P_{25}^{365}}{365^{25}} = 0.4313$$

since 365^{25} is the number of possible sequences of 25 birthdays and

$$P_{25}^{365} = 25!\binom{365}{25}$$

is the number of permutations of 365 objects taken 25 at a time, where repeated objects are not permitted. Therefore,

$$P(\text{At Least 1 Match}) = 1 - P(\text{No Match}) = 1 - 0.4313 = 0.5687$$

William Feller, who first published this birthday matching problem in the days when this kind of computation was not easy, shows a way to get an approximate result using tables of logarithms. Today, statistical software can do the complex calculations easily, and even some statistical calculators can do the numerical computation accurately and with little difficulty.

```
> prod(1 - (0:24)/365)
[1] 0.4313003
> factorial(25)*choose(365, 25)/365^25
[1] 0.4313003
```

Figure 4.6 Two ways to calculate the probability of no matching birthdays among 25 people selected at random.

```
p <- numeric(50)
for (n in 1:50) {
       q <- 1-(0:(n-1))/365
       p[n] <- 1-prod(q) }
plot(p)
```

Figure 4.7 R code to calculate the probability of matching birthdays when the number of people in the room ranges from 1 to 50.

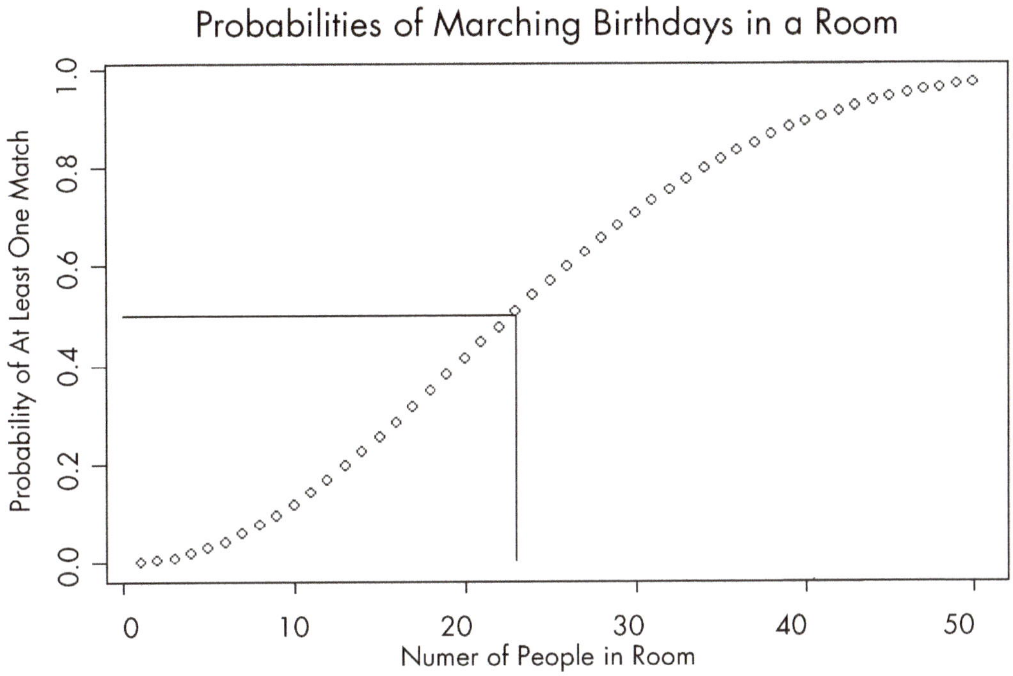

Figure 4.8 Plot from R of the probability of at least one pair of matching birthdays when the number of people in the room ranges from 1 to 50.

In Figure 4.6, we show two ways to use the statistical software R to calculate the probability of no matches.

Of course, different values of n would give different probabilities of a match. With a computer package like R that has built-in procedures for doing probability computations and making graphs, it is easy to loop through various values of n and graph the relationship between n and P(At Least 1 Match). Figure 2 shows the small amount of R code required, and Figure 4.8 shows the resulting plot. (The labels and the reference lines were added later.)

By looking at the plot, we see the probability of at least one match increases from zero to near one as the number of people in the room increases from 1 to 50. We can see that $n = 23$ is the smallest value of n for which P(At Least 1 Match) exceeds 1/2. The computations show the probability for $n = 23$ to be 0.5073. A room with $n = 50$ randomly chosen people is very likely to have at least one match. Indeed, for $n = 50$, the probability is 0.9704.

The solution using simulation

A completely different approach to solving the birthday match problem is by simulation. Simulation is widely used in applied probability to solve problems that are too difficult to solve by combinatorics or other analytical methods. For example, we can use R to build a simulation model to approximate the probability that there are no matching birthdays among 25 people in a room.

This consists of first simulating the birthdays in many rooms, each with 25 people, and then checking to see what percentage of these rooms have matching birthdays. It is a little like taking a public opinion poll where the "subjects" are the rooms. We create the imaginary rooms by simulation, and then we "ask" each room, "Do you have any birthday matches?" If we ask a large number of rooms, the percentage of rooms with no match should be very near the true probability of no match in such a room.

This approach allows us to find the approximate distribution of the number of repeated birthdays (X). From this distribution, we can approximate $P(X = 0)$, which we already know to be 0.4313. As a bonus, we also can approximate $E(X)$, the expected number of matches among 25 birthdays. This expectation would be difficult to find without simulation methods.

Now let's build the simulation model step by step.

Step One: *Simulating birthdays for 25 people in one room*

Programmed into R is a function called "sample" that allows us to simulate a random sample from a finite population. To use this random sampling function, we need to specify three things.

First, we must specify the population from which to sample. For us, this is the 365 days of the year. In R, the notation 1:365 can be used to represent the list of these population elements.

Second, we have to specify how many elements of the population are to be drawn at random. Here, we want 25.

Third, we have to say whether sampling is to be done with or without replacement. Because we want to allow for the possibility of matching birthdays, our answer is "with replacement." In R, this is denoted as repl=T. We put the 25 sampled birthdays into an ordered list called b. Altogether, the R code is

```
b <- sample(1:365, 25, repl=T)
```

Each time R performs this instruction, we will get a different random list b. Below is the result of one run. For easy reference, the numbers in brackets give the position along the list of the first birthday in each line of output. For example, the 22nd person in this simulated room was born on the 20th day of the year, January 20.

```
 [1]  352  364  246  190  143  (272)  (149)
 [8]  206  154  (272)  61   199  357   141
[15]  264  157  42   340  287  166   335
[22]  20   123  214  (149)
```

You can see that there happen to be two matches in this list. The 6th and 10th birthdays both fall on the 272nd day of the year, and the 7th and 25th both fall on the 149th day of the year. Note that we also would have said there are two matches if, for example, the last birthday in the list had fallen on the 272nd day.

Step Two: *Finding the number of birthday matches among 25 people*
In a large-scale simulation, we need an automated way to find whether there are matching birthdays in such a room and, if so, how many repeats there are. In R, we can use the "unique" function to find the number of different birthdays, then subtract from 25 to find the number of birthday matches ("redundant" birthdays):

```
x <- 25 - length(unique(b))
```

For our run above, the list "`unique (b)`" is the same as `b`, but with the 10th and 25th birthdays removed. It is a list of the 23 unique birthdays since its "`length`" is 23. So the value of the random variable X for this simulated room is $X = 25 - 23 = 2$.

Step Three: *Using a loop to simulate X for many rooms*
If we repeat this process for a very large number of rooms, we obtain many realizations of the random variable X, and thus a good idea of the distribution of X. Counting the proportion of rooms with $X = 0$, we get the approximate probability of no match $P(X = 0)$. Taking the average of these realizations of X, we get a good approximation to $E(X)$.

When we simulated 10,000 such rooms, our result was P(No match) $\approx .4338$, which is close to the exact value 0.4313 calculated using combinatorics. We also obtained $E(X) \approx 0.8081$. Additional runs of the program consistently gave values of $E(X)$ in the interval 0.81 ± 0.02.

The histogram in Figure 4.9 shows the approximate distribution of X–the Number of Birthday Matches. Our approximations would have been more precise if we had simulated more than 10,000 rooms, but the results seem good enough for practical purposes.

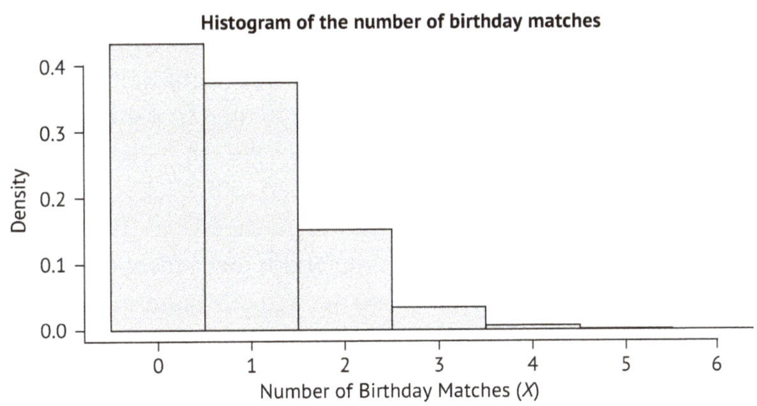

Figure 4.9 The simulated distribution of the number of birthday matches (X) in a room of 25 randomly chosen people.

Testing assumptions
With simulation, it is relatively easy to test the impact of the simplifying assumptions about 365 rather than 366 birthdays and that birthdays are equally likely. Actual 1997–1999 vital statistics for the United States show some variation in daily birth proportions. Monthly averages range from a low of about 94.9% of uniform in January 1999 to a high of about 107.4% in September 1999 (*www.cdc.gov/nchs/products/pubs/pubd/vsus/vsus.htm*). These fluctuations are illustrated in Figure 4.10. Daily birth proportions typically exceed 1/365 from May through September.

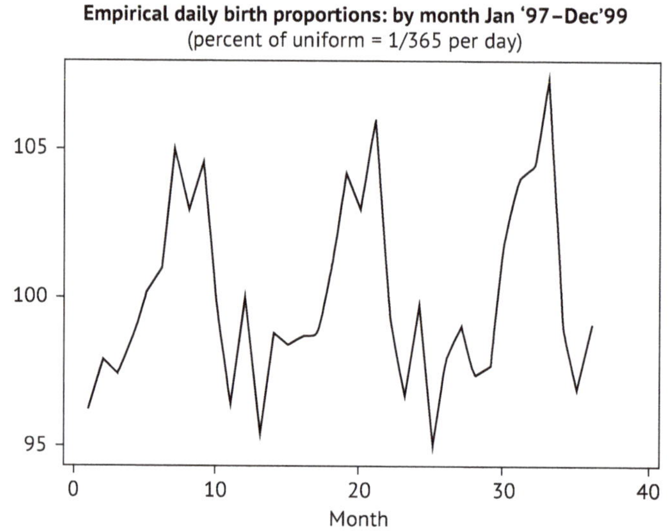

Figure 4.10 Cyclical pattern of birth frequencies in the United States for 36 consecutive months.

For nonuniform birthdays, computing the probability of no matches by analytical methods is beyond the scope of undergraduate mathematics; but using R, it is easy to modify our simulation so that 366 birthdays are chosen according to their true proportions in the United States population—rather than being chosen uniformly. We ran such a simulation and found that within the precision provided by 10,000 simulated rooms (about two decimal places), the results for the true proportions cannot be distinguished from the results for uniformly distributed birthdays. From these and related simulations on birthday matching, we conclude that, although birthdays in the United States are not actually uniformly distributed, it seems harmless in solving the birthday match problem to assume they are. However, important differences in the values of $P(X = 0)$ and $E(X)$ do occur if departure from uniform is a lot more extreme than in the United States population (See Nunnikhoven or Pitman and Camarri).

Using R statistical software

You can download R free of charge online at *www.r-project.org*. The program for doing the birthday matching problem with an explanation of the R code and an elementary tutorial on R are available online at *www.sci.csueastbay.edu/~btrumbo/bdmatch/ index.html*. This website also includes further details of the birthday matching problem in a paper the authors presented at the 2004 Joint Statistical Meetings. Peter Dalgaard provides an introduction to statistics using R in *Introductory Statistics with R*. His book also is available as an electronic book, so check with your library.

Summary comments on simulation

Simulation is an important tool in modern applied probability modeling and in certain kinds of statistical inference. Many problems of great practical importance cannot be solved analytically.

From the birthday matching problem, we can see that—for practical purposes—simulation gives the same answer as does combinatorics for P(No Match) under the simplifying assumption that there are 365 equally likely birthdays. The same simulation provides a value for the expected number of matches, which would be difficult to find by elementary methods. Because this simulation gives what we know to be the correct answer for $P(X = 0)$, credibility is given to the value it gives for $E(X)$.

When we want to drop the uniformity assumption, we enter territory where analytic methods are much more difficult. But a minor modification of the simulation program provides us with values of $P(X = 0)$ and $E(X)$. This allows us to investigate the influence of our simplifying assumptions on results we intend to apply to real life.

In summary, we verified the correctness of a simulation method for an easy problem and then modified it to solve a closely related, but more difficult, problem. This process of building more complex simulation models based upon simpler trusted ones illustrates an important principle for using simulation reliably to solve a wide variety of important practical problems.

References

Peter Dalgaard. *Introductory Statistics with R*, Springer, 2002.

William Feller. *An Introduction to Probability Theory and Its Applications*, Vol. 1, 1950 (3rd ed.), Wiley, 1968.

Thomas S. Nunnikhoven. “A birthday problem solution for nonuniform frequencies.” *The American Statistician*, 46, 270–274, 1992.

Jim Pitman and Michael Camarri. “Limit distributions and random trees derived from the birthday problem with unequal probabilities.” *Electronic Journal of Probability*, 5, Paper 2, 1–18, 2000.

Bruce E. Trumbo, Eric A. Suess, and Clayton W. Schupp. “Using R to Compute Probabilities of Matching Birthdays.” *2004 Proceedings of the Joint Statistical Meetings* [CD-ROM], Alexandria, Virginia: American Statistical Association.

Chapter 5

Probability Models for a Single Discrete Random Variable

- Using probability to make decisions when facing uncertainty is one of the main applications of probability. Consider the following example.

> A florist stocks a perishable flower which costs 50 cents and which is sold at a price of \$1.50 on the first day it is in the shop. Any flowers not sold that first day are worthless and are thrown away. Let X denote the number of flowers that customers order in a randomly selected day. Records of numerous other days in the past reveal that 0 flowers were sold 10% of the days, 1 flower 40% of the days, 2 flowers 30% of the days and 3 flowers 20% of the days.
>
> How many flowers should the florist stock in order to maximize the expected value of the florist's net profit?
>
> **(Goldberg 1960)**

5.1 New representation of a familiar problem

Are you happy with the number and the sex of siblings in your family? Would you have liked to have more or less siblings or siblings of different sex?

We saw in Chapter 4 that if we were to draw at random a family of three siblings, the sample space of this experiment is:

$$S = \{bbb, bbg, bgb, bgg, gbb, gbg, ggb, ggg\},$$

where, for example, *bgg* is a family where the oldest is a boy, the second child is a girl and the third is also a girl. Notice that in Chapter 4 we used $S_1 \times S_2 \times S_3$ where $S_i = \{b, g\}$, to tag that sample space. But once that is learned, we make our life easier from now on by denoting the sample space of the sequence as S, implicitly knowing that it is Cartesian product of the sample space of each of the trials.

The human sex ratio is a very active area of research in biology (Orzack 2016). Science tends to support the notion that $P(b) = P(g) = \frac{1}{2}$, although it all depends on which probability model you assume. Some authors question that assumption, based on observation of families and government data (Carlton and Stansfield 2005). Under the assumption of even chance for each sex, we saw in Chapter 4 that the probabilities of each of those outcomes in S are, respectively:

$$\left(\frac{1}{2}\right)^3, \left(\frac{1}{2}\right)^3, \left(\frac{1}{2}\right)^3, \left(\frac{1}{2}\right)^3, \left(\frac{1}{2}\right)^3, \left(\frac{1}{2}\right)^3, \left(\frac{1}{2}\right)^3, \left(\frac{1}{2}\right)^3$$

If what concerns us is k, the number of boys in a family of 3, we learned in Chapter 4 that this can be easily computed by

$$P(k \ boys|n,p) = \binom{n}{k} p^k (1-p)^{(n-k)}, \quad k = 0, 1, 2, 3.$$

That is a very convenient formula. We can find the probability of k successes in n Bernoulli trials, where the probability of a success is p. We could even make it easy for those scared of formulas and put all the information represented by that formula in Table 5.1. Let Y be the number of boys, and let us represent a particular value of Y by the lower case letter y. *Y is a random variable because its value depends on outcomes of a random experiment represented by S.*

Table 5.1 The first two columns are the probability mass function of Y.

y	$P(Y = y)$	*Events in S*	$P(event) = P(Y = y)$
0	$\left(\frac{1}{2}\right)^3$	{ggg}	$\left(\frac{1}{2}\right)^3$
1	$3\left(\frac{1}{2}\right)^3$	{bgg,gbg,ggb}	$\left(\frac{1}{2}\right)^3 + \left(\frac{1}{2}\right)^3 + \left(\frac{1}{2}\right)^3$
2	$3\left(\frac{1}{2}\right)^3$	{bbg,bgb,gbb}	$\left(\frac{1}{2}\right)^3 + \left(\frac{1}{2}\right)^3 + \left(\frac{1}{2}\right)^3$
3	$\left(\frac{1}{2}\right)^3$	{bbb}	$\left(\frac{1}{2}\right)^3$

We could also represent the probability of each value of Y by using the following formula:

$$P\left(Y = y|n = 3, p = \frac{1}{2}\right) = \binom{3}{y} (1/2)^y (1-1/2)^{(3-y)}, \qquad y = 0, 1, 2, 3.$$

This formula would provide the part of Table 5.1 containing y and $P(Y = y)$. That formula is the formula for the probability mass function of a binomial discrete random variable.

The first two columns of Table 5.1 represent in a table a probability mass function for Y. A probability mass function gives all the unique possible values of the random variable and the probability of the events consisting of all outcomes to which we map that value of the random variable.

How does all that differ from what we have been doing so far? It gives us all possible questions we could ask about the number of boys in a family of three in one single table or formula.

A random variable with a finite number of values (not matter how large) n, whose probabilities can be computed by plugging in y, n, and p in that formula, is called a *Binomial random variable*. If $p = 1/2$ and $n = 3$ then we get the results of Table 5.1.

The binomial probability model is at the heart of the debate as to whether the probability of being a boy equals the probability of being a girl. But it is at the heart of many more things, particularly in statistics when the interest of statisticians is to estimate the p (for example, the proportion of shoppers who responded to sweepstakes for a free car from Toyota).

Using the definition of *expectation*, *variance, standard deviation* and *moment generating function* that we will introduce in this chapter, we will find that for the binomial model those important properties are:

$$E(Y) = np,$$
$$Var(Y) = np(1-p),$$
$$SD(Y) = \sqrt{np(1-p)},$$
$$M_y(t) = (pe^t + (1-p))^n.$$

Knowing this, and the formula or the table, we can conclude that for the family of 3, the expected number of boys is 3/2, the variance is 3/4, and the standard deviation is 0.866. We could also compute the probability of less than 2 boys by extracting the information from the table:

$$P(Y < 2) = P(Y = 0) + P(Y = 1) = \left(\frac{1}{2}\right)^3 + \left(\frac{1}{2}\right)^3 + \left(\frac{1}{2}\right)^3 + \left(\frac{1}{2}\right)^3.$$

We can not deny how convenient it would be to have formulas and tables like these for a host of random phenomena that might step in our way. That is new jargon, but its usefulness will be seen shortly. An advantage of a probability model formula is its usefulness to model our uncertainty about many phenomena of similar nature and help us make predictions and better decisions when faced with uncertainty. The above binomial model could be used to calculate any question that asks: what is the number of successes in n Bernoulli trials? For example,

- The probability of k heads in n tosses of a coin when probability of a head is p.
- The probability of k defective components in a system with n components where the probability of defectives is p.
- The probability of finding k qualified applicants in a pool of n applicants for a job when the probability that an applicant is qualified is p.
- The probability of finding k sparrows in a sample of n birds in a region known to have a proportion p of sparrows.
- The probability that there are k myopic persons in a group of unrelated adults over age 40 when the probability that a person is myopic in this group is p.
- Many other phenomena involving Bernoulli trials, where we are interested in the number of n trials that are successes when the probability of success is p.

Y, the number of boys, is what we call in probability theory a random variable.

Box 5.1

Properties of discrete random variables

If X is a discrete *random variable* with a probability mass function P(X),

$\mu = E(X) = \sum_x xP(X = x)$ is the *expected value* of X,

$\sigma^2 = Var(X) = \sum_x (x-\mu)^2 P(X = x)$ is the *variance* of X,

$\sigma = SD(X) = \sqrt{Var(X)}$ is the standard deviation of X.

The moment generating function of the random variable is

$$M_X(t) = E(e^{tx}) = \sum_x e^{tx} P(X = x),$$

and can be used, by taking derivative of the function with respect to t and evaluating it at $t = 0$, to obtain the expected value. If we take the second derivative and evaluate it at $t = 0$, we get the $E(X^2)$.

5.2 Random variables

We just saw in the previous section that all we want to know about the number of successes in n Bernoulli trials can be conveniently summarized in a probability mass mathematical function, its expectation, variance, standard deviation and moment generating function. The reader may wonder whether we can do that for the other problems we have studied, such as number of successes in trials that are not Bernoulli, for example, or other questions we may have about Bernoulli trials. Indeed, we can do that and more. The concept of random variable and probability mass function are at the core of such possibility.

Definition 5.2.1

A function whose domain is a sample space and whose range is the set of real numbers is called a *random variable*. If the random variable is denoted by X and has as domain the sample space $S = \{o_1, o_2, \ldots\ldots, o_n\}$, then we write $X(\{o_{k,\ldots}, o_j\})$ for the value of X that is shared by each of the elements in the event $\{o_{k,\ldots}, o_j\}$.

Random variables can be *discrete* or *continuous*. A discrete random variable can have a finite or countably infinite number of values. A continuous random variable has an uncountably infinite number of values.

It is conventional to represent a random variable by an upper case letter like *X, Y, W*, etc. and a particular value of the random variable by the corresponding lower case letter, *x, y, w* etc.

5.2.1 The probability mass function of a discrete random variable

Random variables receive their name from the fact that prior to a random draw from the sample space, one has no idea of the value that the random variable function will generate. The values obtained as repeated drawing is carried out will vary randomly as the events in the sample space are randomly chosen.

For each type of experiment, there is a sample space and associated with it several possible random variables.

Example 5.2.1

Consider two fair six-sided dice are tossed. The sample space of this experiment is (see Figure 2.3, Chapter 2):

$$S = \{(1,1), (1,2), \ldots\ldots, (6,6)\},$$

i.e, all the 36 ordered pairs of numbers between 1 and 6.

Let X assign to each outcome (a, b) in S the maximum of its numbers, i.e. $X(a,b) = max(a,b)$. Then X is a random variable that can take the possible values 1 to 6.

The event that X has the value 5, for example, is the subset of S,

$$\{(5,1),(5,2),(5,3),(5,4),(5,5),(1,5),(2,5),(3,5),(4,5),(5,5)\},$$

containing those elements of S, o_k, for which $X(o_k) = 5$. The probability that X = 5 then is the sum of the probabilities of all those outcomes:

$$P(X=5) = P(\{5,1\}) + P(\{5,2\}) + P(\{5,3\}) + P(\{5,4\}) + P(\{5,5\}) + P(\{1,5\}) \\ + P(\{2,5\}) + P(\{3,5\}) + P(\{(4,5)\}).$$

Definition 5.2.2

A *probability mass function* for a discrete random variable X, denoted $P(X)$, is a listing of the unique values of the random variable and the probabilities of these unique values, which are in turn the probability of the event in S consisting of all outcomes that have that value of X. This listing may be given in the form of a table (when X does not have a very large number of values), or in the form of a formula. More technically, the function P whose value for each real number x is given by $P(X = x)$ is called the probability mass function of the random variable X.

Assigning probabilities to the possible values of the random variable requires that we calculate the probability of all the outcomes in the sample space that result in that value of X and add them up.

$$P(X = x) = \sum P(o_k \in S : X(o_k) = x)\cdot$$

For another example, $P(X = 3)$ is the same as the probability of the event

Box 5.2

Random variables are peculiar functions

In Mathematics, a function is specified whenever we are given a set of elements (the domain of the function), together with a rule by which one and only one number is associated with each element of the domain. The number which the rule assigns to an element of the domain is called the value of the function for (or at) that element. The set of all values of a function is called the range of the function. For example, $y = x^2$ determines a function whose domain is the set of all real numbers and whose range is the set of nonnegative real numbers. A random variable, say X, is a very peculiar function. Its domain is events in the sample space S consisting of outcomes that share the same value of X.

When using random variables, the probability of an event is defined in terms of X.

$$P(X = x) = P(\{event\ cointaining\ all\ outcomes\ in\ S\ for\ which\ X = x\})\cdot$$

$A = \{(1,3),(3,1),(2,3),(3,2),(3,3)\}$. Similarly, assuming that each outcome in S is equally likely (which we can assume if the dice are fair and the rolls are independent),

$$P(X = 1) = P(\{1,1)\}) = 1/36,$$
$$P(X = 2) = P(\{(2,1),(1,2),(2,2)\}) = 3/36,$$
$$P(X = 3) = P(\{(1,3),(3,1),(2,3),(3,2),(3,3)\}) = 5/36,$$
$$P(X = 4) = P(\{(4,1),(1,4),(4,2),(2,4),(4,3),(3,4),(4,4)\}) = 7/36,$$
$$P(X = 5) = P(\{(5,1),(5,2),(5,3),(5,4),(5,5),(1,5),(2,5),(3,5),(4,5)\}) = 9/36,$$
$$P(X = 6) = P(\{6,1),(6,2),(6,3),(6,4),(6,5),(6,6),(1,6),(2,6),(3,6),(4,6),(5,6)\}) = 11/36.$$

Once all the work is done to figure out the probability for each unique value of X, we present the information in a discrete probability mass function table like Table 5.2:

Table 5.2

X	**1**	**2**	**3**	**4**	**5**	**6**
P(X = x)	1/36	3/36	5/36	7/36	9/36	11/36

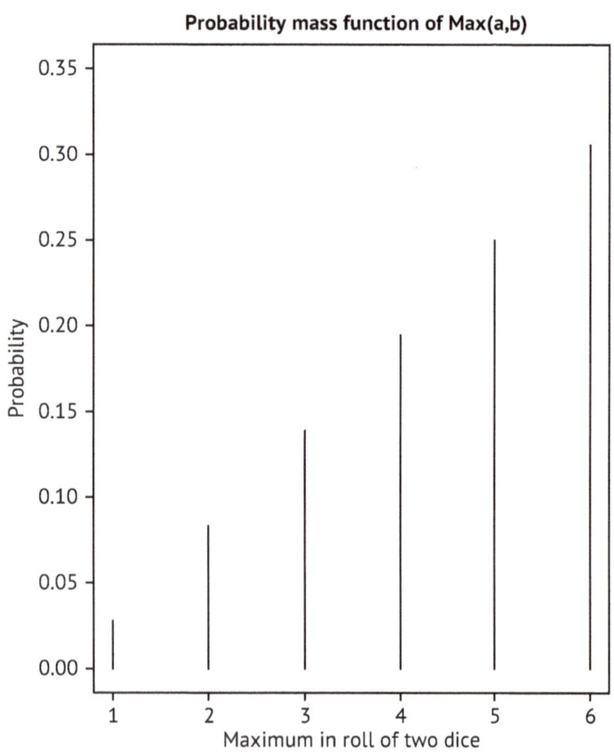

Figure 5.1 Probability mass function plot for the highest number in the roll of two fair six-sided dice.

This is called the *probability mass function (pmf) of X* and is usually given in the form of a table, or a mathematical formula. The sum of the probabilities must be one and each probability must be equal to or greater than 0, in order to satisfy the Axioms introduced in Chapter 3.

This information can also be presented graphically, as in Figure 5.1, where the probability table of the random variable X is drawn. In the graph, values of X are on the horizontal X-axis, and values of $P(X = x)$ on the vertical Y-axis.

Once the probability mass function of X is known, we can compute probabilities on intervals of X.

$$P(x_1 \le X \le x_2) = \sum_x P(\{o_k \in S : x_1 \le X(o_k) \le x_2\}),$$

where x_1, x_2 are specific values of X.

For example, looking at Table 5.2, we can see that

$$Prob(X < 5) = P(X = 0) + P(X = 1) + P(X = 2)$$
$$+ P(X = 3) + P(X = 4) = 1/36 + 3/36$$
$$+ 5/36 + 7/36 = 16/36$$

Box 5.3

Summation operators

The rules of summation are widely used when talking about the probability of a single discrete random variable. We review here:

If x takes n values $x_1, x_2, \ldots, x_n$ then their sum is

$$\sum_{i=1}^{n} x_i = x_1 + x_2 + \ldots, + x_n$$

If a is a constant then

$$\sum_{i=1}^{n} ax_i = a\sum_{i=1}^{n} x_i$$

$$\sum_{i=1}^{n}(x_i - a)^2 = \sum_{i=1}^{n}(x_i^2 + a^2 - 2ax_i)$$

$$= \sum_{i=1}^{n}(x_i^2) + \sum_{i=1}^{n}(a^2) - \sum_{i=1}^{n}(2ax_i) = \sum_{i=1}^{n}(x_i^2) + na^2 - 2a\sum_{i=1}^{n}(x_i).$$

If X and Y are two variables, then the following two equalities hold:

$$\sum_{i=1}^{n}(x_i + y_i) = \sum_{i=1}^{n} x_i + \sum_{i=1}^{n} y_i,$$

$$\sum_{i=1}^{n}(ax_i + by_i) = a\sum_{i=1}^{n} x_i + b\sum_{i=1}^{n} y_i.$$

The following function will be very important when we study the central limit theorem.

$$\bar{x} = \frac{\sum_{i=1}^{n} x_i}{n} = \frac{x_1 + x_2 + \ldots + x_n}{n},$$

and

$$\sum_{i=1}^{n}(x_i - \bar{x}) = 0.$$

Can you prove this last result?

Abbreviated forms of the summation notation:

$$\sum_{i} x_i$$

or

$$\sum_{x} x$$

5.2.2 The cumulative distribution function of a discrete random variable

Definition 5.2.3

Let X be a random variable. The cumulative distribution function F (cdf) of X is the function $F: R -> R$ defined by

$$F(x) = P(X \leq x) = \sum_{x_i < x} P(x_i).$$

That is, it is the sum of the probabilities of all values of X smaller than or equal to specific value x.

If X is a discrete random variable with probability mass function $P(X)$, then F is a step function.

In either case, F is monotonic increasing, i.e, $F(a) \leq F(b)$ whenever $a \leq b$ and the limit of F to the left is 0 and to the right is 1. The cumulative distribution function may also be summarized in a table like Table 5.3 granted that the number of possible values of X is finite and not too many to list.

Example 5.2.2

For the experiment of Example 5.2.1, the cumulative distribution function is given in Table 5.3, where the computations done to arrive to the value of the F(x) are given inside the table for convenience. Usually, those computations would be done outside the table.

Table 5.3 Cumulative probability of the max(a,b) in the roll of two six-sided fair dice

x	1	2	3	4	5	6
F(x)	1/36	3/36 + 1/36 = 4/36	5/36 + 3/36 + 1/36 = 9/36	7/36 + 5/36 + 3/36 + 1/36 = 16/36	9/36 + 7/36 + 5/36 + 3/36 + 1/36 = 25/36	11/36 + 25/36 = 36/36

We can see in Table 5.3 that:

$$F(3) = P(X \leq 3) = 9/36.$$

We can also see that:

$$P(2 \leq X \leq 5) = F(5) - F(1) = \frac{25}{36} - \frac{1}{36} = \frac{24}{36}.$$

Using Table 5.2, we would have computed the same probability as

$$P(2 \leq X \leq 5) = P(X = 2) + P(X = 3) + P(X = 4) + P(X = 5) = \frac{3}{36} + \frac{5}{36} + \frac{7}{36} + \frac{9}{36}$$
$$= \frac{24}{36}.$$

5.2.3 Functions of a discrete random variable

Once a random variable and its probability mass function have been defined, it is straightforward to compute probabilities for functions of the random variable. For example, if X denotes the number of Amber alerts per year in a small community, and X has probability mass function as in Table 5.4,

Table 5.4 Probability mass function of number of amber alerts per year

x	**0**	**1**	**2**
P(X=x)	0.85	0.1	0.05

and if each Amber alert costs the city a value Y, where Y is a function of X, $g(X)$, as follows,

$$Y = 1000 + 200X,$$

then the probability mass function of Y is directly related to that of X, because Y takes, for example, value 1000 when $X = 0$ and that happens with probability 0.85, which makes the probability that $Y = 1000$ equal to 0.85. We can illustrate that with Table 5.5.

Table 5.5 Probability mass function of a function of a discrete random variable

x	**0**	**1**	**2**
y	$1000 + 200(0) = 1000$	$1000 + 200(1) = 1200$	$1000 + 200(2) = 1400$
P(X = x)	0.85	0.1	0.05

Thus,

$$P(Y > 1000) = P(Y = 1200) + P(Y = 1400) = P(X = 1) + P(X = 2) = 0.15.$$

5.2.4 Exercises

Exercise 1. Let Y be a random variable denoting the sum of the roll of two fair six-sided dice. Give a table with the probability mass function of Y in the first two columns, the cumulative probability mass function of Y in the third column and the corresponding sample space members in the last column. Use Tables 5.1 to 5.3 in Section 5.2 as examples.

Exercise 2. (This exercise is based on an exercise in Mansfield (1994).) CycleTheWorld sells bicycles. Based on history of the store, it is known that in May it is equally likely that the store will sell 0, 1, 2, 3, or 4 bicycles in a day. The store has never sold more than 4 bicycles in a day. It is February and CycleTheWorld needs to start planning for the months ahead. (i) Write the table containing the probability mass function for the number of bicycles sold in a random May day. (ii) What is the probability that the number of bicycles sold in a given May day will be less than 3? (iii) CycleTheWorld has only one sales person in the store, whose income depends on the number of bicycles the sales person sells per day. Specifically, there is no commission on the first bicycle sold in a day, a \$20 commission on the second bicycle sold in a day, a \$30 commission on the third, and a \$40 commission on the fourth (thus if there are three bicycles sold in a given day, the commission for that day is $50). The income of this sales person is certainly a random variable because it depends on the random amount sold. Write the probability mass function table for the income that the sales person can make in a May day (assume no deductions or income taxes). Make sure to define the notation you use. For example, Y = income.

Exercise 3. Daily tooth brushing by residents in a remote country was found to follow the probability mass function in Table 5.6, where the random variable Y represents the number of times brushing teeth per day

Table 5.6 Probability mass function of frequency of teeth brushing per day

y	0	1	2	3
P(Y = y)	0.325	0.474	0.15	0.051

The value 0.474 means that 47.4% of the residents brush their teeth once per day. What is the probability that a randomly chosen resident of this country brushes teeth at least twice a day?

5.3 Expected value, variance, standard deviation and median of a discrete random variable

It is convenient to have numeric summaries of the distribution of a discrete random variable. The most widely used summaries are the *expected value* and the *standard deviation* of a random variable. The *variance* must be known before we compute the standard deviation.

These summaries should not be underestimated. In fact, most decision making nowadays is done based on expectations and standard deviations. Consider, for example, a computer scientist that needs to make a decision between several algorithms to sort numbers. The efficiency of the algorithm will depend on the number of comparisons that need to be made before all the numbers are sorted (Sanchez, 2009). The expected number of comparisons will be the right metric to use to judge the efficiency. This is harder than the level of this book requires, but the reader is encouraged to read further in Ross's Probability models for computer scientists (Ross, 2002).

5.3.1 The expected value of a discrete random variable

Example 5.3.1

For the random variable $X = max(a,b)$, the maximum value of the rolls of two fair six-sided dice seen in Examples 5.2.1 and 5.2.2, the expected value of X is, according to Definition 5.3.1,

$$\mu_X = E(X) = 1(1/36) + 2\,(3/36) + 3\,(5/36) + 4\,(7/36) + 5\,(9/36) + 6\,(11/36) = 4.472.$$

As we can see in Figure 5.1, some values of X are above or below the μ but most of them will be around the average. However, for a skewed left distribution like that the expected value is not a very representative metric for average. A more appropriate summary would be the median. We define the median later in this section.

The way in which the values of X are spread around the expected value is described by the variance and the standard deviation. The variance is the expectation of a function of X.

Definition 5.3.1

The expected value of a discrete random variable X with probability mass function $P(X = x)$ is given by

$$\mu_X = E(X) = \sum_x xP(X = x).$$

The sum is over all values of X for which $P(X = x) > 0$. We denote $E(X)$ with the greek letter μ. The $E(X)$ is a characteristic of the probability mass function of X, not of data that we may be observing. We use different notation for observed data. We assume absolute convergence when the range of X is countable; we talk about an expectation only when it is assumed to exist. The expected value of a random variable is the center of gravity of the probability mass function (in the discrete case) or the density function (in the continuous case seen in Part II of this book). It is representative of the average of the distribution only if there are no extreme values skewed to one side of the distribution.

5.3.2 The expected value of a function of a discrete random variable

Example 5.3.2

Consider again Example 5.2.1. The probability mass function was given in Table 5.2. We show in Table 5.7 how we construct the probability mass function of the function of X defined by $Y = g(X) = X^2$

Table 5.7. Probability mass function of nonlinear function of a discrete random variable.

$y = x^2$	1	$2^2 = 4$	$3^2 = 9$	$4^2 = 16$	$5^2 = 25$	$6^2 = 36$
x	1	2	3	4	5	6
P(X = x)	1/36	3/36	5/36	7/36	9/36	11/36

We can see that X^2 takes value 9, for example, when X takes value 3, i.e.,

$$P(Y = 9) = P(X^2 = 9) = P(X = 3) = 5/36.$$

Therefore, the random variable $Y = g(X) = X^2$ has expected value

$$E(Y) = E(X^2) = 1(1/36) + 4\,(3/36) + 9\,(5/36) + 16\,(7/36) + 25\,(9/36) + 36\,(11/36) = 21.97222.$$

Definition 5.3.2

Let $g(X)$ be a function of the random variable X. Then $g(X)$ is also a random variable, with probability mass function directly related to that of X, as we saw in Section 5.2.3. The expected value of a real-valued function of a discrete r.v. X, $g(x)$, is given by

$$\mu_{g(X)} = E(g(X)) = \sum_x g(x)P(X = x)$$

5.3.3 The variance and standard deviation of a discrete random variable

The variance is a very special function of X. It is formed by first measuring the distance of each value of X from its expected value, then squaring the distances.

Definition 5.3.3

Let $g(X) = ((X - \mu))^2$ represent the square distance of a random variable from its expected value. This is a very special function of X. Some values of X will be very far and others very close. Small variance will mean that most of the values are very close to the expected value, and few far.

$$\sigma_X^2 = Var(X) = E[(X - \mu_X)^2] = \sum_x (x - \mu_X)^2 P(x).$$

We denote the variance with the Greek symbol σ^2. The smallest value that σ^2 can take is 0, when all the probability is concentrated at a single point (that is, when X takes on a constant value with probability 1). The variance becomes larger as the points with positive probability spread more.

Definition 5.3.4

The standard deviation of a random variable X is the square root of the variance:

$$\sigma = SD(X) = \sqrt{Var(X)}.$$

The SD is denoted by the greek letter σ. It is more convenient to use the standard deviation as a measure of spread of the values of the random variable around the expected value because the standard deviation has the same units as the random variable whereas the variance has the units squared.

Example 5.3.3

Returning to Example 5.2.1,

$$\begin{aligned}\sigma_x^2 &= Var(X) = (1-4.472)^2\ (1/36) + (2-4.472)^2\ (3/36)\\ &+ (3-4.472)^2\ (5/36) + (4-4.472)^2\ (7/36)\\ &+ (5-4.472)^2 (9/36) + (6-4.472)^2\ (11/36) = 1.9714.\end{aligned}$$

Example 5.3.4

Returning to Example 5.2.1,

$$\sigma_x = SD(X) = \sqrt{Var(X)} = \sqrt{1.9714} = 1.404778.$$

- Large standard deviation means that many values of the random variable are very far from the expected value, and few are close.
- Small standard deviation means that most of the values of the random variable are concentrated around the expected value with very few far.
- The smallest value that the standard deviation can take is 0, when all the probability is concentrated at a single point.

Definition 5.3.5

The moment generating function of a discrete random variable X is defined as the expectation of a nonlinear function of the random variable:

$$M_X(t) = E(e^{tx}) = \sum_x e^{tx} P(x).$$

5.3.4 The moment generating function of a discrete random variable

This function helps us find the expected value of powers of a random variable. The function also is a powerful tool to do many proofs in probability theory that would otherwise be very difficult to do. Its usefulness will become more apparent in the second part of the book.

Example 5.3.5
For example, the probability mass function in Table 5.8

Table 5.8.

x	**1**	**2**	**5**
P(X = x)	1/8	1/4	5/8

has moment generating function:

$$M_x(t) = \frac{1}{8}e^t + \frac{1}{4}e^{2t} + \frac{5}{8}e^{5t}.$$

A nice feature of the moment generating function is that if you take the first derivative with respect to t and evaluate it at $t = 0$, you get the expected value of the random variable. If you take the second derivative and evaluate it at $t = 0$, you get $E(X^2)$ and so on.

5.3.5 The median of a discrete random variable

The median of a discrete random variable X is the value of X such that

$$P(X \leq median) = F(median) = 0.5 = P(X > median) = 1 - F(median).$$

When the random variable is discrete, the median, like the expected value, may take values that are not in the domain. For example, for the random variable $X = max(a,b)$, the maximum value of the rolls of two fair six-sided dice seen in Examples 5.2.1 and 5.2.2,

$$\mu_x = E(X) = 1(1/36) + 2\,(3/36) + 3\,(5/36) + 4\,(7/36) + 5\,(9/36) + 6\,(11/36) = 4.472.$$

However,

$$F(4) = \frac{16}{36} = 0.44444, \quad F(5) = \frac{25}{36} = 0.6944444,$$

which means that the median is between $X = 4$ and $X = 5$.

5.3.6 Variance of a function of a discrete random variable

The variance of a function of a discrete random variable, $g(x)$, is:

$$\sigma^2_{g(X)} = Var(g(x)) = \sum_x (g(x) - E(g(x)))^2 P(x).$$

We will see some examples of this in Section 5.4.

5.3.7 Exercises

Exercise 1. A resident of Boston spends the Summers in the Grand Tetons, Wyoming. Every day there is expectation that a moose may pass in front of the house. Moose are wild animals that live around that area. The daily sighting (number of times seen) of moose has the probability mass function in Table 5.9, where X is the daily number of sightings:

Table 5.9. Daily sighting of moose

x	**0**	**1**	**2**
$P(X = x)$	0.1	0.5	0.4

In a randomly chosen day, what is the expected number of sightings and the standard deviation of sightings? (ii) Find the moment generating function of X. Then take the first derivative with respect to t, evaluate the first derivative at $t = 0$ and compare the result to the expected value of X found with the definition of expected value.

Exercise 2. Let X be a random variable defined as follows:

$$X = number\ \ of\ \ heads - number\ \ of\ \ tails$$

when a coin is tossed three times. Find (i) the probability mass function of X and compute the (ii) expected value of X and the (iii) standard deviation of X. Find also the (iv) moment generating function of X. A few elements of the pmf can be seen in Table 5.10.

Table 5.10. Probability mass function of "number of heads-number of tails" (to complete by reader)

x	**P(X = x)**
−3	$1/8 = 0.125$
−1	$3(1/8) = 0.375$

Exercise 3. You have $1,000 and a certain commodity presently sells for $2 per ounce. Suppose that after one week the commodity will sell for either $1 or $4 an ounce, with these two possibilities being equally likely. If your objective is to maximize the expected amount of money that you possess at the end of the week, what strategy should you employ?

Exercise 4. (This exercise is based on Christensen (2015).) You are the forecaster responsible for hurricane warnings on the southeast coast of the United States in September. The cost of issuing a warning like a hurricane involves people taking shelter, business stopping, the area's economy paused—a moderate cost of C dollars. You also know the preventable loss should a hurricane come and the area be unprepared: property damaged, lives lost—an extremely high loss of L dollars. Your weather forecast indicates hurricane with a probability of p. (i) Should you issue a warning? Use expectations to create a decision rule. (ii) For what kinds of extreme weather warnings is the decision more likely to be "do not warn" than "warn"? Explain.

5.4 Properties of the expected value and variance of a linear function of a discrete random variable

Quite often, we are interested not in a random variable per se, but in a function of a random variable. For example, a business selling a very specialized tool to automobile producers may be more interested in the profits made each week. Of course, the profits depend on how many tools are sold during the week, which is random. But the ultimate goal is to compute the expected profit per week and the give or take, the standard deviation of the profit. The contents of this section will be helpful in computing any linear function of a discrete random variable.

Theorem

Let X be a discrete random variable and let $g(X)$ be a function of X. Then

i. If $g(X) = k$, where k is a constant,

$$E(k) = k \text{ and } Var(k) = 0.$$

ii. If $g(X) = kX$, then

$$E(kX) = kE(X) \text{ and } Var(kX) = k^2 \; Var(X).$$

iii. If a and b are constants and $g(X) = a + b(X)$, then

$$E(a + bX) = a + bE(X) \text{ and } Var(X) = b^2 \; Var(X).$$

We can prove properties *(i)* and *(iii)* using the definitions of expected value and variance given in Sections 5.3.1 and 5.3.3. The reader is asked to prove *(ii)* in the exercises. We always use the definitions to prove results.

Proof

i. If $g(X) = k$, where k is a constant,

$$E(X) = k \text{ and } Var(X) = 0.$$

Using the definition in Section 5.3.1, for $g(X) = k$, and Math Tibdit box 5.3,

$$\mu_{g(X)} = E(g(X)) = \sum_x g(x)P(x) = \sum_x kP(x) = k\sum_x P(x) = k(1) = k,$$

because the sum of the probabilities all the values of X must equal 1.

The variance is

$$\sigma^2_{g(X)} = Var(g(x)) = \sum_x (g(x) - E(g(x)))^2 P(x) = \sum_x (k - E(k))^2 P(x) = \sum_x (k - k)^2 P(x) = 0.$$

ii. If a and b are constants and $g(X) = a + b(X)$, then

$$E(a + bX) = a + bE(X) \text{ and } Var(X) = b^2 \; Var(X).$$

$$\mu_{g(X)} = E(g(X)) = \sum_x g(x)P(x) = \sum_x (a+bx)P(x) = \sum_x (aP(x) + bxP(x)) =$$

$$\sum_x (aP(x)) + \sum_x (bxP(x)) = a\sum_x P(x) + b\sum_x xP(x) = a(1) + bE(X) = a + bE(X).$$

The variance is

$$\sigma^2_{g(X)} = Var(g(x))$$

$$= \sum_x (g(x) - E(g(x)))^2 P(x) = \sum_x (a + bx - (a + bE(x)))^2 P(x)$$

$$= \sum_x (b(X - E(X))^2 P(x) = \sum_x b^2(X - E(X))^2 P(x)$$

$$= b^2 \sum_x (X - E(X))^2 P(x) = b^2 Var(X).$$

5.4.1 Short-cut formula for the variance of a random variable

As we said in section 5.3.2, if X is a random variable,

$$\sigma^2 = Var(X) = E[(X-\mu)^2] = \sum_x (x-\mu)^2 P(x).$$

We will use that definition now to prove that:

$$\sigma^2 = Var(X) = E(X^2) - \mu^2.$$

Proof

$$\sigma^2_x = \sum_x (x-\mu)^2 P(x) = \sum_x (x^2 + \mu^2 - 2\mu x)P(x)$$

$$= \sum_x (x^2)P(x) + \sum_x (\mu^2)P(x) - \sum_x (2\mu x)P(x) =$$

$$= E(X^2) + \mu^2 \sum_x P(x) - (2\mu)\sum_x xP(x) = E(X^2) + \mu^2 - 2\mu^2 = E(X^2) - \mu^2.$$

Example 5.4.1

Daily sales records for a car dealership is a random variable with probability mass function given in Table 5.11. (i) Consider three independent days. What is the probability that in less than 2 of those days the number of sales will be at least 2? (ii) The price of a car is \$3000 and total daily revenue is given by $T = 3000X^2$, where X is the number of cars sold. What is the expected total daily revenue?

Table 5.11 Probability mass function of daily sales

x	**0**	**1**	**2**	**3**
$P(X = x)$	0.5	0.3	0.15	0.05

(i) $P(X \geq 2) = P(X = 2) + P(X = 3) = 0.15 + 0.05 = 0.2.$

Thus the probability that in a randomly chosen day the dealership will sell at least 2 cars is 0.2.

Let Y be a new random variable representing the number of days, out of three, in which a success (sell at least two) occurs. P(success) = 0.2 = p. The event in question is

$$A = \{fff, sff, ffs, fsf\},$$

where s is success (sell at least two).

$$P(A) = P(Y < 2) = P(Y = 1) + P(Y = 0) = 3(0.8^2)0.2 + 0.8^3 = 0.896.$$

For part (ii),

$$E(T) = 3000E(X^2) = 3000(\sigma_X^2 + \mu_X^2) = 4050.$$

5.4.2 Exercises

Exercise 1. Prove that If $g(X) = kX$, then

$$E(kX) = kE(X) \text{ and } Var(kX) = k^2\ Var(X)$$

Exercise 2. The manager of a cosmetic products stand in a department store knows that the daily demand for the most expensive item in the stand, the "dramatically beautifying moisturizing lotion" has the probability mass function in Table 5.12:

Table 5.12 Probability mass function of daily demand for expensive cosmetic item

Quantity demanded	0	1	2
Probability	0.1	0.5	0.4

Suppose that the bonus is $10 each time an item is bought. (i) What is the expected daily bonus of selling the expensive item? How much is this bonus expected to vary from one day to another? (ii) What is the $E(X^3)$?

Exercise 3. Weekly downtime of internet services from an internet service provider (in hours) has expected value 0.5 and variance 0.25. Based on past experience, the data scientist of a retailer store has calculated the loss function to the store from the downtime as

$$C = 30X + 2X^2$$

where X is the amount of weekly downtime and C is cost. Find the expected cost.

Exercise 4. The number of calories burnt by a biker on a biking day depends on the number of hours biking plus the fixed amount burnt by the regular functioning of the body to stay alive. Based on past experience it is known that the calories burnt by a biker follows this function

$$Calories = 1000 + 200X$$

where X is the number of hours biking. If X is a random variable with expected value 5 and variance 5, what is the expected number of calories burnt and the standard deviation of the calories?

5.5 Expectation and variance of sums of independent random variables

A fair six-sided die is rolled 100 times. How can we calculate the expected sum and the variance of the sum of the numbers obtained?

If we assume that the dice are of similar quality, then we can calculate the expected sum of the 100 rolls and the variance of the sum of the 100 rolls as follows:

$$E(S_{100}) = E\left(\sum_{i=1}^{100} X_i\right) = \sum_{i=1}^{100} E(X_i) = \mu + \mu + \ldots\ldots + \mu = n\mu = 100\left(\frac{7}{2}\right) = 350$$

$$Var(S_{100}) = Var\left(\sum_{i=1}^{100} X_i\right) = \sum_{i=1}^{100} Var(X_i) = \sigma^2 + \sigma^2 + \ldots\ldots + \sigma^2 = 100(2.916667) = 291.6667$$

Dice are not the only context in the applications of probability where we are interested in sums of random variables. For example,

- The study of the number of cyberattacks to a computer network might require knowledge of the sum of the random number of cyberattacks each hour of the day. The number of cyberattacks of each hour is a random variable and the sum of the independent numbers of cyberattacks of each of the 24 hours of the day is a sum of random variables giving us the total number of cyberattacks per day.
- The random total cost of a building project can be studied as the sum of the random costs for the major independent components of the project.
- The random size of an animal population can be modeled as the sum of the random sizes of the independent colonies within the population.
- At the end of the summer the total weight of seeds accumulated by a nest of seed-gathering ants will vary from nest to nest. We may be interested in the sum of the total weights of seeds of all nests.
- The total weight of people riding an elevator is important to know to prevent overloading the particular elevator.
- An insurance company may want to know the total yearly claim by all the automobile policy holders.

Let's define the sum of n random independent and identically distributed (same mean, same variance) random variables as follows:

$$S_n = X_1 + X_2 + \ldots.. + X_n = \sum_{i=1}^{n} X_i,$$

where S is for sum and n for how many random variables we are adding. Then, without proving it, we claim that:

$$E(S_n) = E\left(\sum_{i=1}^{n} X_i\right) = \sum_{i=1}^{n} E(X_i) = \mu + \mu + \ldots\ldots + \mu = n\mu,$$

$$Var(S_n) = Var\left(\sum_{i=1}^{n} X_i\right) = \sum_{i=1}^{n} Var(X_i) = \sigma^2 + \sigma^2 + \ldots\ldots + \sigma^2 = n\sigma^2.$$

Sums of independent and identically distributed random variables are of central importance in both Probability and Statistics.

The proof of this result will be done in Chapter 6.

Example 5.5.1

A stockbroker recommends three stocks (ORCL, AAPL and Microsoft and GOOGL) to his firm's clients. His bonus will be \$50,000, \$30,000, \$10,000, or nothing, depending on whether the prices of 3, 2, 1, or none of them go up next year. Suppose that the probability that each stock goes up next year is 1/2, and each stock's price behavior is independent of the price behavior of the other stocks. (i) Construct a table with the probability mass function of the bonus that the stockbroker will receive next year. (ii) If the stock broker recommends the same stocks each year, and all other factors are constant, what is the expected total bonus received after five years of predictions; what is the standard deviation?

We will write Table 5.13 with the probability mass function of the stockbroker's bonus.

Table 5.13 Probability mass function of stockbroker bonus

Number of stocks going up next year	**0**	**1**	**2**	**3**
Bonus next year (X)	0	10000	30000	50000
P(X = x)	1/8	3/8	3/8	1/8

$$E(X) = \text{Expected bonus} = \$21250,$$
$$Var(X) = 235937500,$$
$$SD(X) = 15360.26.$$

Let S_5 denote the total bonus after five years. Then,

$$E(S_5) = E\left(\sum_{i=1}^{5} X_i\right) = \sum_{i=1}^{5} E(X_i) = 5(21250) = \$106250,$$

$$Var(S_5) = Var\left(\sum_{i=1}^{5} X_i\right) = \sum_{i=1}^{5} Var(X_i) = 5(235937500) = 1179687500,$$
$$SD(S_5) = \$34346.58.$$

Example 5.5.2

A biologist has determined that the expected number of genetically modified crops to be found in a randomly chosen farm of a remote region is 4 and the standard deviation is 1. If three farms are independent and randomly chosen, what is the expected value and variance of the total amount of genetically modified crops produced by the three farms together, respectively?

$$E(S_3) = E\left(\sum_{i=1}^{3} X_i\right) = \sum_{i=1}^{3} E(X_i) = \mu + \mu + \mu = 3\mu = 3(4) = 12,$$

$$Var(S_3) = Var\left(\sum_{i=1}^{3} X_i\right) = \sum_{i=1}^{3} Var(X_i) = \sigma^2 + \sigma^2 + \sigma^2 = 3(16) = 48.$$

Example 5.5.3

We said earlier that the proof for the results regarding sums of independent random variables would be in Chapters 6 and 8. However, we can informally prove the results with an example.

Consider random variable X which denotes the number of candy that a typical student at College Bliss eats in a typical day. Table 5.14 describes the probability mass function of X:

Table 5.14 Probability mass function of amount of candy eaten per day

x	**0**	**1**	**2**
P(X = x)	1/4	1/2	1/4

We find the expected value of X and the variance of X to be:

$$\mu_X = 1,\ \sigma^2 = 0.5\cdot$$

Consider two randomly chosen unrelated students from this college. Let X_1 denote the amount of candy consumed by student I and X_2 the amount consumed by student II.

We now list all possible values of the sum $X_1 + X_2$, the total amount of candy eaten by the two next Monday, where the two random variables have the probability mass function given above. The table below shows the possible values of the consumption of the two students and the sum and the probability of that sum, which is obtained using the product rule for independent events. For example, $P(X_1 = 0, X_2 = 1) = P(X_1 = 0)P(X_2 = 1) = \left(\frac{1}{4}\right)\left(\frac{1}{2}\right) = \frac{1}{8}$.

X_1, X_2	**(0,0)**	**(0,1)**	**(0,2)**	**(1,0)**	**(1,1)**	**(1,2)**	**(2,0)**	**(2,1)**	**(2,2)**
$S_2 = X_1 + X_2$	0	1	2	1	2	3	2	3	4
$P(S_2)$	1/16	1/8	1/16	1/8	1/4	1/8	1/16	1/8	1/16

We can regroup the results to obtain Table 5.15 of the probability mass function for the unique values of the sum. We will make use of the addition rule for mutually exclusive events.

Table 5.15 Probability mass function of sums of two random variables denoting amount of candy eaten by two students.

S_2	0	1	2	3	4
$P(S_2)$	1/16	4/16	6/16	4/16	1/16

$$E(S_2) = \sum_{S_2=0}^{4} S_2 P(S_2) = 2,$$

$$Var(S_2) = \sum_{S_2=0}^{4} (S_2 - 2)^2 P(S_2) = 1,$$

and we have shown that:

$$E(S_n) = n\mu,$$
$$Var(S_n) = n\sigma^2,$$

where n = 2, $\mu = 1$ and $\sigma^2 = 0.5$.

5.5.1 Exercises

Exercise 1. A painter purchases two types of paints, paint A and paint B. The amount of paint A purchased per week, X_1, has $E(X_1) = 40$ gallons and $Var(X_1) = 4$. The amount of paint B purchased, X_2, has $E(X_2) = 65$ gallons and $Var(X_2) = 8$. Paint A costs \$3 per gallon, whereas paint B costs \$5 per gallon. How much should the firm expect to spend next week in these two types of paint? What is the standard deviation of the amount spent? Assume that X_1 and X_2 are independent.

Exercise 2. A group of 50,000 loan requests made through the web site of a bank report an average income of $\mu = \$37,000$ with a standard deviation of $\sigma = \$20,000$. Furthermore, 20% of the requests report a gross income over \$50,000. A group of 900 requests is chosen at random to check their accuracy. Find the expected value and standard deviation of the total income of the 900 loan requests.

5.6 Named discrete random variables, their expectations, variances and moment generating functions

Probability models play a very important role as mathematical abstractions of the uncertainty we have about different measurements. Although we can reach an answer to our probability questions with the methods we studied in Chapters 2 and 3, some experiments are so similar in nature, and happen so often, that, for the sake of economy of time, it is worth having all the information we need about a given experiment in an equation and methodology like that presented in this chapter to extract from that equation everything we need. Many years of research by probabilists and applied researchers went into arriving at those equations. We call them families of distributions that have been given names. In this section, we introduce the most common ones, but the reader should be aware that there are many more. Each area of application has its own models and it is very likely that even though the reader has seen a lot of models in probability books, the reader will find that the model used in a future job is different.

In Sections 5.7 to 5.11, we will present models that are based on the assumption of independence of a sequence of Bernoulli trials. Section 5.12 presents the case where the trials are not independent. The discussion we had in Chapter 4 that distinguished between these two scenarios will be helpful for those sections. The reader may want to review Chapter 4 before engaging in the study of Sections 5.7 to 5.11. Finally, section 5.14 presents the Poisson random variables and in Section 5.15 we mention other distributions less widely studied in an Introductory Probability book.

The reader should be aware that there are relations between these types of random variables under some conditions. Some of the exercises will require doing some research to establish and prove those relationships.

The last chapter of this book is dedicated to case studies where these, and the continuous random variables, considered together, help us solve complex problems in a host of applications.

5.7 Discrete uniform random variable

A random variable X has a discrete uniform distribution with N points, where N is a positive integer, and possible distinct values x_i, $i = 1,2,\ldots,N$, if its probability mass function is given by:

$$P(x_i) = \frac{1}{N}, \; i = 1,2,\ldots N$$

Box 5.4

Binomial expansion and Pascal's triangle

The Binomial probability mass function is called binomial because it is related to the binomial expansion

$$\sum_{0 \le k \le n} \binom{n}{k} p^k q^{n-k} = (p+q)^n = 1$$

Where q = 1 − p

Pascal's triangle gives the values of $\binom{n}{k}$ for any n and k.

1
1 2 1
1 3 3 1
1 4 6 4 1
....

Example 5.7.1

The roll of a six sided fair die can be represented by this distribution, where N = 6.

5.8 Bernoulli random variable

We talked in Chapter 4 about Bernoulli random variables. A Bernoulli random variable X has the following probability mass function:

$$P(X = x) = \begin{cases} 1-p, & x = 0 \\ p, & x = 1 \end{cases}$$

where p is the probability of a success. The Bernoulli probability mass function is a model for a Bernoulli trial with probability of success p.

As we saw in Chapter 4, a Bernoulli trial is the equivalent of tossing a biased or unbiased coin. It can be used in modeling communication channel errors, for example, or the defective and non-defective status of a component in a system, or the smoking or nonsmoking status of an individual, and many other binary choice problems. Using the definition of expectation and variance of Sections 5.3.1 and 5.3.3, we can see that:

$$E(X) = 0(1-p) + 1(p) = p$$
$$Var(X) = (0-p)^2(1-p) + (1-p)^2(p) = p(1-p)$$

Example 5.8.1

A genetic counselor is conducting genetic testing for a genetic trait. Previous research suggests that this trait is found in 1 out of every 8 people. He examines one randomly chosen person and observes whether it has the genetic trait or not. The pmf is:

$$P(X = x) = \begin{cases} 1 - \frac{1}{8}, & x = 0 \\ \frac{1}{8}, & x = 1 \end{cases}$$

$$E(X) = \frac{1}{8}, \quad Var(X) = \left(\frac{1}{8}\right)\left(\frac{7}{8}\right) = \frac{7}{64}.$$

5.8.1 Exercises

Exercise 1. Find the moment generating function of a Bernoulli random variable using the definition of moment generating function given in Section 5.3.4.

Exercise 2. In 2008, the city of Quebec, a Canadian city, reported that 10% of Quebec students are victims of acts of bullying at least once a week. We choose a student at random. (i) Write the Bernoulli model formula appropriate for this context. What is the expected value of this Bernoulli random variable? (ii) Define what the random variable represents. (iii) Give the expected value and variance of this random variable. (iv) Do you think that the model you wrote applies equally to young and old students? Write your thoughts about this, and if you think the models might be different, write your proposed models. (v) What variables may be most important in discriminating groups of bullied students?

5.9 Binomial random variable

A Park Ranger working in a large National Park is in charge of overseeing groups of ten visitors to describe to them the regulations of the park. There is a probability of 1/3 that a visitor will ask the park ranger a question. The next group of 10 has arrived. What is the probability that at most two people ask a question?

Imagine an experiment that consists of observing a sequence of fixed n independent Bernoulli trials to determine how many successes are there in the n trials. Suppose that on each trial there is success with the same probability p in all trials, failure with probability $q = 1 - p$, and assume the trials are independent. The random variable X that gives the number of successes in n Bernoulli trials with the same probability of success p is a Binomial random variable. A Bernoulli r.v. is just a Binomial with n = 1 (one trial).

Let p be the probability of a success and let $q = 1 - p$. Let n be the number of trials. For $x = 0,1,\ldots,n$, the probability of obtaining any particular ordered sequence of n items containing exactly x successes and $n - x$ failures is $p^x q^{n-x}$. Since there are $\binom{n}{x}$ different ordered sequences of this type, it follows that the probability of finding x successes in n independent trials can be found as we described in Chapter 4, namely

$$P(X = x) = \binom{n}{x} p^x q^{n-x}, \quad x = 0,1,2,\ldots,n.$$

$$E(X) = \sum_x xP(X = x) = \sum_x x \binom{n}{x} p^x q^{n-x} = np,$$

$$Var(X) = \sum_x (x - \mu_X)^2 \binom{n}{x} p^x q^{n-x} = npq.$$

The moment generating function of a binomial random variable is

$$M_X(t) = E(e^{tX}) = \sum_x e^{tX} \binom{n}{x} p^x q^{n-x} = (pe^t + 1 - p)^n$$

Linear functions of a Binomial random variable:

$$E(a + bX) = a + b(np),$$
$$Var(a + bX) = b^2 np(1 - p),$$

There is a linear function of the binomial which is of particular interest in Statistics. This is:

$$Y = \frac{X}{n}.$$

By the linearity of expectations,

$$E(Y) = \frac{E(X)}{n} = p.$$

$$Var(Y) = \frac{Var(X)}{n^2} = \frac{p(1-p)}{n}.$$

Why these results? How did we use the properties of expectations to obtain them? The reader will answer that in one of the exercises.

The parameters of the Binomial distribution, p and n, determine the shape of the binomial distribution. As n stays fixed and p changes the pmf goes from being right to left skewed. As p stays fixed and n increases, the distribution becomes more bell shaped.

Example 5.9.1

In a Mexican town, there are 25% PRI members (event P) and 75% in other party affiliations (event O). What is the probability that a random sample of 3 residents of this town contain less than two PRI members?

First we check the conditions for a Binomial experiment. Because all residents come from the same large population, we can assume that each of the n = 3 draws is an independent Bernoulli trial with the same probability of success (success being a PRI member, because that is what we are asking about). We are assuming with this that the voters, being chosen randomly, are unrelated. However, because they come from the same town they share the same probability of being PRI supporters.

$$n = 3$$
$$p = 0.25$$
$$X = \text{“}number\ of\ PRI\ members\text{”}$$

$$P(X<2) = P(X=1)+P(X=0) = \binom{3}{1}0.25^1(0.75)^2 + \binom{3}{0}0.25^0(0.75)^3 = 0.843750$$

The sample space of this experiment is S = {PPP,PPO, POP, POO, OPP, OPO. OOP, OOO}.

Example 5.9.2

Suppose that a large lot of fuses contains 10 percent defectives. If four fuses are randomly sampled from the lot, find the probability that at least one fuse in the sample of four is defective.

p = probability of defective = 0.1,

n = number of fuses sampled = 4 .

Let X = "number of defective fuses in $n = 4$."

$$P(X \geq 1) = 1 - P(X=0) = 1 - \binom{4}{0}0.1^0(0.9)^3 = 0.271.$$

Example 5.9.3

A sequence of 4 bits is transmitted over a channel with a bit error rate of 0.2. What is the probability that the number of erroneous bits is 2?

p = probability that a bit is erroneous,

n = number of bits transmitted,

X = number of erroneous bits in $n = 4$,

$$P(X=2) = \binom{4}{2}(0.2)^2(0.8)^2 = 0.1536.$$

Example 5.9.4

Suppose that a large manufacturer of earbuds produces 10% defectives. Suppose that four earbuds sampled from the lot were shipped to a customer, before being tested, on a guarantee basis. Assume that the cost of making the shipment good is given by $C = 3X^2$, where X denotes the number of defectives in the shipment of four. Find the expected repair cost.

$$n=4,\quad p=0.1,\ X = \text{"}number\ of\ defectives.\text{"}$$

$$E(C) = 3E(X^2) = 3[Var(X)+(E(X))^2] = 3npq + 3(np)^2 = 3(4)(0.1)(0.9) + 3((4)(0.1))^2 = \$1.56.$$

Example 5.9.5

On a multiple choice exam with three possible choices for each of the five questions, what is the probability that a student would get four or more correct answers just by guessing?

Succeeding in a MC question means that the correct answer is chosen. The probability of choosing the correct answer in a question is 1/3. Think of each question as a Bernoulli trial, assuming that all the questions are unrelated. There are five Bernoulli trials if we have five questions. Let X be the number of questions answered correctly.

$$P(X \geq 4) = P(X = 4) + P(X = 5) = \binom{5}{4}\left(\frac{1}{3}\right)^4\left(\frac{2}{3}\right) + \left(\frac{1}{3}\right)^5 = \frac{11}{243} = 0.04526749.$$

5.9.1 Applicability of the Binomial probability mass function in Statistics

Statisticians sample randomly from populations to observe variables of interest. If the variable is dichotomous and the sample size is small compared to the size of the population, the assumptions of the Binomial model follow, namely the probability of finding a success in one trial is not altered because we drew the sample. Thus, the trials, which are the observed elements of the sample can be assumed to come from the independent trials model that the Binomial represents.

After having collected the sample, statisticians will use the ratio of the number of successes in the sample and the total number of elements in the sample as an estimate of p. With that estimate, they will assume that the Binomial model applies, and future probabilities and planning will be made based on that model.

In biological studies, for example, the populations are so large that the independence of the elements in the sample is not jeopardized by extracting some elements. However, as Samuels et al., (2016, 114) points out, the phenomenon of contagion invalidates the condition of independence between trials. If we sample from a population to estimate the prevalence of tuberculosis then if there is some tuberculosis cases in that population the possibility of contagion exists and the binomial model will underestimate the probability of observing a given number of cases.

5.9.2 Exercises

Exercise 1. Consider the following scenarios:

SCENARIO 1: A system with two independent components is such that the system fails if at least one of the individual components fails. The probability that a component fails is 0.04.

SCENARIO 2: A system with three independent components is such that the system fails if at least one of the individual components fails. The probability that a component fails is 0.04.

SCENARIO 3: A system with four independent components is such that the system fails if at least one of the individual components fails. The probability that a component fails is 0.04.

Write down explicitly the outcomes in the sample space of each scenario. Three sample spaces total. For notation, you may represent "fail" by f and "doesn't fail" by s (success).

(i) For each scenario, provide the probability mass function in the form of a table indicating the values of the random variable X = number of failing components in one table, and the probability of that X in the other column. Add a third column in each table, indicating which events are used to compute the probabilities. In each of the scenarios double check whether the formula:

$$P(X = x) = \binom{n}{x} p^x q^{n-x}, \quad x = 0,1,2,\ldots,n,$$

could be used to compute the probabilities, where x is a value of X, n is the number of components in the system, and p is the probability that a component fails.

(ii) Add a column to your table indicating the computation done using the binomial formula. Check that, in each of the scenarios,

$$\sum_x xP(X = x) = E(X) = np, \text{ and}$$

$$\sum_x (x - E(X))^2 P(X = x) = Var(X) = npq.$$

(iii) What is the value of $E(X^2)$ when $n = 3$?

Exercise 2. According to the Center for Disease Control and Prevention, 1 of every 4 deaths are caused by heart disease. If we choose 10 deaths from the registry at random, what is the probability that the sample would contain 3 heart disease deaths?

Exercise 3. The Physella Zionis, also known as wet-rock physa is a snail found in Zion Canyon, Utah, and Orderville Canyon, along the North Fork of the Virgin River in Utah. This species is believed to have mutated to adapt to the environment in the Canyons. The National Park has a small information post at the entrance of the area where the snail lives indicating the history. What is the probability that in a random sample of 10 visitors, at least one stopped to read the post about the Physella Zionis, if the probability that one will stop is 0.2? This information may help the park rangers decide whether perhaps the post should be moved to a more visible part of the trail.

Exercise 4. In some parts of Africa the prevalence of albinism is as high as 1 in 1,000. How large a sample of individuals from these parts would we have to draw in order to sight at least one person with albinism?

Exercise 5. Outbreaks of cholera, a highly contagious disease, are possible after a natural disaster. For example, after the 2010 earthquake in Haiti, there was a cholera outbreak. If a natural disaster such as that in Haiti were to occur in the future, and you were in charge of sampling the population to screen for cholera, would you use the binomial model?

Exercise 6. Suppose that a large lot of fuses contains 10% defectives. (i) If four fuses are randomly sampled from the lot, find the probability that at least one fuse in the sample of four is defective. (ii) If the four fuses were shipped to a customer, before being tested, on a guarantee basis and X is the number of defectives in the shipment, find the expected repair cost. Assume that the cost of making the shipment good is given by $C = 3X^2$, where X denotes the number of defectives in the shipment of four.

Exercise 7. The probability that the life length of a certain type of battery exceeds four hours is 0.135. If three such batteries are in use in independent operating systems, what is the probability that only one of the batteries will last more than four hours?

Exercise 8. Suppose that X is a binomial random variable with parameters p and n. Define $Y = \frac{X}{n}$. (i) Find the $E(Y)$ and the Standard deviation of Y. (ii) What is the $E(Y^2)$ equal to?

Exercise 9. In the old times, when there were no computers and calculators were not so good, students had access to tables for the distributions. For example, for the Binomial distribution, for any given n, the table would give the probability of $X = 0, X = 1, \ldots, X = n$ for values of p of 0.01,0.05, 0.1, 0.2, 0.3, 0.4, 0.5, 0.6, 0.7, 0.8, 0.9, 0.95. Suppose n is 3, for which value of those p's is $P(X = 1)$ highest?

5.10 The geometric random variable

- What experiment would the sample space given in the box below represent? Describe the experiment (note that the dots (.....) mean that the sequences continue ad infinitum forever following similar pattern).

H TTTTTH
TH TTH
TTTH TTTTH
TTTTTTTH
TTTTTTTTH ....

Suppose we have a situation, like the coin toss, where there is only one of two possible outcomes. Let $X = 1$ if the event of interest happens and $X = 0$ otherwise. We keep trying until the event appears. Suppose that the trials are independent and of constant probability of the event. So we have just described a sequence of Bernoulli trials. The trial in the sequence at which the first occurs is often of interest. For example, we are going to conduct interviews to hire a candidate proficient in Zulu to work for the government of South Africa. How many interviews does it take to find the first qualified Zulu speaker? Let that interview (could be the first, the second, the 20th, or other) be denoted by a random variable Y. We say that Y is a geometric random variable, with probability mass function

$$P(Y=k)=P(x_1=0,x_2=0,..,x_k=1)=(1-p)^{k-1}p, \quad y=1,2,\ldots.$$

We allow the experiment to go on for a countable infinite number of trials. To show that P(Y) is a distribution, we must show that the probabilities add up to 1. In order to do that, we must allow for the random value to take any possible nonnegative value because only then

$$\sum_{k=1}^{\infty}(1-p)^{k-1}\ p=p\sum_{k=1}^{\infty}(1-p)^{k-1}=1$$

By computing the probabilities this way, we have counted all the probability of all possible outcomes in the sample space. The sample space has a probability of 1. See Math tidbit 5.5 to review the geometric progression formula needed to reach this conclusion.

Using similar mathematics, we can prove that:

$$E(Y)=\frac{1}{p}.$$

Similarly, we can show that:

$$Var(Y)=\frac{1-p}{p^2}.$$

The moment generating function of the geometric random variable is

$$M_X(t)=\frac{pe^t}{1-(1-p)e^t}.$$

Example 5.10.1

It is presumed, based on past OECD information, that 7% of Koreans younger than 17 years old live below the poverty level. Finding people below poverty level is hard, but the government is determined to find them and help them. What is the probability that the first Korean below poverty level found by government is the fourth person interviewed? What is the expected number of interviews needed to find the first Korean below poverty level?

$$P(Y=4)=(1-0.07)^3\ 0.07=0.05630499.$$

$$E(Y)=\frac{1}{0.07}=14.2857.$$

Box 5.5

Geometric series

In the study of the properties of the geometric distribution, it helps to remember the formulas for the sum of a geometric series.

$$1+q+q^2+q^3+\ldots.=\frac{1}{1-q}$$

or

$$q+q^2+q^3+\ldots.=\frac{q}{1-q}$$

for $-1<q<1$.

The sum of a finite number of terms in the geometric distribution,

$$\sum_{k=1}^{n}pq^{k-1}=\frac{p}{1-q}$$

Example 5.10.2

(This problem is from Trivedi (2002, page 83).) A mischievous student wants to break into a computer file, which is password-protected. Assume that there are n equally likely passwords, and that the student chooses passwords independently and at random and tries them. Let N_n be the number of trials required to break into the file. Determine the pmf of N_n (i) if unsuccessful passwords are not eliminated from further selections, and (ii) if they are.

(i) If the password is not eliminated, then each trial is independent of another and has the same probability of success, 1/n, since each password is equally likely. Then the probability that the number of trials required to break into the file equals some number c is given by a geometric distribution with parameter p = 1/n:

$$P(N_n = c) = (1 - 1/n)^{c-1}\,\frac{1}{n}.$$

(ii) If the password is not replaced, then each trial is not independent of the other, and the probability of hitting the right password changes as we eliminate passwords. Therefore,

$$P(N_n = c) = \left(1-\frac{1}{n}\right)\left(1-\frac{1}{n-1}\right)\cdots\cdots\left(1-\frac{1}{n-c-1}\right)\left(1-\frac{1}{n-c}\right).$$

5.10.1 Exercises

Exercise 1. The National Health and Nutrition Examination Survey (NHANES) examines about 5000 persons per year https://www.cdc.gov/nchs/nhanes/index.htm and constantly screens diabetic people to detect people with type I diabetes. Let X be a random variable measuring the number of diabetic people it takes to find the first Type I diabetic person. Assume that the probability of a diabetic person having type I diabetes is 0.05. (i) List five or six outcomes of the sample space. Use as notation: 1 for type I and 0 for not type I diabetic. Also provide the probability of each of these outcomes and write a small table containing the first values of X and P(X). (ii) Granted that populations are finite. But would a finite number of outcomes in this sample space suffice to guarantee that all the axioms of probability hold? Explain why or why not.

Exercise 2. A Statistics department finds that 20% of applicants for the Ph.D. have had prior statistics courses. Applicants are reviewed at random from the pool. (i) Find the probability that the first applicant having had statistics prior to applying is the sixth applicant. (ii) Suppose that the first applicant who has had statistics courses prior to applying is accepted in the program, and the applicant visits the department. Suppose each interview costs $100 and there are $350 in travel expenses. Find the expected value and standard deviation of the cost of reviewing applicants until the first qualified applicant is found and visits.

Exercise 3. The probability of being able to access a disaster zone area by air after a hurricane is 0.7. A helicopter keeps trying until access is achieved. (i) What is the probability that access

probability that access will occur in any of the next four attempts? (iii) What is the average number of attempts that will have to be made to gain access?

Exercise 4. The probability that a watch produced by a factory is defective is 0.01. In a recent audit of the factory, products are inspected until the first defective is found. The first 10 trials have been found to be free of defectives (that is, it takes more than 10 trials to find a defective). Calculate the probability that the first defective will occur in the 15th trial given that it took more than 10 trials.

5.11 Negative Binomial random variable

What experiment would the sample space given in the box below represent? Describe the experiment (Note that (....) means that the sequences continue ad infinitum forever following similar pattern.)

HH	TTTHTTH
THH	THTH
HTTTH	TTTHTH
TTHTTTTTH	
TTTTTTTHTTTTH	
....	

Let Y denote the number of the trial on which the *rth* success occurs in a sequence of independent Bernoulli trials where the probability of success is p. For example, when do we reach the first two heads?

$$P(Y=2)=p(p)=P(HH),$$
$$P(Y=3)=2(1-p)p(p)=P(\{THH, HTH\}),$$
$$P(Y=4)=3(1-p)^2 p(p)=P(\{HTTH, THTH, TTHH\}),$$
$$\cdots\cdot$$

$$P(Y=y)=P[\{\text{ first } (y-1) \text{ trials contain } (r-1) \text{ successes and the last trial is a success}\}]$$
$$=P[\text{first } (y-1) \text{ trials contain } r-1 \text{ successes}]\, P\,[y\text{th trial is a success}].$$

$$P(Y=y)=\binom{y-1}{r-1}p^{r-1}(1-p)^{(y-1)-(r-1)}p=\binom{y-1}{r-1}p^{r}(1-p)^{(y-r)} \quad y=r, r+1, \ldots\ldots.$$

The first probability statement is identical to the one that results in a binomial model for the probability of obtaining $r-1$ successes in $y-1$ trials. The last statement is just p. When combining the two, we get the final formula for the negative binomial.

Box 5.6

Binomial expansion with negative exponent

The name negative binomial distribution results from its relationship to the binomial series expansion with negative exponent, −r.

$$\sum_{i=0}^{\infty}\binom{i+r-1}{r-1}q^{i}(1-q)^{-r}$$

The formulas for the expectation and variance of a negative binomial random variable Y are:

$$E(Y)=\frac{r}{p},$$

$$Var(Y)=\frac{r(1-p)}{p^2}.$$

The moment generating function of the negative binomial random variable is

$$M_X(t)=\left[\frac{pe^t}{1-(1-p)e^t}\right]^r.$$

Example 5.11.1

(This example is from Bain (1987, 69).) Team A plays team B in a seven-game world series. That is, the series is over when either team wins four games. For each game, $P(A \text{ wins}) = 0.6$, and the games are assumed independent. What is the probability that the series will end in exactly six games?

We have $X = 6$, $r = 4$, and $p = 0.6$. The random variable X is the number of games it takes A to win 4 games.

The series will end in exactly 6 games if team A wins 4 games by game 6 or if B wins 4 games by game 6. We calculate the terms separately first.

$$P(\text{"}A \text{ wins series in } 6\text{"}) = P(X=6) = \binom{5}{3}(0.6)^4(0.4)^2 = 0.20736.$$

We have $Y = 6$, $r = 4$, and $p = 0.4$. The random variable Y is the number of games it takes B to win 4 games.

$$P(\text{"}B \text{ wins series in } 6\text{"}) = \binom{5}{3}(0.4)^4(0.6)^2 = 0.09216.$$

$$P(\text{"series goes 6 games"}) = P(\text{"}A \text{ wins in } 6\text{"} \cup \text{"}B \text{ wins in } 6\text{"}) = 0.20736 + 0.09216 = 0.29952.$$

Example 5.11.2

(This example is from Scheaffer (1995).) 30% of the applicants for a certain position have received advanced training in computer programming. Suppose that three jobs that require advanced programming training are open. Find the probability that the third qualified applicant is found on the fifth interview, if the applicants are interviewed sequentially and at random. In the example of the candidates for the jobs, how many interviews should we expect to conduct to find the three experienced candidates?

$3/0.3 = 10$ is the expected value, but we could need 4.804 more or less interviews than that (4.8304) is the standard deviation).

5.11.1 Exercises

Exercise 1. Down syndrome (DS) is a chromosomal condition that occurs in about 1 in 1000 pregnancies. How many pregnancies should we expect to observe in order to find 3 Down syndrome pregnancies?

Exercise 2. On a hunting trip, the probability that a lion will find suitable prey is 0.8. What is the probability that it takes 10 trips to find three suitable prey?

5.12 The hypergeometric distribution

The following quote illustrates the use of a probability model to assess danger.

> *Several cases of anthrax including the deadly inhalation form of the disease were associated with an envelope containing anthrax spores addressed to Senate Majority Leader Thomas Daschle of South Dakota and found in a mailroom serviced by the U.S. Postal Service distribution in Brentwood, MD, in the Autumn of 2001. In addition to the mailroom serving Senator Daschle's office, the Brentwood facility services approximately 3200 other mailrooms in the district of Columbia and surrounding area. Once the letter was discovered and cases of the disease emerged, a logical step for consideration in control of the disease is to inspect a sample of the approximately 3200 other mailrooms served by the Brentwood facility for the presence of anthrax spores.*
>
> *Under simple random sampling without replacement, the conditional probability that k of the n sampled mailrooms are contaminated with anthrax spores is given by the hypergeometric distribution.*
>
> **(Levy, Hsia, Jovanovic, and Passaro 2002, 19)**

Statisticians use all kinds of probability models like those seen in this book to set up their statistical hypothesis testing. The statistics calculation is complicated, and requires that the reader learn inferential statistics to do it. However, after studying this section, the reader is challenged to think: what in the hypergeometric model is the unknown that the statistician is after in this anthrax case?

When trials are not independent Bernoulli trials, how can we calculate the probability of k successes in n trials? We learned to do that in Chapter 4, Section 4.2.1. In the spirit of Chapter 5, where we try to find a formula that summarizes all possible values and their probabilities, a name was given to the formula we used in Chapter 4 to solve this problem. The name given is the hypergeometric probability mass function.

The hypergeometric probability mass function is applicable to any situation in which a population of N units can be considered to contain two mutually exclusive and exhaustive subpopulations: subpopulation 1 having K units and subpopulation 2 containing N−K units. Let Y be the number of units from subpopulation 1. The probability that a sample of n distinct units from this population contains k units from subpopulation 1 and n−k from subpopulation 2 is:

$$P(Y=k\mid N,n,K)=\frac{\binom{K}{k}\binom{N-K}{n-k}}{\binom{N}{n}},\quad k=0,1,\ldots,K.$$

The reader will recognize the above formula as the one arrived at when trying to figure the probability of obtaining n draws from an urn in such a way that k elements are from population 1 and $n-k$ from population 2. In the language of trials used in Chapter 4, Y represents the number of successes in n trials that are not independent and are obtained from a box containing M items, M_s of which are successes and $M-M_s$ are failures. The reader will realize that the formula we got in Chapter 4, namely,

$$P(Y=k)=\frac{\binom{M_s}{k}\binom{M-M_s}{n-k}}{\binom{M}{n}},\quad k=0,\ 1,\ldots,\ M_s$$

does the same job as the one we wrote in terms of N and K.

Example 5.12.1

A department store has ten discounted espresso coffee machines, four of which are defective. A customer randomly selects five of the machines for purchase, to give to relatives for the holidays. What is the probability that all five of the machines are non-defective?

Then

$$P(Y=5)=\frac{\binom{6}{5}\binom{4}{0}}{\binom{10}{5}}=\frac{6}{252}=0.02380952,\quad k=0,1,\ldots,M_s$$

5.12.1 Exercises

Exercise 1. A supermarket sells boxes containing five chirimoyas, two of which are spoiled inside. If we were to randomly pick two chirimoyas from a box, what is the probability that no more than one chirimoya is spoiled inside?

Exercise 2. Proposition 134 is on the ballot for the next election. In a small town of 50 people, 30 favor the proposition and 20 do not. A committee of 4 people is selected from this town. Answer the following questions: (i) What is the probability that there will be no one in favor of Prop. 134 in the sample? (ii) What is the probability that there will be at least one person in favor? (iii) What is the probability that exactly one pro-Proposition 134 person will appear in the sample? What is the probability that the majority of the sample will vote for Prop. 134?

Exercise 3. A museum curator collects paintings from local artists in sets of size 10 sent by an intermediary. It is the curator's policy to inspect 3 paintings randomly from a set and to accept the set if all 3 satisfy the standards of the museum. If 30 percent of the sets have 4 unacceptable paintings and 70 percent have only 1, what proportion of the sets does the curator reject?

Exercise 4. There are 20 data scientists and 23 statisticians attending the Annual Data Science conference. Three of these 43 people are randomly chosen to take part in a panel discussion. What is the probability that at least one statistician is chosen?

Exercise 5. In a small town of 20 people in Ecuador, South America, 4 favor Proposition 27 to make Quechua the official language and 16 are against. A committee of 5 people is selected at random from this town to represent the town in front of the national government of Ecuador. What is the probability that there will be no one in favor of proposition 27 in the committee?

5.13 When to use binomial, when to use hypergeometric? When to assume independence in sampling?

This is a question that always arises when talking about the hypergeometric probability mass function. Binomial is for independent Bernoulli trials and for draws with replacement. Hypergeometric is for draws without replacement. Where is the gray area? Consider the following example.

Suppose we have a box that contains 1000000 red and 1000000 yellow balls. We are going to draw two balls at random without replacement. What is the probability of getting two red balls? Let R_1 be the first ball drawn that is red, and let R_2 be the second red ball. Then:

$$P(R_1R_2) = P(R_1)P(R_2 \mid R_1) = \left(\frac{1000000}{2000000}\right)\left(\frac{999999}{1999999}\right) = \left(\frac{1}{2}\right)(0.4999997) = 0.2499999.$$

If we had drawn the two balls with replacement, the probability would have been

$$P(R_1R_2) = P(R_1)P(R_2 \mid R_1) = \left(\frac{1000000}{2000000}\right)\left(\frac{1000000}{2000000}\right) = \left(\frac{1}{2}\right)^2 = 0.25.$$

The results are very close. Thus, when the box (or urn, or population) is very large, drawing with or without replacement does not make much difference, independence may be assumed and the binomial would be a good approximation.

Consider now a small box with 10 Red and 10 Yellow balls, the same experiment of drawing two balls and the same question, what is the probability of two reds?

Without replacement,

$$P(R_1R_2) = P(R_1)P(R_2 \mid R_1) = \left(\frac{10}{20}\right)\left(\frac{9}{19}\right) = \left(\frac{1}{2}\right)(0.4736842) = 0.2368421.$$

With replacement,

$$P(R_1R_2) = P(R_1)P(R_2 \mid R_1) = \left(\frac{10}{20}\right)\left(\frac{10}{20}\right) = \left(\frac{1}{2}\right)^2 = 0.25.$$

Now there is an appreciable difference. Thus, if the box (urn, population) is small, drawing with and without replacement makes a difference in the results. The Binomial approximation is not appropriate in this case. We cannot assume independence if we are not drawing with replacement. But we can assume independence if the box we are drawing from is so large that the composition of the box does not change noticeably.

5.13.1 Implications for data science

As we have said in several occasions, in data science, sampling to learn about populations is done without replacement. A statistical random sample is a sample without replacement. When the sample is drawn from a large population, statisticians do not make corrections. But when the population is small, statisticians make corrections to their estimates and computations, which are done assuming that the observations they obtained are independent.

5.14 The Poisson random variable

- What would be the distribution of the number of sparrows landings across a landscape if the landing is completely random?

If (a) the probability that a sparrow lands at a given point on the landscape is the same everywhere, and (b) every point in the landscape is equally likely to be a landing point, then the number of sparrows landing per square mile, as in Figure 5.2, is a Poisson random variable. Those assumptions are equivalent to assuming that there is a large number of independent sources affecting the landing or not landing. Alternatively, they are equivalent to assuming that the landing is random.

The Poisson random variable is used to model the number of occurrences of a certain event, in a unit of space or time, when the occurrences happen independently of each other and occur with equal probability at every point in time and space. Examples are:

- the number of calls to an office telephone during business hours
- the number of atoms decaying in a sample of some radioactive substance

- the number of visits to a web site per hour
- the number of customers entering a store per hour
- the number of mutations carried per individual in a population
- the number of seeds successfully germinated per mother plant
- the number of seizure counts per patient undergoing anti-convulsant therapy.
- the number of failures per power plant pump.
- the number of mutations per plate assaulted by phage under the directed mutation hypothesis (Zheng, 2010)
- the random emission of electrons from the filament of a vacuum tube, or from a photosensitive substance under the influence of light
- the spontaneous decomposition of radioactive atomic nuclei
- demand for a service per unit of time, such as demand for runway to takeoff by a plane
- the number of calamities per unit of time
- the number of α-particles discharged per 1/8-minute from a film of plutonium
- the number of minute corpuscles to be found in sample drops of liquid

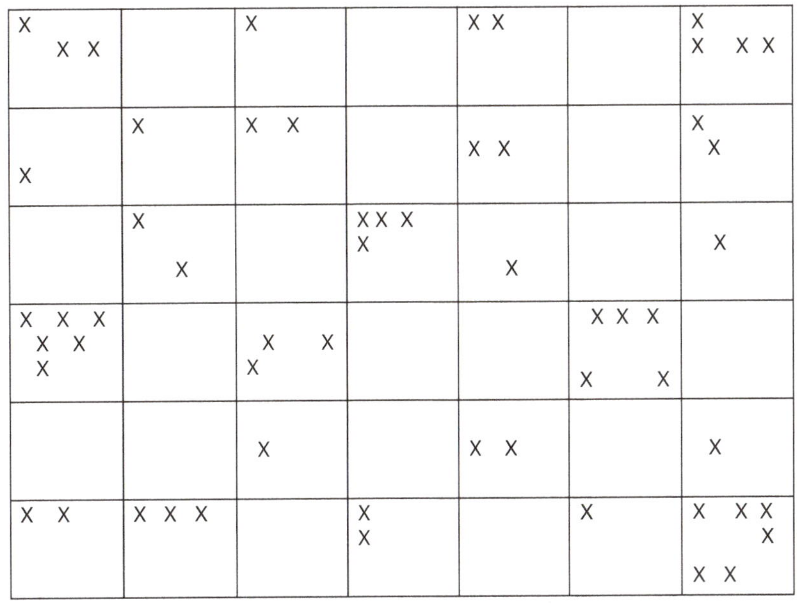

Figure 5.2 Random landing of sparrows in a landscape.

> *For the biologist, the Poisson distribution is just a model for how successes may be distributed in time and space in nature. Life gets interesting when the model doesn't fit the data, because then we learn that one or more of the main assumptions is false, hinting at the existence of interesting biological processes. (For example, some individuals may be more prone to mutations than others, some fishers may be better catchers than others, or some plants may produce better quality seeds.* (Whitlock and Schluter 2009)

The alternative to a Poisson model is counts that are clumped together, meaning that one occurrence (e.g., a sparrow landing) increases the probability of another occurrence nearby (another sparrow landing nearby), as when there are contagious diseases. The distribution of counts is more clustered than what is expected by chance. Another alternative is when the occurrences are more dispersed than expected, as when for example, seeing a territorial animal decreases the probability of seeing another animal nearby. There is more dispersion than would be expected by chance.

A random variable X is said to be a Poisson r.v. with parameter λ if

$$P(X=k)=\frac{\lambda^k e^{-\lambda}}{k!}, \quad k=0,1,2,\ldots\ldots$$
$$E(X)=\lambda, \quad Var(X)=\lambda.$$

The moment generating function of X is:

$$M_X(t)=e^{\lambda(e^t-1)}$$

The Poisson is a discrete random variable. The parameter λ is larger than 0. Parameters are assumed constant. Different values of the parameter characterize different members of the Poisson family.

Example 5.14.1

As part of a National Parks survey, an aerial photograph of a region where cars are not allowed could be taken by the National Parks Service. A grid of four hundred 100-meter squares could be superimposed over the photograph to enable a study to be made of the distribution of cars. The number of 100 meter squares containing 0,1,2,3, up to 6 cars could be recorded. Suppose the following information was found:

Number of cars	**0**	**1**	**2**	**3**	**4**	**5**	**6**
Number of squares	278	92	25	4	0	0	1

If cars can be considered to be random occurrences in space, the probability of finding any specified number of cars in a 100-meter square can be calculated using the Poisson. A Poisson model with a parameter 0.4 could be fit to these data to see if the number of squares predicted by the Poisson model for each number of cars is closed to the number observed. Statisticians then use other probability models for tests they conduct to determine how reliable is the hypothesis that the Poisson model fits these data.

Example 5.14.2

The reader is encouraged to read further into uses of the Poisson distribution, in particular to model rare phenomena, such as for example "horse-kick" deaths or death by surgery in the wrong location. The following web site https://onlinelibrary.wiley.com/doi/full/10.1111/anae.13261 dwells on this application.

Example 5.14.3

Modern statistical software packages give us the power to perform many computations with a few clever keystrokes and mouse clicks. However, this computing power does not come without drawbacks. Specifically it has become all too easy to overlook data-entry errors and/or nonsense codes meant to represent missing data, with many researchers including these values in their final computations. Codes such as 9999 or −1 for missing data were often used to indicate that a value was missing like systolic blood pressure (which should be in the vicinity of 100–150 mm Hg.) While programs such as SAS, SPSS, and Stata are now capable of handling missing data merely by leaving the cell blank, the use of dummy codes is still a common practice. Statistical software programs do not know that a blood pressure of 9999 is a nonsense value.

Suppose we are about to receive the 2019 annual Survey of Health Measures data. Throughout the years it has been learned that, on average, the annual survey contains 30 dummy codes for missing values. What is the probability that the 2019 survey will contain less than 10 missing data dummy codes?

$$P(X<10)=\sum_{k=0}^{9}\frac{30^k e^{-30}}{k!}\approx 0.$$

Example 5.14.4

The number of emergency calls to 911 for health reasons per day in a certain road has a Poisson distribution with expected value 4. There are only three locally owned ambulances which are dispatched to the first three emergency calls locations, after that ambulances from the neighboring town are called. On any given day, what is the probability of having to call ambulances from the nearest town?

Let X be the number of fatal accidents per day.

$$P(X>3)=1-P(X\leq 3)=1-[P(X=0)+P(X=1)+P(X=2)+P(X=3)]$$
$$=1-\left[e^{-4}+4e^{-4}+\frac{16e^{-4}}{2}+\frac{64e^{-4}}{6}\right]=0.56653.$$

Example 5.14.5 Epidemics, contagious diseases

Under normal circumstances, the number X of monthly mump cases in Iowa has approximately the Poisson distribution with mean 0.1. If we calculate the probability that in a given month there is at most one mumps case in Iowa we find, using the Poisson formula,

$$P(X\leq 1)=\frac{0.1^0 e^{-0.1}}{0!}+\frac{0.1^1 e^{-0.1}}{1!}=0.9953.$$

So the probability of more than one case is 0.0047. That is, we would consider seeing two or more mump cases in a month a very unlikely event if mump cases are random and independent. More than two cases per month would be almost impossible to observe.

Mumps cases, like all cases of rare diseases, must be reported to CDC and are published in the Morbidity and Mortality Weekly Report. In January 2006, Iowa reported 4 cases of mumps. What is the probability of getting 4 or more cases of mumps under the model assumed? We find, again using the Poisson model that

$$P(X\geq 4)=0.000004.$$

Thus, assuming that cases of mumps are random and independent, we would expect to almost never see 4 or more cases of mumps in Iowa in any given month. The unusually high count of mumps in January 2006 pointed to a substantial departure from the Poisson model, for instance, because of a contagious outbreak. In a contagion model, knowing that one person is infected increases the chance that another person is also infected. Therefore, contagious events are not independent. These first four cases were indeed the beginning of a major mumps outbreak in Iowa in 2006. You may read about this epidemic in https://www.cdc.gov/mmwr/preview/mmwrhtml/mm5513a3.htm

Example 5.14.6

(This problem is from Ross (2010).) The number of times that a person contracts a cold in a given year is a Poisson random variable with parameter $\lambda = 6$. Suppose that a new wonder drug (based on largē quantities of vitamin C) has just been marketed that reduces the Poisson parameter to $\lambda = 3$ for 75% of the population. For the other 25 percent of the population the drug has no appreciable effect on colds. If an individual tries the drug for a year and has 2 colds in that time, how likely is it that the drug is beneficial for him or her?

Let B denote beneficial and X = the number of colds

The probability sought is

$$P(B \mid X=2) = \frac{P(X=2 \mid B)P(B)}{P(X=2 \mid B)P(B) + P(X=2 \mid B^c)P(B^c)}$$

$$= \frac{\frac{3^2 e^{-3}}{2!}\left(\frac{3}{4}\right)}{\frac{3^2 e^{-3}}{2!}\left(\frac{3}{4}\right) + \frac{6^2 e^{-6}}{2!}\left(\frac{1}{4}\right)} = 0.9377.$$

5.14.1 Exercises

Exercise 1. Almost every year, there is some incidence of volcanic activity on the island of Japan. In 2005 there were 5 volcanic episodes, defined as either eruptions or sizable seismic activity. Suppose the mean number of episodes is 2.4 per year. Let X be the number of episodes in the next two years. (i) What model might you use to model X? (ii) What is the expected number of episodes in the next two years period according to your model? (iii) What is the probability that there will be no episodes in the next two years? (iv) What is the probability that there are more than three episodes in this period?

Exercise 2. In a town with two public libraries, people borrow books from public library A at a rate of one for every two minutes and, independently, from library B at a rate of two for every two minutes. People tend to prefer borrowing from library A, with 60% of library patrons preferring A, and 40% preferring B. Nobody is known to borrow books from both libraries. The two public libraries are open all day. (i) What is the probability that there are no books borrowed between 12:00 and 12:05? (ii) What is the probability that at least four books are borrowed during that time?

Exercise 3. Prove that $\sum_{i=0}^{\infty} \frac{e^{-\lambda}\lambda^{x}}{x!} = 1.$

Exercise 4. Prove that the expected value of a Poisson random variable with parameter λ is λ. The following result from Mathematics may pop up during the computations: $\sum_{k=0}^{\infty} \frac{k^{n}}{n!} = e^{k}$. Use it as needed.

Exercise 5. A random variable Y is Poisson with parameter λ. Find

$$E(20Y + 10Y^{2} - 3Y^{3} + e^{tY}).$$

Simplify as much as possible.

Exercise 6. Some online sellers allow free examination of products for a month. The customer can return the product within a month and get a full refund. In the past, an average of 2 of every 10 products sold by a seller are returned for a refund. Using the Poisson probability distribution formula, find the probability that exactly 6 of the 40 products sold by this company on a given day will be returned for a refund.

Exercise 7. Suppose that X and Y are independent Poisson random variables with parameters λ_1 and λ_2, respectively. (i) What is the expected value of the sum of these two random variables? (ii) What is the variance of the sum of these two random variables?

5.15 The choice of probability models in data science

The Poisson, the Binomial and many of the common discrete distributions, are such that for large values of the random variable the distribution decays exponentially. These models find wide applicability in many areas of science and engineering. Side Box 5.7 contains an example in genomics. The reader is encouraged to read Nolan and Speed (2000) where the problem is presented in detail.

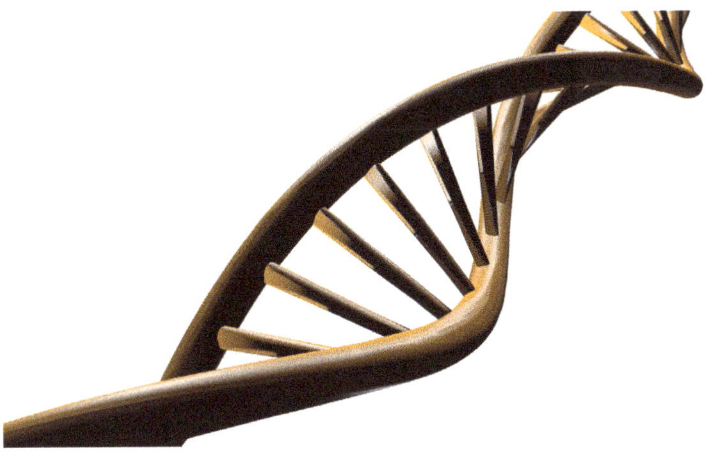

Figure 5.3 Palindromes appear in DNA sequences.

Box 5.7

Genomics

DNA is a long, coded message made from a four letter alphabet: A, C, G, and T. It is believed that some patterns may flag important sites on the DNA, such as the area on a virus' DNA that contains instructions for its reproduction. A particular type of pattern is a complementary palindrome, a sequence that reads in reverse as the complement of the forward sequence. Palindromes were found in the area of replication of several viruses of the Herpes family. Nolan and Speed (2000) studied a DNA sequence the human cytomegalovirus member of the herpes virus family. Places with clusters of palindromes were found along the DNA sequence and they were believed to be suspect of being the origin of replication. Those clusters were located. The locations are the numbers assigned to the pairs in the DNA sequence, which has 229354 pairs of letters.

A question of interest posed by Nolan and Speed (2000) is whether the Poisson model fits these data well. The Poisson model can help determine whether there are clusters or not. Would you like to think how?

Solution: Divide the 229,354 locations into equal intervals. Count the number of palindromes per interval. Create a table that shows X (the number of palindromes) and P(X) the number of intervals containing that number of palindromes divided by the total number of intervals. Calculate the average number of palindromes per interval and calculate the Poisson probabilities for X, using a Poisson with the data average. If the numbers are systematically off, we can say that the Poisson does not fit.

However, the exponential decay at the tail, this type of behavior, has been found not to be very appropriate for random variables describing the internet. Even before the internet, it was found that they are not very useful to model several hydrolic variables such as river runoff.

5.15.1 Zipf laws and the Internet. Scalability. Heavy tails distributions.

Consider the ranked web pages of a particular institution. Lower number in the rank means more frequently visited. Assign rank = 1 to the page with the highest frequency. The probability of requesting the rth ranked page is a power law called Zipf's law.

The assumption made about the distribution of the popular web pages has a lot of implications for web cache replacement algorithms.

A discrete power law distribution with coefficient $\gamma > 1$ is a distribution of the form

$$P(X = k) = Ck^{-\gamma} \quad k = 1, 2, \ldots$$

This equation describes the behavior of the random variable X for sufficiently large values of X. The distribution for small values of X may deviate from the expression above.

Power law distributions decay polynomially for large values of the random variable. That is, the distribution decays polynomially as $k^{-\gamma}$ for $\gamma > 1$. This means that in a power law distribution, rare events are not so rare.

To detect a power law in a log-log plot (log of probability vs log of X) we should see a line with a slope determined by the coefficient γ.

In Web applications, the distribution of web pages ranked by their popularity (frequency of requests in a large set of web pages) is believed to be a power law distribution known as Zipf's law (Zipf 1949).

Zipf's law states that the request for the kth ranked of n web pages is a power law with $\gamma = 1$, where

$$C \cong \frac{1}{\log(n) + 0.577}$$

Sanchez and He (2009) has a discussion about how to fit a probability model to ranked web pages.

Paxson and Floyd (1995) and Willinger and Paxson (1998) talk more specifically about the failure of the Poisson model to capture internet network data's random behavior.

Box 5.8

Changing probabilistic nature of telecommunication networks.

The static nature of traditional public switched telephone networks (PSTN) contributed to the popular belief in the existence of universal laws governing voice networks, the most significant of which is the Poisson nature of call arrivals at links in the network where traffic is heavily aggregated, such as interoffice trunk groups. The average number of calls predicted well the performance of the networks and variability was limited.

But data traffic through the internet is much more variable, with individual connections ranging from extremely short to extremely long and from extremely low rate to extremely high rate. The term to describe that is bursty (rollercoaster) or highly variable. The Poisson model does not hold. Distributions with thick tails, power laws are more useful. (Willinger and Paxson 1998, 961–70)

5.16 Mini quiz

Question 1. (This problem is from Moore (1996, 352).) Joe reads that one out of four eggs contains salmonella bacteria, so he never uses more than three eggs in cooking. If eggs do or do not contain salmonella independently of each other, the number of contaminated eggs when Joe uses three chosen at random has the distribution

a. binomial with n = 4 and p = 1/4
b. binomial with n = 3 and p = 1/4
c. binomial with n = 3 and p = 1/3
d. Hypergeometric with $M = 4$, $M_s = 2$

Question 2. A student of Probability proposed a discrete random variable Y which can take the possible values 1, 2, and 3. The student claims that the following function is a probability mass function for Y:

$$P(Y = y) = \frac{\theta^{2y}}{\theta^2 + \theta^4 + \theta^6}, \quad y = 1,2,3; \quad \theta > 0$$

Is P(Y) a probability mass function? Why? (circle one and explain)

(A) YES (B) NO

Question 3. Which of the following does NOT equal the variance of X?

$$(a) \sum_x x^2 P(x) + \sum_x (E(X))^2 P(x) - \sum_x 2xE(X)P(x)$$

$$(b) \sum_x x^2P(x) + \sum_x (\mu)^2 P(x) - \mu \sum_x 2xP(x)$$

$$(c)\ E(\mu^2) + (E(X))^2 - 2(E(X))^2$$

Question 4. A supermarket chain reports that 40% of all mineral water purchases are "sparkling water." Consider the next 20 mineral water purchases made. Suppose Y is the number of those 20 bottle purchases that are "sparkling water." What distribution model does Y follow?

a. Bernoulli
b. Binomial
c. Poisson
d. Negative Binomial

Question 5. Which of the following cannot be a probability mass function?

a. $P(x) = \frac{x}{4},\ x = 1,2,3$

b. $P(x) = \frac{x^2}{8},\ x = 1,2,3$

c. $P(x) = \frac{x}{3},\ x = -1,+1,+3$

d. $P(x) = \frac{x}{3},\ x = -2,-1,+2$

Question 6. The mean number of baby deliveries by obstetricians at a large hospital per day is 5. If we assume that the deliveries are random, independent events, the count of daily baby deliveries by obstetricians at this hospital follows approximately

a. A binomial distribution with mean 5 and standard deviation 2.
b. A Poisson distribution with mean 5 and standard deviation 5.
c. A Poisson distribution with mean 5 and standard deviation 2.236
d. A geometric distribution with mean 0.5 and standard deviation 0.2

Question 7. Suppose that 65% of the American public approves of the way the President is handling the economy. A random sample of 8 adults is taken and Y is made to represent the number who approve in that sample, a Binomial random variable with $n = 8$ and $p = 0.65$. A student of survey sampling theory proposed looking instead at $X = 8 - Y$, another random variable. The probability model for X is

a. Poisson($\lambda = 8$)
b. Geometric($p = 0.65$)
c. Binomial($n = 8$, $p = 0.35$)
d. Negative Binomial

Question 8. A biologist is examining frogs for a genetic trait. Previous research suggests that this trait is found in 1 out of every 8 frogs. She collects and examines 150 frogs chosen at random. How many frogs with the trait should he expect to find?

a. 20
b. 18.75
c. 45
d. 8

Question 9. An oil exploration firm is to drill ten wells, each of which has probability 0.1 of successfully striking recoverable oil. It costs \$10000 to drill each well so there is a total fixed cost of \$100000. A successful well will bring oil worth \$500000. The expected value and standard deviation of the firm's gains are, respectively

a. \$500000, \$500000
b. \$500000, \$0.9
c. \$450000, \$474.34
d. \$400000, \$474341.6

Question 10. 20% of the applicants for a certain sales position are fluent in both Chinese and Spanish. Suppose that four jobs requiring fluency in Chinese and Spanish are open. Find the probability that two unqualified applicants are interviewed before finding the fourth qualified applicant, if the applicants are interviewed sequentially and at random.

a. 0.87
b. 0.2
c. 0.0124
d. 0.541

5.17 R code

It is possible to use R as a calculator of discrete probabilities. The discrete cumulative probability calculators in R are functions beginning with "p" to compute Prob(X less than or equal to x), with "d" to compute Prob(X = x), with "q" to compute the value of X such that Prob(X less than or equal to x) is some value, and with "r" to generate random numbers (data) that follow the distribution. Those letters are followed by the short name of the distribution. Here is a list of functions computing probability for known distributions.

Binomial distribution-Computing probabilities with cumulative distribution:

pbinom(x, size, prob)
size = number of trials (n).
prob = probability of success on each trial (p).

Example: To calculate P(X <= 10), given X follows a binomial distribution with n = 30, p = 0.4, you can use the command:

```
  pbinom(10, 30, 0.4)
#will give you
  [1] 0.2914719
```

Example: To calculate P(X > 13), given X follows a binomial distribution with n = 30, p = 0.4, you can use the command, since P(X > 13) = 1 − P(X <= 13),

```
  1-pbinom(13, 30, 0.4)
#will give you
  [1] 0.2854956
```

Example: To calculate P(10 < X < 13), given X follows a binomial distribution with n = 30, p = 0.4, you can use the command, since P(10 < X < 13)= P(X <= 12) − P(X <= 10),

```
  pbinom(12, 30, 0.4)-pbinom(10, 30, 0.4)
  will give you
[1] 0.1473752
```

Poisson distribution:

ppois(x, lambda)

This computes the probability that given the average rate of hits per unit of time (lambda) there are x hits or less during a given unit of time.

Example: calculate P(X <= 2) given that the average rate per unit of time is 3 (lambda).

```
x
ppois(2,3)
# will give you
[1] 0.4231901
```

Example: Likewise, if you were interested in calculating the exact probability that there were 4 hits in one unit of time if the average is 3 hits per unit of time, you could use:

```
 ppois(4,3)-ppois(3,3)
# will give you
[1] 0.1680314
```

Example: What happens if we want to calculate the probability that we get more than four hits if the average amount of successes is three per unit of time? Remember that if you sum up all of the probabilities for a distribution they equal 1. Thus if you calculate the probability of having four or less hits and then take one minus that probability you get the probability of having more than four hits. In R, it looks like:

```
1-ppois(4,3)
#will give you
[1] 0.1847368
```

You can find more information by typing a command like "?ppois" in R.

You may want to try on your own the other distributions.

Notice the geometric distribution in R works a little different from the above and from what we defined in this chapter.

Setting up and plotting a generic discrete random variable

1. A neat way to plot the probability mass function of a discrete random variables is a line plot. Write by hand the probability mass function and cumulative probability given below and then plot it with R with the following commands. Plot also the cumulative probability.

```
stacksize=0:7
probability=c(0.05,0.10,0.25,0.2,0.2,0.10,0.05,0.05)
plot(stacksize,probability,xlab="Stack
  size",ylab="probability",type="h")
cumprob=c(0.05,0.15,0.4,0.6,0.8,0.9,0.95,1)
plot(stacksize,cumprob,xlab="stack  size",ylab="Cumulative
  probability",type="S")
```

Finding expected value and variance of a generic discrete random variable

```
expectation=sum(stacksize*probability)
variance=sum((stacksize-expectation)^2 * probability)
standarddeviation=sqrt(variance)
```

Sampling (generating random numbers) from families of discrete random variables.
Example: Drawing random numbers from the Poisson distribution

```
sample= rpois(1000, 1) # 1000 random numbers from Poisson with
lambda=1
plot(table(sample)/1000,,xlab="X=Poisson Random variable",
ylab="Proportion in the sample",type="h",main="Distribution of
random numbers in sample from Pois(lambda=1)")
```

Example: Drawing random numbers from the Binomial distribution

```
sample= rbinom(1000, 20, 0.3) # 1000 random numbers, Bin(n=20,p=0.3)
plot(table(sample)/1000,,xlab="X=Binomial random variable",
ylab="Proportion in the sample",type="h",main="Distribution of
random numbers in sample from Bin(0.3,20)")
```

5.18 Chapter Exercises

Exercise 1. In a digital commutation system, bits are transmitted over a channel in which the bit error rate is assumed to be 0.0001. The transmitter sends each bit five times, and a decoder takes a majority vote of the received bits to determine what the transmitted bit was. Determine the probability that the receiver will make an incorrect decision.

Exercise 2. Prove that the moment generating function of the Binomial Distribution is

$$M_x(t) = (pe^t + 1 - p)^n$$

Exercise 3. A large corporation is interested in reconsidering its retirement policy and proceeds to conduct a survey of a random sample of its employees. The survey asks each of the surveyed employees whether they favor or not the current retirement policy. In a random sample of 10 employees, what is the probability that all of them favor the current retirement policy if overall 65% of all employees favor it?

Exercise 4. Jury deliberation in the jury room sometimes takes a long time. Jurors are randomly selected members of the population. In a population where 90% of the population thinks that someone that robs a convenience store at gun point is guilty, what is the probability that 5 out of 12 jurors will find the defendant of such a robbery case guilty?

Exercise 5. Show that the cumulative distribution function of a geometric random variable equals

$$P(Y \leq y) = 1 - (1 - p)^y$$

Exercise 6. Coltron has an unfair coin (it's unfair because it only comes up heads 40% of the time). He bets his older brother Garrett that if he flips the coin 10 times he can't get at least 6 heads. Garrett, unaware of the nature of the coin agrees. What is the chance that Garrett defeats Coltron and flips at least 6 heads with the unfair coin. Do this problem theoretically and also with R. Compare the answers.

Exercise 7. Zandree Rose is a car salesperson. On average she sells two brand new BMWs every week. She gets a bonus if she sells at least three BMWs in any given week. Given a seven-day span, what is the probability that she gets the bonus? What is the probability she sells exactly three? Do this problem theoretically and also with R. Compare the answers.

Exercise 8. Let X be a random variable representing the roll of a fair six-sided die. Write the probability mass function that corresponds to the roll of a fair die, find the expected value, the standard deviation, the cumulative distribution function and the moment generating function.

Exercise 9. Suppose that you and a friend are matching balanced coins. Each of you tosses a coin. If the upper faces match, you win one dollar; if they do not match, you lose one dollar (your friend wins one dollar). The probability of a match is 0.5. Let X = your winnings.

a. Draw the probability mass function of your winnings.
b. On the average, how much will you win per game over the long run? Is this a fair game?
c. Now payday has arrived and you and your friend up the stakes to 10 dollars per game of matching coins. You now win -10 or +10 with equal probability. What is your expected winning per game? Is the game still fair?
d. You and your friend decide to complicate the payoff picture in the coin-matching game by agreeing to let you win one dollar if the match is tails and two dollars if the match is heads. You still lose one dollar if the coins do not match. What are your expected winnings now? Is this a fair game?
e. You compensate for that by agreeing to pay your friend $1.50 if the coins do not match. Compute the expected winnings.
f. What is the difference between the last game in part (e) and the first one in part (a)? Display the probability mass function of this new game. (g) Compute the variance and standard deviation of the winnings for the first game (part (a)) and the last game of part (e). What game would you rather play?

Exercise 10. A coin weighted so that P(H) = 2/3 and P(T) = 1/3 is tossed three times.

a. Write down the outcomes in the sample space S and the probability of each of the individual outcomes.
b. Let X be the random variable which assigns to each point in S the largest number of successive heads which occurs. Write the values of X under each of the outcomes in the sample space S. For example, in a different context, if we observe, say, 10 boxes of cereal in a supermarket in sequence to see if they contain a coupon inside, we could observe cccnnncncc. In this case, the largest number of successive coupons is 3. We observe successive coupons cc and c and ccc. The largest sequence is ccc.
c. Write the probability distribution of X in tabular form.

Exercise 11. (This problem is from Petrucelli, Nandram and Chen (1999, 211).) Ming's Seafood Shop stocks live lobsters. Ming pays $6.00 for each lobster and sells each one for $12.00. The demand X for these lobsters in a given day has the following probability mass function.

x	**0**	**1**	**2**	**3**	**4**	**5**
P(X = x)	0.05	0.15	0.30	0.20	0.20	0.10

a. What is the expected demand?
b. What is the expected profit on a given day if Ming has a stock of exactly three lobsters that day.

Exercise 12. A computer store has purchased three computers at $500 apiece. It will sell them for $1000 apiece. The manufacturer has agreed to repurchase any computers still unsold after a specified period at $200 apiece. Let X denote the number of computers sold, and suppose that the probability mass function is

x	0	1	2	3
$P(X = x)$	0.1	0.2	0.3	0.4

Let g(X) represent the profit associated with selling X units.

a. Write the formula g(X) function.
b. Compute the expected value of the g(X) function.
c. Compute the standard deviation of the g(X) function.

Exercise 13. You have $1,000 and a certain commodity presently sells for $2 per ounce. Suppose that after one week the commodity will sell for either $1 or $4 an ounce, with these two possibilities being equally likely. If your objective is to maximize the expected amount of the commodity that you possess at the end of the week, what strategy should you employ?

Exercise 14. A typical slot machine has 3 dials, each with 20 symbols (cherries, lemons, plums, oranges, bells, and bars). A typical set of dials is shown below. According to this table, of the 20 slots on dial 1, 7 are cherries, 3 are oranges, and so on. A typical payoff on a 1-unit bet is as shown in below. Compute the player's expected winnings on a single play of the slot machine. Assume that each dial acts independently.

Slot Machine Dial Set up

	Dial 1	Dial 2	Dial 3
Cherries	7	7	0
Oranges	3	7	6
Lemons	3	0	4
Plums	4	1	6
Bells	2	2	3
Bars	1	3	1
	20	20	20

Typical payoff on a 1-unit bet

Dial 1	Dial 2	Dial 3	Payoff
Bar	Bar	Bar	\$60
Bell	Bell	Bell	20
Bell	Bell	Bar	18
Plum	Plum	Plum	14
Orange	Orange	Orange	10
Orange	Orange	Bar	8
Cherry	Cherry	Anything	2
Cherry	No Cherry	Anything	0
Anything else			−1

Exercise 15. Prove that $E(X) = np$ and $Var(X) = np(1 - p)$ if X is a binomial random variable.

Exercise 16. The number alpha particles emitted by a radioactive substance has expected value of 12 per square centimeter. If two 1-square centimeters are independently selected, find the probability that the two received 4 alpha particles. How many 1-square-centimeter samples should be selected to establish a probability of approximately 0.95 that at least one will contain one or more alpha particles?

Exercise 17. A department store classifies its charge customers as either high-volume or low-volume purchasers. Ten percent are high-volume purchasers. If a sample of four customers is randomly selected, what is the chance that none of them is a high-volume purchaser?

Exercise 18. Assume that 13% of people are left-handed. If we select five people at random, find the probability of each outcome below:

a. The first lefty is the fifth person chosen
b. There are some lefties among the five people
c. The first lefty is the second or third person
d. There are exactly three lefties in the group
e. There are no more than three lefties in the group

Exercise 19. Suppose we randomly select five cards without replacement from an ordinary deck of playing cards. What is the probability of getting exactly two red cards?

Exercise 20. Let X be the number of bacterial colonies per cubic centimeter, a Poisson random variable with expected value 3. (i) What is the probability that there is at least one bacterial colony in a randomly chosen cubic centimeter? (ii). What is the probability that in five randomly chosen cubic centimeters there is at least one cubic centimeter where the event of part (i) occurs? (iii) How many cubic centimeters must be observed for the probability of observing at least one satisfying the event in (i) to be 0.95?

Exercise 21. Show that if X is a random variable having a discrete uniform distribution with N points, then

$$E(X) = \frac{1}{N}\sum_{i=1}^{N} x_i, \quad E(X^r) = \frac{1}{N}\sum_{i=1}^{N} x_i^r, \quad M_x(t) = \frac{1}{N}\sum_{i=1}^{N} e^{tx_i}$$

Exercise 22. A student member collecting money for the Probability Club thinks that the probability of collecting money from a person is 0.01 if nothing is offered for the money in return, but it is 0.4 if doughnuts from the Sinking Donuts store is given to the person donating. If this week Sinking Donuts has offered to give a doughnut to everyone that donates money, how many people will the student have to address to collect money?

Exercise 23. (This problem is based on Marchette and Wegman (2004, 13).) Cybersecurity is of major concern these days. It is rare the large institution that has not received a virus or is attacked on the internet. A network attack may occur by compromising multiple systems which are then used as platforms to mount a distributed attack. The victims send their responses to the spoofed addresses of the compromised system, not to the attacker. The compromised systems are chosen at random by the attacker. Assume that an attacker is selecting spoofed IP addresses from N total addresses. Further, assume that a network sensor monitors n IP addresses. The attacker sends k packets to the victim. A packet is the basic data part of the internet. (i) What is the probability of detecting an attack? (ii) What is the probability of seeing j packets?

Exercise 24. In a town, the probability that the air quality is good is p. If we choose n days at random, what is the probability of finding 4 with good air quality?

Exercise 25. A certain water purification system contains five filters. Each one of the five filters of the water-purification system functions independently with probability of 0.95. For good water quality, at least 4 of the filters should be functioning. What is the probability that the quality of the water is good?

Exercise 26. Consider the following random variables and identify each type.

a. Weekly counts of new typhoid fever cases in the world
b. The number of biology majors in a class of 50 students
c. The number of people interviewed it takes to find a qualified candidate

Exercise 27. A homework assignment has 12 problems. A quiz is designed containing a random selection of 4 of these problems. If a student has figured out how to do 6 of the problems in the homework, what is the probability that the student will answer correctly more than 2 questions on the quiz?

Exercise 28. Suppose that 16% of cars fail pollution tests in California. What is the probability that an entire fleet of seven cars pass the test?

Exercise 29. Do extinctions occur randomly through the long fossil's record of Earth's history?, or are there periods in which extinction rates are unusually high ("mass extinctions") compared with background rates? Whitlock and Schluter (2009) give data on the number of extinctions of marine invertebrate families in 76 blocks of time of similar duration.

0 extinctions happened in none of the blocks
1 extinction happened in 13 blocks
2 extinctions happened in 15 blocks
3 extinctions happened in 16 blocks
4 extinctions happened in 7 blocks
5 extinctions happened in 10 blocks
6 extinctions happened in 4 blocks, 7 extinctions happend in 2 blocks, 8 in 1 block, 9 in 2 blocks and greater or equal to 10 in 6 blocks.

Compute the expected number of blocks for the number of extinctions given above and compare with the observed. Is there much difference between the two?

Exercise 30. The moment-generating function of a random variable X is given below.

$$M_X(t) = \left(\frac{1}{2}\right)^{10} (e^t + 1)^{10}$$

Find $P(X = 4)$.

5.19 Chapter References

Bain, Lee J., and Max Engelhardt. 1987. *Introduction to Probability and Mathematical Statistics.* Duxbury Press.

Carlton, Matthew A., and William D. Stansfield. 2005. "Making Babies by the Flip of a Coin." *The American Statistician* 59, no. 2 (May): 180–182.

Christensen, Hannah. 2015. "Outlook for the weekend? Chilly, with a chance of confusion." *Significance* (October): 4–5.

Goldberg, Samuel. 1960. *Probability. An Introduction.* New York. Dover Publications, Inc.

Levy, Paul S., Jason Hsia, Borko Jovanovic, and Douglas Passaro. 2002. "Sampling Mailroom for Presence of Anthrax Spores: A Curious Property of the Hypergeometric Distribution under an Unusual Hypothesis Testing Scenario." *Chance* 15, no. 2: 19–21.

Mansfield, Edwin. 1994. *Statistics for Business and Economics: Methods and Applications.* Fifth Edition. W.W. Norton & Company.

Marchette, David J., and Edward J. Wegman. 2004. "Statistical Analysis of Network Data for Cybersecurity." *Chance* 17, no 1: 8–18.

Moore, David S. 2010. *The Basic Practice of Statistics*. W.H. Freeman and Company.

Nolan, Deborah, and Terry Speed. 2000. *Stat Labs: Mathematical Statistics through Applications*. Springer Verlag.

Orzack, Steven Hecht. 2016. "Old and New Ideas About the Human Sex Ratio." *Significance* 13, no. 1: 24–27.

Paxson, Vern, and Sally Floyd. 1995. Wide-Area Traffic: The Failure of Poisson Modeling. *IEEE/ACM Transactions in Networking*, 3(3) pp.226–244.

Petrucelli, Joseph D. Balgovin Nandram, and Minghui Cheni. 1999. *Applied Statistics for Engineers and Scientists*. Prentice Hall, New Jersey.

Ross, Sheldon M. 2002. *Probability Models for Computer Science*. Harcourt/Academic Press.

Ross, Sheldon. 2010. *A First Course in Probability*. 8th Edition. Pearson Prentice Hall.

Samuels, Myra L., Jeffrey A. Witmer, and Andrew A. Schaffner. 2016. *Statistics for the Life Sciences*. Fifth Edition. Pearson.

Sanchez, J. (2003). "Data Analysis Activities and Problems for the Computer Science Major in a Post-Calculus Introductory Statistics Course." *2003 Proceedings of the American Statistical Association. Statistical Education Section,* Alexandria, VA: American Statistical Association.

Sanchez, Juana, and Y. He. 2005. "Internet Data Analysis for the Undergraduate Statistics Curriculum." *Journal of Statistics Education* 13, no. 3.

Scheaffer, Richard L. 1995. *Introduction to Probability and its Applications*. Duxbury Press.

Trivedi, Shridharbhai. 2002. *Probability and Statistics with Reliability, Queueing, and Computer Science Applications*. John Wiley and Sons, Inc.

Willinger, Walter, and Vern Paxson. 1998. "Where Mathematics Meets the Internet." *Notices of the AMS,* 45, no. 8 (1998): 961–970.

Whitlock, Michael C., and Dolph Schluter. 2009. *The Analysis of Biological Data*. Roberts and Company Publishers.

Zheng, Qi. 2010. The Luria-Delbrück Distribution. *Chance* 23, no. 2: 15–18.

Zipf, G. K. 1949. *Human Behavior and the Principle of Least Effort*. Cambridge, MA: Addison-Wesley.

Chapter 6

Probability Models for More Than One Discrete Random Variable

> Do you think people can tell if their significant other (SO) were cheating on them? How would people feel if they accused their SO of cheating and found out they were wrong? Do people themselves get away with cheating? How would you conduct an investigation to answer these questions?
>
> **(Jane M. Watson 2011)**

6.1 Joint probability mass functions

We are often interested in more than one characteristic of the outcome of an experiment. If a conservation measure is implemented to determine its environmental effect on plants and wild animals in the Santa Monica mountains, California, this experiment could produce a wide array of outcomes, each outcome involving a metric on plant abundance and animal abundance. Most of the time we would be interested in how those are related. Moreover, before we implement the measure, we would like to know the probability of the logically possible outcomes, to prevent the implementation of a measure with devastating effects. Probability theory for the relation between events, which we saw in Chapter 3 when we studied conditional probability and independence, can be used to this end. Chapter 6 contains the alternative random variable representation of metrics related to outcomes in the sample space. We focus in two metrics in this chapter, vector random variables such that the vector has two elements.

The reader is encouraged to review Sections 3.3 and 3.4 of this textbook before embarking on the study of Chapter 6.

Definition 6.1.1

Let a sample space $S = \{o_1, o_2, o_3 \ldots, o_N\}$ be given together with an acceptable assignment of probabilities to its simple events or outcomes of the experiment, $o_1, o_2, o_3, \ldots, o_N$. Let X and Y be two random variables defined on S. Then the function P whose value at the point $(X = x, Y = y)$ is given by:

$$P(x,y) = P(X = x, Y = y)$$
$$= \sum P(\{o_i\} \in S \mid X(o_i)$$
$$= x \text{ and } Y(\{o_i\}) = y),$$

is called the joint probability mass function of the random variable X and Y. (The domain of the function is the set of all ordered pairs of real numbers, although P has nonzero values for only a finite number of such pairs.) If the number of values of X and Y is not large, we may represent the $P(x, y)$ in a two-way table, as in examples 6.1.1 and 6.1.2.

The joint probability mass function must satisfy the axioms, namely the joint probabilities must add to 1, joint probabilities are between 0 and 1 and the probability of the union of mutually exclusive events is the sum of the probabilities.

Example 6.1.1

It is customary at American universities to employ students to do many of the services. A campus ticket office employs three work-study students on Mondays. The students are unrelated, and come from different parts of campus. The first student works 10 a.m.–12 p.m., the second 12–2 p.m. and the third 2–4 p.m. Sometimes, a student gets sick, or does not show up. Suppose each student is, independently, equally likely to show up or not. The manager of the office must have a backup plan in the event that some of the students do not show up to work that day. It would be convenient to have a model that gives the probability of all possible events next Monday.

Let W indicate whether a work-study student shows up to work on a given Monday, and A(for absent) if the student does not show up. As usual, we refer to the sample space of the experiment of observing the attendance of the three students on a randomly chosen Monday.

$$S = \{WWW, WWA, WAW, WAA, AWW, AWA, AAW, AAA\}.$$

The simple event *WWW*, for example, denotes the outcome where the first student shows up to work, the second student shows up to work and the third shows up to work. The probability of this simple event is $\frac{1}{8} = \left(\frac{1}{2}\right)\left(\frac{1}{2}\right)\left(\frac{1}{2}\right)$, if we are willing to assume that all outcomes are equally likely. Similarly, we can compute the probability of the other simple events, which is 1/8 for each of them.

Define the following random variables:

$$X = \begin{cases} 0 \text{ if the first student in the shift is absent } (A) \\ 1 \text{ if the first student in the shift works } (W) \end{cases}$$

$$Y = \text{the total number of student employees appearing that monday}, Y = 0,\ 1,\ 2,\ 3.$$

Of interest in this example is whether the fact that the student in the first shift is absent has some effect on how many students show to work that day.

We want to determine not only the possible pairs of values of X and Y, but also the probability with which each such pair occurs. To say, for example, that the event consisting of X taking value 0 and Y the value 1 occurs is to say that the event $\{AWA, AAW\}$ occurs. The probability of this event is therefore 2 / 8 or 1 /4. We write

$$P(X = 0, Y = 1) = P(\{AWA, AAW\}) = \frac{1}{8} + \frac{1}{8} = 1/4,$$

adopting the usual convention in which a comma is used in place of $\cap$ to denote the intersection of the two events $X = 0$ and $Y = 1$.

We similarly find:

$$P(X = 0, Y = 0) = P(\{AAA\}) = \frac{1}{8}$$

$$P(X = 1, Y = 0) = P(\phi) = 0.$$

etc. In this way, we obtain the probabilities of all possible pairs of values of X and Y. These probabilities are conveniently arranged in Table 6.1, in order to see the association between the outcomes of the sample space and the values of the random variables.

Note that when we define random variables in this way, the equality sign is used as a shorthand for "is the random variable whose value for any outcome (element of S) is". The distinction between the random variable and the value of the random variable should be kept clearly in mind when, as here, the customary notation is somewhat misleading.

We list in Table 6.1 the values of these two random variables for each element of the sample space S.

Table 6.1 Two random variables defined on the same sample space.

Outcome *o*	**WWW**	**WWA**	**WAW**	**WAA**	**AWW**	**AWA**	**AAW**	**AAA**
P(*o*)	1/8	1/8	1/8	1/8	1/8	1/8	1/8	1/8
X value	1	1	1	1	0	0	0	0
Y value	3	2	2	1	2	1	1	0

From the information in Table 6.1, and Definition 6.1.1, we obtain then the joint probability mass function table of X and Y.

The resulting joint or bivariate distribution of X and Y, containing all values of $P(X = x,\ Y = y)$, can be seen in Table 6.2.

Table 6.2 Joint probability mass function of the random variables X and Y of Table 6.1. The numbers in the cells are the joint probabilities of values of X and Y.

x \y	**0**	**1**	**2**	**3**
0	P(X = 0,Y = 0) = 1/8	P(X = 0,Y = 1) = 2/8	P(X = 0,Y = 2) = 1/8	P(X = 0,Y = 3) = 0
1	P(X = 1,Y = 0) = 0	P(X = 1,Y = 1) = 1/8	P(X = 1,Y = 2) = 2/8	P(X = 1,Y = 3) = 1/8

We can also represent these results graphically as in Figure 6.1, where we draw a three dimentional chart in which $P(X = x, Y = y)$ is the height of a vertical line drawn above the point (x, y) in the horizontal x, y plane, the height being the value of the joint probability at that point.

Once the joint probability mass table is constructed, we may answer questions relevant to the problem at hand. For example, what is the probability that the student in the first

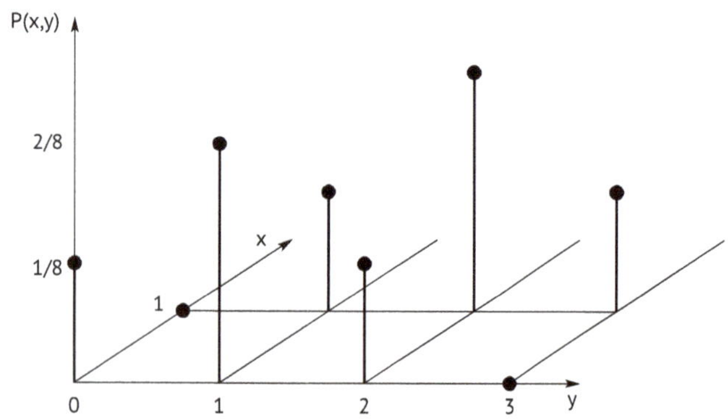

Figure 6.1 Graphical display of the joint probability mass function in Table 6.2.

shift misses work and less than two students show up to work?

$$P(X=0,\ Y<2)=\frac{1}{8}+\frac{2}{8}=\frac{3}{8}.$$

Example 6.1.2

In trying to determine whether the employee in charge of checkout register 1 at a large department store is much more desirable to customers than the employee in charge of checkout register 2, a big department store conducts a study of the 6 PM rush hour over many days that these two employees work. Let X be the number of customers in register 1 and Y the number in register 2. The joint probability mass function is given below.

x \y	0	1	2	3
0	0.08	0.07	0.03	0.3
1	0.02	0.01	0.02	0.2
2	0.01	0.02	0.03	0.2
3	0	0	0.01	0.0

What is the probability that the employee in checkout register 1 has more customers than the employee in register 2? We add the probabilities of the mutually exclusive events where this happens.

$$\begin{aligned}P(X>Y)&=P(X=1,\ Y=0)+P(X=2,\ Y=0)+P(X=2,\ Y=1)+P(X=3,\ Y=0)\\&\quad+P(X=3,\ Y=1)+P(X=3,\ Y=2)\\&=0.02+0.01+0.02+0+0+0.01\\&=0.06.\end{aligned}$$

6.1.1 Exercises

Exercise 1. Two aeronautics companies (I, II,) bid for contracts for space in a satellite navigation system. A company that bids for a contract gets funded for their contract by the European Union. Past information shows that firm I and firm II get each one contract with probability 1/9, firm I and firm II can each get two contracts with probability 1/9 and firm I and firm II can each get 3 contracts with probability 1/9. But any other distribution of the contracts between the two companies have also 1/9 probability of happening. There are then a total of 9 outcomes. The Sample space would be represented by:

$$S=\{(1,1),\ (1,2),\ (1,3),\ (2,1),\ (2,2),\ (2,3),\ (3,1),\ (3,2),\ (3,3)\}.$$

The first component of the pair indicates how many contracts company I gets and the second how many contracts company II gets. Consider random variable X denoting the number of contracts granted to firm I and Y the number of contracts granted to firm II. (i) Construct the joint probability mass function of X and Y. (ii) What is is the probability that I gets the same number of contracts as II? (iii) What is the probability that a contract goes to company one?

Exercise 2. Two species, A and B, affected by the same environmental factors, are being studied to see if there is association between them. The species live in fruits. The random variable X measures the number of species A per fruit, and the random variable Y measures the number of species B per fruit. The joint probability mass function $P(X, Y)$ is given by the following table.

x \y	**0**	**1**	**2**
0	0.40	0.1	0.1
1	0.1	0.1	0.02
2	0.1	0.02	0.03
3	0.01	0.01	0.01

What is the probability that the number of species B is larger than the number of species A?

6.2 Marginal or total probability mass functions

Suppose that in example 6.1.1 we are interested only in Y yet have to work with the joint distribution of X and Y. Specifically, suppose we are interested in the event Y = 2, which is a vertical slice of Table 6.2. Of course, as we learned in Chapter 3, by law of total probability (Section 3.5),

$$P(Y=2)=P(Y=2,\ X=0)+P(Y=2,\ X=1)=\frac{1}{8}+\frac{2}{8}=\frac{3}{8}.$$

That is, to obtain the total probability of the event Y = 2, we sum all the joint probabilities events in which Y takes the value 2.

Example 6.2.1. Example 6.1.1 (continued)

Consider Table 6.3. It is a copy of Table 6.2, but we have added a column and a row.

Definition 6.2.1

Knowing the joint probability mass function of two random variables X and Y, we can construct the total probability mass function of X, $P(X)$, and the total probability mass function of Y, $P(Y)$ as follows:

$$P(x)=P(X=x)=\sum P(\{o\_i\}\in S \mid X(o\_i)=x)$$

$$P(y)=P(Y=y)=\sum P(\{o\_i\}\in S \mid Y(o\_i)=y)$$

Table 6.3 The marginal probabilities of X and Y obtained by summing rows and columns of the joint distribution.

x \y	**0**	**1**	**2**	**3**	**P(X = x)**
0	1/8	2/8	1/8	0	1/2
1	0	1/8	2/8	1/8	1/2
P(Y = y)	1/8	3/8	3/8	1/8	1

The event $Y=0$ is the union of the mutually exclusive events $\{X=0, Y=0\}$ and $\{X=1, Y=0\}$. Hence,

$$P(Y=0)=\sum_x P(x,\ 0)=P(X=0,\ Y=0)+P(X=1,\ Y=0)=\frac{1}{8}+0=1/8.$$

In Table 6.3, this probability is obtained as the sum of the entries in the rows headed $Y=0$. By adding the entries in the other columns, we similarly find:

$$P(Y=1)=\sum_x P(x,\ 1)=\frac{3}{8},\quad P(Y=2)=\sum_x P(x,\ 2)=3/8,\quad P(Y=3)=\sum_x P(x,\ 3)=1/8.$$

In this way, we obtain the total or marginal probability function of the random variables Y from the joint probability table of X and Y. Since values of this probability function are written in the lower margin of the joint table, the function is commonly called the marginal probability of Y, in spite of the fact that the adjective "marginal" is redundant. By adding across the columns in the joint table, one similarly obtains the (marginal) probability mass function of X. Marginal probabilities are total probabilities.

It may be more familiar to the reader to have the marginal probability mass functions in the usual table format that we used in Chapter 5. Table 6.4 contains them. This is the way they should be presented. Table 6.3 is just to illustrate where they come from.

Table 6.4 Marginal pmfs of X and Y give us total probabilities of X and Y.

x	**P(X = x)**
0	1/2
1	1/2

y	**P(Y = y)**
0	1/8
1	3/8
2	3/8
3	1/8

Once the marginal probabilities have been found, it is not hard to go back to Chapter 5 and review how to compute the expected value, variance, moment generating function and other functions of the single random variable. For example,

$$\mu_X=E(X)=\frac{1}{2},\quad \mu_Y=E(Y)=\frac{12}{8},\quad E(X^2)=\frac{1}{2},\ E(Y^2)=\frac{24}{8},$$

which allow us to compute the variance of X and the variance of Y as we did in Chapter 5, namely

$$\sigma_X^2=Var(X)=\frac{1}{2}-\left(\frac{1}{2}\right)^2=\frac{1}{4},\quad \sigma_Y^2=Var(Y)=\frac{24}{8}-\left(\frac{12}{8}\right)^2=\frac{3}{4}$$

6.2.1 Exercises

Exercise 1. (This problem is from Goldberg (1960, 211, problem 4.8).) The joint probability mass function of X and Y is given by

$$P(X=x,\ Y=y)=\frac{1}{32}(x^2+y^2), \quad x=0,\ 1,\ 2,\ 3 \quad \text{and} \quad y=0,\ 1\cdot$$

Show that the marginal probability mass function of X is given by

$$P(X=x)=\frac{1}{32}(2x^2+1), \quad x=0,\ 1,\ 2,\ 3.$$

Show that the marginal probability mass function of Y is given by

$$P(Y=y)=\frac{1}{16}(2y^2+7), \quad y=0,\ 1.$$

Exercise 2. In exercise 1, section 6.1.1, calculate the expected number of contracts going to company I and the standard deviation of contracts going to company I.

6.3 Independence of two discrete random variables

Example 6.3.1. Example 6.1.1 (continued)
Let's check whether X and Y in Table 6.2 are independent. We ask whether it holds, for example, when $X=0$ and $Y=0$. In other words, is it true that:

$$P(X=0,Y=0)=P(X=0)P(Y=0)?$$

The answer is no, because, using Table 6.3,

$$P(X=0)P(Y=0)=(1/2)(1/8)=1/16,$$

but

$$P(X=0,Y=0)=1/8,$$

and clearly,

$$\frac{1}{16}\neq\frac{1}{8}.$$

We got all the information we needed from the joint probability mass function table.

Because $P(X=0,Y=0)\neq P(X=0)P(Y=0)$ we can conclude that X and Y are not independent random variables. This was intuitively obvious by the way we obtained these two random variables from the sample space.

Definition 6.3.1

Two discrete random variables X and Y are independent if the events $(X=x)$ and $(Y=y)$ are independent, i.e., if

$$P(X=x,Y=y)=P(X=x)P(Y=y), \quad \textit{for all } x \textit{ and } y$$

or, put differently, if the joint probabilities are the product of the respective marginal probabilities.

For example, for the distribution in Table 6.2, are X and Y independent? For independence, the condition must hold for every (x,y) combination.

Notice that all we need to show is that the condition is not satisfied in one cell of the table. If it had been satisfied we would have had to continue trying until we were convinced that the condition is not true in all the cells.

6.3.1 Exercises

Exercise 1. Suppose that X and Y have the following joint probability mass function $P(X, Y)$.

x\y	0	2	4	6
10	0.04	0.08	0.08	0.05
15	0.12	0.24	0.24	0.15

Are X and Y independent?

Exercise 2. (This exercise is from Wonnacott and Wonnacot (1990).) A salesman has an 80% chance of making a sale on each call. If three calls are to be made next month, let

$X =$ total number of sales
$Y =$ total profit from the sales

Where the profit Y is calculated as follows: Any sales on the first two calls yields a profit of \$100 each. By the time the third call is made, the original product has been replaced by a new product whose sale yields a profit of \$200. Thus, for example, the sequence (sale, no sale, sale) would give $Y =$ \$300. (i) List the sample space. (ii) Tabulate and graph the bivariate probability mass function of X and Y. (iii) Calculate the marginal distribution of X and Y. (iv) What is the expected value of X and the expected value of Y? (v) Are X and Y independent?

Exercise 3. Consider the planning of a 3 day vacation. The cost of the vacation R depends upon both the number of days that the person spends the vacation outside the home (Y) and whether the last day of the three was spent outside the home (X)

$P(X, Y)$ is given by the following table

x \y	0	1	2	3
0	0.04	0.08	0.08	0.05
1	0.12	0.24	0.24	0.15

1. What is the expected cost? (ii) Compute the marginal pmf of X and Y. (iii) Are X and Y independent?

6.4 Conditional probability mass functions

▶ Consider again the probability mass function of Table 6.2. If we are interested only in the probability of Y when X = 1, how would we go about finding the distribution of Y for that case when all we have to work with is the joint distribution of X and Y? Do you think that the probability mass function of *Y* will be different when $X = 2$?

We add one final observation about the random variables *X* and *Y* in Table 6.2. It is clear from the meaning of *X* and *Y* that knowing the value of *X* changes the probability that a given value of *Y* occurs. For example, $P(Y = 2) = 3/8$. But if we are told that the value of *X* is 1, then the conditional probability of the event $Y = 2$ becomes $1/2$. For, by the definition of conditional probability of two events, the conditional probability of the event $Y = 2$ given that $X = 1$ is

$$P(Y = 2 | X = 1) = \frac{P(X = 1, Y = 2)}{P(X = 1)} = \frac{1/4}{1/2} = \frac{1}{2}.$$

We can also think of this as follows: knowing that $X = 1$ reduces the sample space to the four outcomes where $X = 1$ ({*WWW, WWA, WAW, WAA*}). In this reduced sample space, only two outcomes have $Y = 2$, namely {*WWA, WAW*}, and the probability of this last event is $(1/8) + (1/8) = 1/2$ by axiom 3, since these two outcomes are mutually exclusive.

As we expect, the events $X = 1$ and $Y = 2$ are not independent: knowing that the first student in the shift works on a Monday increases the probability of 2 students working on a Monday.

We introduce now a new type of univariate probability mass function: the conditional distributions. There are four conditional distributions of *X*, one for each value of *Y*. And there are two conditional distributions of *Y*, one for each value of *X*. We will illustrate the extraction of conditional probability mass functions from a joint probability mass function using the latter.

Example 6.4.1. Example 6.1 (continued)

Table 6.5 Conditional pmfs of Y given a value of X. Obtained from Table 6.3.

y	$\mathbf{P(Y = y \mid X = 0)}$
0	$P(Y = 0 \mid X = 0) = \frac{P(X = 0, Y = 0)}{P(X = 0)} = \frac{1/8}{1/2} = 1/4$
1	$P(Y = 1 \mid X = 0) = \frac{P(X = 0, Y = 1)}{P(X = 0)} = \frac{2/8}{1/2} = 2/4$
2	$P(Y = 2 \mid X = 0) = \frac{P(X = 0, Y = 2)}{P(X = 0)} = \frac{1/8}{1/2} = 1/4$
3	$P(Y = 3 \mid X = 0) = \frac{P(X = 0, Y = 3)}{P(X = 0)} = \frac{0}{1/2} = 0$

y	$P(Y = y \mid X = 1)$
0	$P(Y=0\mid X=1)=\dfrac{P(X=1,Y=0)}{P(X=1)}=\dfrac{0}{1/2}=0$
1	$P(Y=1\mid X=1)=\dfrac{P(X=1,Y=1)}{P(X=1)}=\dfrac{1/8}{1/2}=1/4$
2	$P(Y=2\mid X=1)=\dfrac{P(X=1,Y=2)}{P(X=1)}=\dfrac{2/8}{1/2}=2/4$
3	$P(Y=3\mid X=1)=\dfrac{P(X=1,Y=3)}{P(X=1)}=\dfrac{1/8}{1/2}=1/4$

These pmfs are univariate probability mass functions. Thus, their expectation and variance must be computed using the same methodology used for the marginal probability mass functions and for a univariate mass function as seen in Chapter 5. However, the notation must change to allow for the fact that each expectation depends on the value of the other variable and the pmf is the conditional pmf. We will illustrate this idea by computing the conditional expectation and variance of Y when X is 1. We obtain what we need from Table 6.5. We will use the table for $P(Y \mid X = 1)$.

$$\mu_{Y|X=1} = E(Y|X=1) = (0)(0) + 1\left(\frac{1}{4}\right) + 2\left(\frac{2}{4}\right) + 3\left(\frac{1}{4}\right) = 2.$$

$$\mu_{Y^2|X=1} = E(Y^2|X=1) = (0)(0) + 1\left(\frac{1}{4}\right) + 4\left(\frac{2}{4}\right) + 9\left(\frac{1}{4}\right) = 9/2.$$

$$\sigma^2_{Y|X=1} = E(Y^2|X=1) - (E(Y|X=1))^2 = \frac{9}{2} - 4 = 1/2.$$

With this particular example understood, we can now proceed to discuss the general case of any two random variables defined on the same sample space.

6.4.1 Exercises

Exercise 1. Find the following conditional probability mass functions obtained from Table 6.2 and then compute the conditional expectations and conditional variances in each case.

$$P(X \mid Y=0);\ P(X \mid Y=1);\ P(X \mid Y=2);\ P(X \mid Y=3).$$

What happens to the expected value of X as Y increases?

6.5 Expectation of functions of two random variables

- Consider the planning of undergraduate education for all three kids that a young couple plans to have. The costs depend on the number of kids that go to college (X) and the number of years each takes to complete the degree (Y). We denote the cost by C. Thus

$$C = g(X,Y).$$

What will be the expected cost, μ_c? By how much could the actual cost deviate from the expected cost?

We could tabulate the distribution of C and calculate:

$$E(C) = \sum CP(C).$$

Or we could use the joint distribution of X and Y, $P(X,Y)$, directly:

$$E[g(X,Y)] = \sum_x \sum_y g(x,y)p(x,y).$$

Example 6.5.1

Suppose the joint probability mass function of X and Y for the planning couple is as follows

x \y	**3**	**4**	**5**
1	0.05	0.05	0
2	0.07	0.3	0.03
3	0	0.4	0.1

Suppose the cost function is (in tens of thousands)

$$C = g(X,Y) = 10XY.$$

We can calculate the mean of C directly from $P(X,Y)$. We will first use the joint probability mass table as a tool to do the computations directly in each cell. We calculate in each cell $10xyP(X = x, Y = y)$.

x \y	**3**	**4**	**5**
1	(10xy)0.05 = 1.5	(10xy)0.05 = 2	0
2	(10xy) 0.07 = 4.2	(10xy)0.3 = 24	(10xy) 0.03 = 3
3	0	(10xy)0.4 = 48	(10xy) 0.1 = 15

$$E(C) = E[g(X,Y)] = \sum_x \sum_y g(x,y)p(x,y) = 1.5 + 2 + 4.2 + 24 + 3 + 48 + 15 = 97.7.$$

We can also find the same result by first finding the probability mass function of C, also obtained from the table, as follows, and then compute the expected value of this univariate discrete random variable.

c	**P(C = c)**
$10(1)(3) = 30$	0.05
$10(1)(4) = 40$	0.05
60	0.07
80	0.3
100	0.03
120	0.4
150	0.1

$$E(C) = \sum CP(C) = 97.7.$$

We shall see in this section that if two random variables X and Y are defined on a sample space S, then there are automatically many other random variables also defined on S. In particular, the sum $X+Y$ and the product XY turn out to be especially important random variables.

$$g(X,Y) = X+Y$$

is the sum of the random variables X and Y.

$$g(X,Y) = (X-\mu_x)(Y-\mu_y)$$

is the product of the deviations of X and Y from their respective means.

Example 6.5.2

Consider the random variables X and Y of Table 6.2. The possible values of X and Y, together with their joint probabilities, are given in that table. Let $g(X,Y) = X+Y$.

From the joint probability Table 6.2, we can determine the possible values of random variable $U = X + Y$ as well as the probabilities with which each value occurs. For example,

$$P(U=2) = P(X=0, Y=2) + P(X=1, Y=1) = 1/8 + 1/8 = 1/4.$$

In this way, we obtain the entries in the following probability table for the random variable $U = X + Y$.

u	**0**	**1**	**2**	**3**	**4**
P(U = u)	1/8	1/4	1/4	1/4	1/8

From this table, we calculate the mean of U:

$$E(U) = E(X+Y) = 0\left(\frac{1}{8}\right) + 1\left(\frac{1}{4}\right) + 2\left(\frac{1}{4}\right) + 3\left(\frac{1}{4}\right) + 4\left(\frac{1}{8}\right) = 2.$$

From the marginal probability functions of X and Y, given in Table 6.3, we find that:

$$E(X) = 0\left(\frac{1}{2}\right) + 1\left(\frac{1}{2}\right) = \frac{1}{2};\ E(Y) = 0(1/8) + 1(3/8) + 2(3/8) + 3(1/8) = 3/2.$$

Observe that $E(X+Y) = E(X) + E(Y)$, a result that we will soon establish for all random variables X and Y.

Example 6.5.3

If we define $g(X,Y)$ as the product rather than the sum of X and Y, then $V = XY$ is a random variable whose probability table is similarly found from Table 6.2:

v	0	1	2	3
P(V = v)	1/2	1/8	1/4	1/8

Now we compute the mean of V,

$$E(V) = 0\left(\frac{1}{2}\right) + 1\left(\frac{1}{8}\right) + 2\left(\frac{1}{4}\right) + 3\left(\frac{1}{8}\right) = 1.$$

Observe that $E(XY) \neq E(X)E(Y)$ in general. The equality will only be true when the two random variables are independent, which is not the case in this exercise.

You should note that what we do to determine the probability function of $g(X,Y)$ is collect all possible pairs of X and Y values that lead to the same value of $g(X,Y)$ and add their probabilities. But to compute the mean of $g(X,Y)$, we could just take a short-cut and use the joint probability table as a computational tool.

Theorem 6.1

Let X and Y be random variables with joint probability function P and let g be a function of X and Y. Then

$$E[g(X,Y)] = \sum_x \sum_y g(x,y)P(x,y)$$

In words, we find $E[g(X,Y)]$ by moving from cell to cell in the joint probability table of X and Y, multiplying the value of $g(X, Y)$ corresponding to each cell by the probability appearing in that cell, and then adding these products for all cells.

Example 6.5.4

Consider Example 6.1.1 and let us illustrate the use of the last formula by recalculating the mean of X + Y and XY. We find directly from Table 6.2, moving across the first row and then the second,

$$E(X+Y)=0\left(\frac{1}{8}\right)+1\left(\frac{1}{4}\right)+2\left(\frac{1}{8}\right)+3(0)+1(0)+2\left(\frac{1}{8}\right)+3\left(\frac{1}{4}\right)+4\left(\frac{1}{8}\right)=2,$$

as before.

There is of course no need to write down terms that have zero factors. Indeed, any cell in the joint probability table for which $g(x_{j,}y_k) = 0$ can be skipped in computing $E[g(X,Y)]$. Hence we skip these and find

$$E(XY)=1\left(\frac{1}{8}\right)+2\left(\frac{1}{4}\right)+3\left(\frac{1}{8}\right)=1.$$

Theorem 6.1 enables us to prove the following extremely important and often-used results.

$$E(X+Y)=E(X)+E(Y)$$

In words, the expected sum of two random variables is equal to the sum of their means.

$$E(X+Y)=\sum_x x\sum_y P(x,y)+\sum_y y\sum_x P(x,y)=\sum_x xP(x)+\sum_y yP(y)$$

This result generalizes to

$$E(aX+bY)=aE(X)+bE(Y)$$

Still more general is the following theorem

Theorem 6.2

Let n be any positive integers. If $X_1, X_2, X_3, \ldots, X_n$ are any random variables defined on a sample space S, and if $a_1, a_2, a_3, \ldots, a_n$ are any constants, then

$$E(a_1X_1+a_2X_2+,\ldots,+a_nX_n)=a_1E(X_1)+a_2E(X_2)+,\ldots,+a_nE(X_n).$$

Proof:

The result is true for n = 1 and n = 2 by formula earlier. The theorem is proved by mathematical induction) as soon as we show that if the theorem is true for any positive integer, say n = k, then it is also true for the next integer, n = k + 1. Let us therefore assume that the last statement is true for n = k. That is, letting $Y=a_1X_1+a_2X_2+,\ldots,+a_kX_k$, we are assuming:

$$E(a_1X_1+a_2X_2+,\ldots,+a_kX_k)=a_1E(X_1)+a_2E(X_2)+,\ldots,+a_kE(X_k).$$

The key idea of the proof is the observation that the sum of k + 1 random variables can be thought of as the sum of two random variables to which the two variable results can be applied. In particular,

$$
\begin{aligned}
E(a_1X_1 + a_2X_2 +,\ldots,+ a_kX_k +\ldots,+ a_{k+1}X_{k+1}) &= E(Y + a_{k+1}X_{k+1}) \\
&= E(Y) + a_{k+1}E(X_{k+1}) \\
&= a_1E(X_1) + a_2E(X_2) +,\ldots,+ a_kE(X_k) + a_{k+1}E(X_{k+1}).
\end{aligned}
$$

But this last equality shows that theorem 6.2 is true for n = k + 1, and so the proof is complete.

Theorem 6.3

Let X and Y be independent random variables defined on a sample space *S*. Then

$$E(XY) = E(X)E(Y).$$

In words, the mean of the product of two independent random variables is equal to the product of their means.

Proof

The proof needs you to recall that if two variables are independent, $P(X,Y) = P(X)P(Y)$

$$E(XY) = \sum_x \sum_y xyP(x,y) = \sum_x \sum_y xyP(x)P(y) = \sum_x xP(x) \sum_y yP(y) = E(X)E(Y).$$

It is very important to note that the converse of Theorem 6.3 is false. As the following example shows, it is possible to find this last result, (E(XY) = E(X)E(Y)), to be true for random variables that are dependent.

Example 6.5.5

Suppose X has probability table

x	**−1**	**0**	**1**
P(X = x)	1/4	1/2	1/4

Let $Y = X^2$. Then *X* and *Y* are surely dependent, since the value of *X* determines the value of *Y*. This dependence is obvious from the joint probability of *X* and *Y*, as given below

		x		
		−1	0	1
y	0	0	1/2	0
	1	1/4	0	1/4

$E(XY) = 0$; $E(X) = 0$; $E(Y) = 1/2$. So $E(XY) = E(X)E(Y)$ but the two random variables are not independent. We know that because, for example,

$$P(X = 0, Y = 0) = 1/2;$$

$$P(X = 0) = 1/2;\ P(Y = 0) = 1/2\ .$$

So $P(X = 0)P(Y = 0) = \frac{1}{4}$.

Obviously, $P(X = 0, Y = 0) \neq P(X = 0)P(Y = 0)$, so the two random variables are not independent, and yet

$$E(XY) = E(X)E(Y).$$

6.5.1 Exercises

Exercise 1. In the example of the planning family, suppose that

$$C = X^2 + Y^2$$

instead of the function given for C in Example 6.5.1.

(i) Find the expected value of C by first finding its distribution $P(C)$ and then calculating the expectation. (ii) Find expected value of C by using $E[g(X,Y)] = \sum_x \sum_y g(x,y)p(x,y)$. (iii) Find the following expected values: (a) $E[(X-2)(Y-2)]$; (b) $E[(X-2)^2]$; (c) $E[(4X+2Y)]$

6.6 Covariance and Correlation

The *covariance* is used to measure how two variables X and Y vary together, using the familiar concept of $E[(g(X,Y)]$, where $g(X,Y) = (X - \mu_X)(Y - \mu_Y)$. To measure how two variables vary together, we start with the deviations, we multiply them and then take the expectation.

When two variables are independent, the covariance is 0. But the reverse is not true in general.

6.6.1 Alternative computation of the covariance

The covariance can also be calculated using the following short-cut formula, which follows directly from the definition of covariance, using techniques already seen in Section 6.5.

$$\sigma_{X,Y} = Cov(X,Y) = E(XY) - \mu_x \mu_y$$

6.6.2 The correlation coefficient. Rescaling the covariance

The covariance depends upon the units in which X and Y are measured. If X, for example, was measured in degrees centigrade instead of degrees Fahrenheit, the covariance would be different.

Definition 6.6.1

Let X and Y be random variables defined on a sample space S. The covariance of X and Y, denoted by $Cov(X,Y)$, is defined as the number given by

$$\sigma_{X,Y} = Cov(X,Y) = E\left[(X - E(X))(Y - E(Y))\right]$$
$$= \sum_x \sum_y \left[(X - E(X))(Y - E(Y))\right] P(X = x, Y = y).$$

When large values of X and Y occur together, the deviations are both positive, and so their product is positive. Similarly, when small values of X and Y occur together, both deviations are negative, so their product is positive. If most of the joint probabilities were in this situation, then the covariance would be positive and would summarize the positive relation between the two variables.

When one deviation is positive and the other is negative, the calculated product is negative. If there is high joint probabilities in those cases, the covariance would be negative.

The *correlation coefficient*, a number between −1 and 1 takes care of eliminating the units, thus giving us a unitless measure of the direction and strength of a linear relation between the two random variables. See Figure 6.2.

$$\rho = \frac{\sigma_{X,Y}}{\sigma_X \sigma_Y}$$

This expression neutralizes any change in the scale of X and Y. Correlation is independent of the scale in which either X or Y is measured. Correlation is also always bounded.

$$-1 \leq \rho \leq +1$$

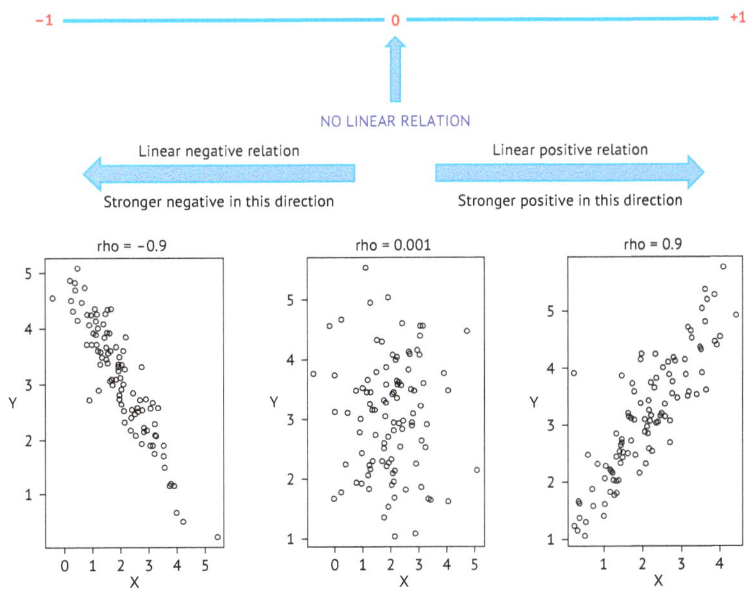

Figure 6.2 Correlation coefficient interpretation.

Example 6.6.1

x \y	0	1	2	3
0	(0 – 1/2)(0 – 3/2)1/8	(0 – 1/2)(1 – 3/2)2/8	(0 – 1/2)(2 – 3/2)1/8	0
1	0	(1 – 1/2)(1 – 3/2)1/8	(1 – 1/2)(2 – 3/2)2/8	(1 – 1/2)(3 – 3/2)1/8

Let X and Y be random variables with joint probability as given in Table 6.2. In Example 6.2.1, we found that $\mu_x = \frac{1}{2}, \mu_y = \frac{3}{2}$. In Example 6.5.3, we saw that $E(XY) = 1$.

Using the short-cut formula,

$$Cov(X,Y) = E(XY) - \mu_x\mu_y = 1 - \left(\frac{1}{2}\right)\left(\frac{3}{2}\right) = \frac{1}{4}$$

And we know that X and Y are positively correlated because the covariance is positive. We now use the calculation we made in Example 6.2.1, namely,

$$\sigma_X^2 = \frac{1}{4}, \quad \sigma_Y^2 = \frac{3}{4},$$

$$\rho(X,Y) = \frac{\frac{1}{4}}{\sqrt{\frac{1}{4}}\sqrt{\frac{3}{4}}} = 0.57773503.$$

Notice that we could obtain the same number for the value of the covariance if we add the values of the cells given in the table at the top of this example. That is,

$$\sigma_{X,Y} = Cov(X,Y) = \sum_x\sum_y \left[(X - E(X))(Y - E(Y))\right] P(X = x, Y = y) = \frac{1}{4}.$$

The correlation coefficient of 0.57773503 means that there is a positive linear association between random variables X and Y, but it is not very strong. It is not very weak either.

6.6.3 Exercises

Exercise 1. For the joint distribution in Example 6.5.1, calculate the covariance and correlation using the definition and the short-cut formula. Is there a strong positive relation between the number of kids that go to college and how long it takes to complete the degree?

Exercise 2. Let X be a random variable with the following probability mass function:

x	–2	–1	1	2
P(X = x)	1/4	1/4	1/4	1/4

Let $Y = X^2$

(i) Find the joint probability mass function of X and Y. (ii) Determine $E(XY)$ and the value of the correlation. (iii) Are the variables independent?

6.7 Linear combination of two random variables. Breaking down the problem into simpler components

Now that we know the concept of covariance, we can study a new important result concerning the variance of the sum of two random variables. Consider the following function of two random variables

$$g(X,Y) = X + Y.$$

$$\begin{aligned} Var(X+Y) &= E(((X+Y)-E(X+Y))^2) \\ &= \sum_x \sum_y ((X+Y)-E(X+Y))^2 P(X=x,Y=y) \\ &= \sum_x \sum_y ((X-\mu_x)+(Y-\mu_y))^2 P(X=x,Y=y) \\ &= \sum_x \sum_y ((X-\mu_x)^2+(Y-\mu_y)^2+2(X-\mu_x)(Y-\mu_y))P(X=x,Y=y). \end{aligned}$$

By bringing the summation operators into each term, we obtain the desired result:

$$Var(X+Y) = Var(X) + Var(Y) + 2Cov(X,Y).$$

Similarly, we can prove that

$$Var(aX+bY) = a^2Var(X) + b^2Var(Y) + 2abCov(X,Y).$$

When X and Y are uncorrelated,

$$Var(aX+bY) = a^2Var(X) + b^2Var(Y).$$

Example 6.7.1

Let's revisit the sum of the rolls of two dice. What would be the expected sum and the variance?

$$E(X+Y) = E(X) + E(Y) = 3.5 + 3.5 = 7,$$

$$Var(X+Y) = Var(X) + Var(Y) = 2.92 + 2.92 = 5.84.$$

And we can see that the formulas are satisfied, knowing that the correlation between the roll of two dice is 0.

6.7.1 Exercises

Exercise 1. Complete the steps needed to arrive from

$$
\begin{aligned}
Var(X+Y) &= E((X+Y)-E(X+Y)^2) \\
&= \sum_x\sum_y ((X+Y)-E(X+Y))^2 P(X=x,Y=y) \\
&= \sum_x\sum_y ((X-\mu_x)+(Y-\mu_y))^2 P(X=x,Y=y) \\
&= \sum_x\sum_y (X-\mu_x)^2+(Y-\mu_y)^2+2(X-\mu_x)(Y-\mu_y)(P(X=x,Y=y)
\end{aligned}
$$

to

$$Var(X+Y) = Var(X)+Var(Y)+2Cov(X,Y).$$

Exercise 2. Prove that

$$E(aX+bY) = aE(X)+bE(Y), \text{ and}$$

$$Var(aX+bY) = a^2Var(X)+b^2Var(Y)+2abCov(X,Y).$$

using, respectively, the definitions:

$$E(aX+bY) = \sum_x\sum_y (ax+by)P(X=x,Y=y) \text{ and}$$

$$Var(aX+bY) = E\left(((aX+bY)-E(aX+bY))^2\right) = \sum_x\sum_y ((aX+bY)-E(aX+bY))^2 P(X=x,Y=y).$$

Exercise 3. A machine produced by plant A of a company is expected to weigh 1000 pounds with standard deviation 200 pounds. A machine produced by plant B is expected to weigh 1,500 pounds with standard deviation of 170 pounds. The weights of the machines produced in the two plants are correlated, as some parts of the machines in plant B are produced by plant A. This correlation is 0.4. When shipped to a customer, the mailing cost is a function of the total weight, in this way:

$$C = 0.04\ (X+Y),$$

where C is cost, X is the weight of the machine from plant A and B is the weight of the machine from plant B. Find the expected cost and the variance of the cost.

6.8 Covariance between linear functions of the random variables

We can prove the following:

$$Cov(a+X, Y) = Cov(X,Y),$$

$$Cov(aW + bX, cY + dZ) = acCov(W,Y) + bcCov(X,Y) + adCov(W,Z) + bdCov(X,Z),$$

$$Cov\left(a + \sum_{i=1}^{n} b_i X_i, c + \sum_{j=1}^{m} d_j Y_j\right) = \sum_{i=1}^{n}\sum_{j=1}^{m} b_i d_j cov(X_i, Y_j).$$

The proofs will be left as exercises.

6.9 Joint distributions of independent named random variables. Applications in mathematical statistics

Australia's Highway 1 is considered to be the longest national highway in the world at over 14,500 km or 9,000 miles and runs almost the entire way around the continent. There are 16 major intersections of this highway and therefore there are 16 major exit ramps. Suppose the average number of cars per minute in each of these major ramps is 10. What is the joint probability mass function of the number of cars for all ramps?

We denote by X_i, $i = 1, 2, \ldots, 16$, the number of cars exiting each of the 16 ramps. Assuming that $P(X_i)$ is Poisson with parameter $\lambda = 10$, we can compute the joint probability mass function as follows:

$$P(X_1, X_2, \ldots\ldots, X_{16}) = P(X_1)P(X_2)\ldots\ldots P(X_{16}) = \frac{10^{\sum_{i=1}^{16} x_i} e^{-10}}{\prod_{i=1}^{16} x_i!}$$

In Probability, we could show, although it is too advanced for this book, where we only prove it for two random variables, that

$$P(X_1 < 2, X_2 < 2, \ldots\ldots, X_{16} < 2) = P(X_1 < 2)P(X_2 < 2)\ldots\ldots P(X_{16} < 2)$$

$$= \left(\frac{10^0 e^{-10}}{0!} + \frac{10^1 e^{-10}}{1!}\right)^{16} = (e^{-10}(1+10))^{16}$$

The job of statisticians is very different from that of the probabilist. In the problem discussed above, a typical statistical problem would start not knowing the value of the parameter λ. Then the above joint distribution would be looked at not as a function of the random variable, but rather as a function of the parameter lambda. The function will then be called "likelihood function." To estimate λ the statistician will maximize the likelihood with respect to λ. The result will be called the "maximum likelihood estimator." The statistician will then proceed to use expectations and variances rules learned in probability to determine the properties of the estimator and the distribution of the estimator. Decisions about whether the estimator is an accurate value, close to the true λ, will be made based on those properties, but, given uncertainty, probability theory plays a very important role in that decision.

Another typical problem encountered by statisticians is, for example, to estimate the average number of customers appearing in a store the first hour of the day. Why? Say a store owner is struggling and wants to make better personnel management. Perhaps it is not necessary to have so many clerks during the first hour. It is very common to contract a statistician to help make that decision. A statistician will ask the store owner to bring the number of people entering in a randomly selected number of days, say 250 days.

The statistician then will assume a model for

Y = the number of customers entering the store in the first hour of a random day.

The statistician thinks that there is the same distribution for Y in every one of the n = 250 days. But there is independence, so the joint distribution of all the observed number of customers, $Y_1, Y_2, \ldots, Y_n$, is:

$$P(y_1, y_2, \ldots, y_n) = \prod_{i=1}^{n} \frac{\lambda^{y_i} e^{-\lambda}}{y_i!}.$$

The statistician then will design methods to estimate the λ, which is the only unknown in that equation. Probability has contributed the model assumption, the procedure to find the joint distribution, since independence implies you may multiply the marginal. But that is it. The statistician will provide the store owner with the best estimate of the average number of customers, but also the standard error of the estimate, so that the uncertainty can be taken into account in decision making.

6.10 The multinomial probability mass function

The multinomial distribution is a generalization of the binomial distribution to the situation where we have $k > 2$ categories to which to classify objects. For example, if we are interested in whether in a group of 100 people, 20 are from California, 30 are from other states of the United States, 40 are from European countries and 10 are from other countries in the world, and if we know the probabilities that a randomly chosen person is from any of those, then we can calculate

$$P(X_c = 20, X_{USAother} = 30, X_{Europe} = 40, X_{other\ countries} = 10)$$
$$= \binom{100}{20,30,40,10}(0.1208)^{20}(0.0092)^{30}(0.08)^{40}(0.79)^{10} = 0.$$

In general, if there are n objects that fall in one and only one of k categories, and we denote by X_i, i = 1,...,k, the number of objects in category i, and by p_i, $i = 1, \ldots, k$, the probabilities of the object being in category i, then

$$P(X_1 = x_1, \ldots\ldots\ldots, X_k = x_k) = \binom{n}{x_1, \ldots, x_k}(p_1)^{x_1} \ldots\ldots(p_k)^{x_k}$$

gives the probability of a particular given allocation to k categories by chance.

The probabilities must add to one, and the sum of the X's must add to n.

6.10.1 Exercises

Exercise 1. The demographic profile of Ecuador in 2018 is

- 27.08% of the population are 0–14 years old
- 18.35% are 15–24 years old
- 39.59 are 25–54 years old
- 7.53% are 55–64 years old
- 7.45% are 65 years and older.

In a random sample of 20 people, what is the probability that 2 are 65 years or older, 5 are 55–64 years old, 6 are 25–54 years old, 4 are 15–24 years old and 3 are 0–14 years old?

https://www.indexmundi.com/ecuador/demographics_profile.htm

Exercise 2. (Scheaffer 1995, 254) With the recent emphasis on solar energy, solar radiation has been carefully monitored at various sites in Florida. Among typical July days in Tampa, 30% have total radiation of at most 5 calories, 60 % have total radiation of at most 6 calories, and 100% have total radiation of at most 8 calories. A solar collector for a hot water system is to be run for 6 days. Find the probability that 3 days will produce no more than 5 calories each, 1 day will produce between 5 and 6 calories, and 2 days will produce between 6 and 8 calories. What assumptions must be true for your answer to be correct?

6.11 Mini quiz

Question 1. A diagnostic test for the presence of a disease has two possible outcomes: 1 for disease present and 0 for disease not present. Let X denote the disease state of a patient, and let Y denote the outcome of the diagnostic test. The joint probability mass function of X and Y is given by

$$P(X=0, Y=0) = 0.800; P(X=1, Y=0) = 0.050; P(X=0, Y=1) = 0.025;$$

$$P(X=1, Y=1) = 0.125.$$

Calculate the variance of the outcome of the diagnostic test for those with the disease.

a. 0.13
b. 0.15
c. 0.2
d. 0.51
e. 0.71

Question 2. The joint probability mass function of random variables X and Y is given as follows:

		X	
		1	2
Y	1	1/8	2/8
	2	2/8	1/8
	3	1/8	0
	4	0	1/8

The $Var(Y \mid X = 2)$ is:

a. 2/9
b. 2/3
c. 2
d. 1.5
e. 1/3

Question 3. For the joint probability mass function of question 2, calculate $Cov(2+3X, 4-2Y)$.

a. 3
b. 6
c. 3/2
d. 0

Question 4. The joint probability mass function of two random variables X and Y is as follows:

$P(X=1,Y=1)=0.1$; $P(X=2, Y=2)=0.35$; $P(X=2, Y=1)=0.05$; $P(X=1,Y=2)=0.5$

The probability

$$P(1 \leq X \leq 2, Y \leq 2)$$

is:

a. 0.6
b. 0.65
c. 0.56
d. 1

Question 5. Consider the planning of undergraduate education for all three kids that a young couple plans to have. The costs depend on the number of kids that go to college (X) and the number of years each takes to complete the degree (Y). We denote the cost by C. Thus

$$C = g(X,Y).$$

Suppose the joint probability mass function of X, and Y for the planning couple is as follows.

x\y	3	4	5
1	0.05	0.05	0
2	0.07	0.3	0.03
3	0	0.4	0.1

Suppose the cost function is (in tens of thousands):

$$C = g(X,Y) = 10X + 20Y.$$

What is the expected cost C?

a. 21
b. 1506
c. 104.2
d. 193

Question 6. Consider the planning of undergraduate education for all three kids that a young couple plans to have. Use the information we used in Question 5. What is the variance of the cost C?

a. 419.78
b. -31.004
c. -661.64
d. 143.96

Question 7. In the joint probability mass function of Question 5, calculate the probability that X is larger than or equal to 3 and Y is larger than or equal to 4.

a. 1/2
b. 1/4
c. 3/7
d. 1/7

Question 8. Suppose that 25% of the people attending a popular gym live within 5 miles of the gym, 55% live between 5 and 10 miles from the gym and 20% live more than 10 miles from the gym. Suppose that 30 people are selected at random from the members of the gym. What is the probability that 10 live within 5 miles, 10 live between 5 and 10 miles, and the other 10 live more than 10 miles from the gym?

a. 0.001373087
b. 0.33145
c. 0.7814
d. 0.999

Question 9. Consider the following joint probability mass function of X and Y.

x\y	0	1	2	3
0	1/8	2/8	1/8	0
1	0	1/8	2/8	1/8

The $E(XY)$ is

a. 0.5
b. 0.007
c. 1
d. 0.11

Question 10. In the same joint probability mass function of Question 9, what is the $Cov(X,Y)$?

a. 0.25
b. 0.577
c. 0.75
d. 0.214

6.12 Chapter Exercises

Exercise 1. This problem is from Di Cook's lecture notes http://homepage.stat.uiowa.edu/~rdecook/stat2020/notes/ch5_pt1.pdf

Suppose that 2 batteries are randomly chosen without replacement from the following group of 12 batteries: 3 new, 4 used (working) and 5 defective. Let X denote the number of new batteries chosen. Let Y denote the number of used batteries chosen.

(i) Find the joint probability mass function.

(ii) Find $E(X)$.

Exercise 2. Suppose that 15% of the families in a certain community have no car, 20% have 1 car, 35% have 2, and 30% have 3. Suppose, further, that in each family, each car is equally likely (independently) to be a foreign or a domestic car. Let F be the number of foreign cars and D the number of domestic cars. (i) Find the joint probability mass function of F and D, showing your work. (ii) Write the marginal distribution for the number of foreign cars and find expected number of foreign cars per family and the standard deviation. (iii) Write the marginal distribution for the number of domestic cars and find expected number of domestic cars per family and the standard deviation.

Exercise 3. Suppose that a surveyor is trying to determine the areas of a rectangular field, in which the measured length X and the measured width Y are independent random variables that fluctuate widely about the true values, according to the following probability distributions

x	**8**	**10**	**11**
P(X = x)	1/4	1/4	1/2

y	**4**	**6**
P(Y = y)	1/2	1/2

The calculated area $A = XY$ is a random variable. What is the expected area?

Exercise 4. Calculate $Cov(2X + 3Y - Z, 5M)$, where X, Y, Z, M are random variables

Exercise 5. Daily sales records for a car dealership show that it will sell 0, 1, 2, or 3 cars, with probabilities as listed

X = Number of sales	**0**	**1**	**2**	**3**
P(X = x)	0.5	0.3	0.15	0.05

(i) Find the probability distribution for the total number of sales in a 2-day period assuming that the sales are independent from day to day. (ii) Find the probability that two or more sales are made in the next two days.

Exercise 6. If individuals always married significant others that are 2 years younger than themselves, what would be the correlation between the age of the two individuals in the married couple?

Exercise 7. (Ibe 2014, Example 5.2) The joint probability mass function of two random variables X and Y is given by

$$P(X = x,\ Y = y) = k(2x + y), \quad x = 1,\ 2; \quad y = 1,\ 2,\ 3$$

where k is a constant. (i) What is the value of k? (ii) Find the marginal probability mass functions of X and Y. (iii) Are X and Y independent?

Exercise 8. (Bain 2014, Example 5.7) For two random variables, X and Y, we know the following:

$$P(X \le 1, Y \le 1) = \frac{1}{8}; \quad P(X \le 1, Y \le 2) = \frac{5}{8}; \quad P(X \le 2, Y \le 1) = \frac{1}{4}; \quad P(X \le 2, Y \le 2) = 1$$

Determine the joint probability mass function of X and Y. (ii) Determine the marginal probability mass function of X. (iii) Determine the marginal probability mass function of Y.

Exercise 9. (Based on Page (1989, 130).) One 4-ohm resistor and two 8-ohm resistors are in a box. A resistor is randomly drawn from the box and inspected; then it is replaced in the box and a second is drawn. We will denote by X the resistance of the first one drawn from the box and Y the resistance of the second. Since this is sampling with replacement, the result of one test has no influence on the result of the other test. (i) Construct the joint probability mass function of X and Y. (ii) What is the probability that $X = Y$?

6.13 Chapter References

Bain, Lee J., and Max Engelhardt. 1987. *Introduction to Probability and Mathematical Statistics.* Duxbury Press.

Goldberg, Samuel. 1960. *Probability: An Introduction.* New York: Dover Publications, Inc.

Ibe, Oliver C. 2014. *Fundamentals of Applied Probability and Random Processes.* 2nd edition. Elsevier.

Page, Lavon B. 1989. *Probability for Engineering with Applications to Reliability.* Computer Science Press.

Scheaffer, Richard L. 1995. *Introduction to Probability and its Applications*, Second edition. Duxbury Press.

Watson, Jane M. 2011. "Cheating partners, conditional probability and independence." *Teaching Statistics. An International Journal for Teachers* 33, no. 3 (Autumn): 66–70.

Wonnacott,Thomas H., and Ronald J. Wonnacott. 1990. *Introductory statistics for Business and Economics*, fourth edition. Wiley and Sons.

Part II

Probability in Continuous Sample Spaces

Now that we have learned the main subjects of probability theory, differential and integral calculus background allows the reader to enter the study of similar concepts in continuous sample spaces—that is, sample spaces that consist of intervals of the real line. For example, a sample space that describes the outcomes of an experiment measuring time to death. There are some practical problems with such a sample space. First, an exact measurement would not be a measurement to the nearest 10s but to an infinite number of decimal places, a mathematician would say. With a not-very-precise ruler, that is impossible to do, but nonetheless, the value of the time to death is such a number. A second difficulty is deciding how to measure probability on an interval in the real line, which is challenging. There is an infinite amount of points in an interval in the real line. Assigning a positive probability to each of them would give infinite probability for the whole sample space, not possible. Assigning each of them a value of 0 makes probability of the sample space 0. As the reader may expect, to keep the axioms and the logic of probability intact under this dilemma, a solution to the dilemma was found with the help of calculus.

With continuous random variables, probabilities are assigned to intervals rather than to individual values. Probability is defined as an area under a graph of a function of a continuous variable, **f(x)**, called the density function. What in the discrete case was the individual probability **P(x)**, is replaced by the infinitesimal probability **f(x)dx** of the event $X \in dx$, the event that X falls in an infinitesimal interval of length **dx** near **x**. Probabilities of events are then areas under a continuous function called the **density function**. Events are intervals in the real line. Expectations are obtained using integration (albeit preserving the linearity of the expectation operator). The mathematics needed to find the area under a continuous curve is

integration of functions of one variable. This makes this part of the book accessible to readers with a good background in differential and integral calculus of one and several variables. Supplementary sidebars with review of some of the mathematics, and references to resources to refresh your calculus are provided. References will be made to the sections of Part I where the concepts were first introduced. Numerous references to authors who have written about probability theory at the accessible level of this part can be found throughout the chapters. As in the discrete case, the reader should be aware that notation varies by authors, vocabulary for the same thing is different across the disciplines, but the probability theory method may be exactly the same in all of them. Probability theory is not a bag of different tricks to solve problems but a very condensed set of a few tools to solve a bag of very different and contextually unrelated problems.

Chapter 7

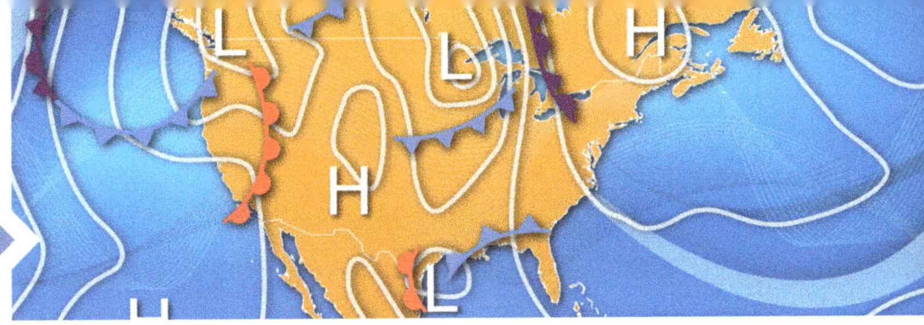

Infinite and Continuous Sample Spaces

Probability Models for a Single Continuous Random Variable

- Pick up a ruler similar to the one drawn below, or larger, and ask 10 different persons to measure the length of a string like that in Figure 7.2. Make them write the number on paper, keeping their record confidential. They all should use the same system of measurement: decimal system or United States customary units.

Figure 7.1 Ruler.

Figure 7.2 String.

When all measurements are done, unveil the measurements made. Enter them in this table, one measurement obtained by the 1st person, another by the 2nd person, and so on.

1st	2nd	3rd	4th	5th	6th	7th	8th	9th	10th

What do you expect to happen? Is that what happened? What is the exact length with this measurement instrument, or would you require an instrument with greater precision? Can you measure the exact length of the line?

7.1 Coping with the dilemmas of continuous sample spaces

Up to this point, we have restricted attention to discrete sample spaces and discrete random variables. But life is not always discrete. There are many problems in which a random variable can have any value in an interval of the Real Line, meaning that there are an infinite number of possible outcomes. For example, concentration of pollutant in some media is a continuous random variable of interest in environmental monitoring. The amount by which a radioactive particle decays is also a continuous random variable (Harris, 2014); the decay can take place at any time and is caused by a nuclear reaction whose behavior is only predictable statistically. As Harris explains, it makes no sense to associate a finite nonzero probability with each instant in time (there are an infinite number of such instants). It is more useful to recognize that for a short time interval Δt the decay probability will have some value proportional to Δt, and that the relevant quantity is the decay probability per unit time (computed in the limit of small Δt). Since this probability of decay per unit time may differ at different times, a general description for the decay within a time interval dt at time t must be of the form $f(t)\,dt$, where f can be called a density function. Then the overall probability of decay during a time interval from times t_1 and t_2 will be given by

$$\textit{Probability of decay between times } t_1 \textit{ and } t_2 = \int_{t_1}^{t_2} f(t)dt.$$

In other words, to deal with continuous random variables, we have to shift our thinking from specific outcomes to intervals of outcomes in the real line. Intervals break up a continuous scale into a set of discrete chunks, and we use these discrete chunks the way we previously used discrete outcomes. Thus, instead of thinking of events such as $\{X = 2\}$, we focus on events such as $\{1 \leq X \leq 2\}$, and ask: what is the probability that a random variable will lie within this range of values instead of at some particular point? (Denny and Gaines (2000, 69))

Harris presents an interesting example that illustrates this point.

Example

(This example is from Harris (2014, 684).) "A particle of unit mass moves (assuming classical mechanics) subject to the potential $V = \frac{x^2}{2}$, where x is the particle position, with a total energy $E = \frac{1}{2}$ (dimensionless units). These conditions correspond to motion with kinetic energy $T = \frac{1-x^2}{2}$, so that when the particle is at x its velocity can be found from $\frac{v^2}{2} = \frac{1-x^2}{2}$ leading to $v(x) = \pm\sqrt{1-x^2}$. We see from this form for $v(x)$ that the particle will move back and forth between turning points at $x = \pm 1$, and that it will move fastest at $x = 0$ and momentarily become stationary at $x = \pm 1$. The probability density for the particle's position will be proportional to the time spent in each element dx of its range, which in turn is proportional to $\frac{1}{|v(x)|}$. The probability density function of X is then

$$f(x)=\frac{1}{\pi\sqrt{1-x^2}}, \quad -1<x<1$$

The reader may be tempted to deduce from the discussion we are leading so far that knowing physics and every single discipline is needed to understand this second part of the book. But that is not the case. This example was presented here to illustrate how these new entities that we are about to learn about in this chapter had their genesis in some application. Each area of science has its own problems and its corresponding probabilistic solutions to those problems. There are many different density functions, but the reader does not need to know them all. In this chapter, we hope to convey the methodology needed to handle all of them once they have been given to us, and we study the ones that are most widely used in a wide array of fields, i.e., the ones that have wide applicability as models for a wide range of random phenomena. We will explore what we can do with them, and how we can use the concepts learned in Part I.

For the purposes of this chapter, the reader needs only have very present that the concept of a continuously distributed random variable is an idealization which allows for calculus to be used as a technical tool. This gives models for chance phenomena involving continuous random variables (Pitman (2005, 259).)

In fact, once we accept the convention used to measure probability in sample spaces for continuous random variables, the methodology will be the same as in the discrete case, but, instead of using summation operators in the calculations, we will use integrals. And regardless of which area of science we move in, we will be able to use the methodology, knowing that only the complexity of the context, and the mathematical complexity of the probability model can stop us.

7.1.1 Event operations for infinite collection of events

An infinite sample space raises the question of whether we are going to have an infinite amount of events. Indeed that is the case, although the reader will not see operations with infinite collections of events in this book. Suffice to know that we can extend the event operations of union and intersection of events to an infinite collection of events. If $A_1, A_2, \ldots$.. is a n infinite collection of events, all defined on a sample space S, then the union of the infinite collection of events consists of all simple outcomes s of S that are in at least one of the events, namely:

$$\cup_{i=1}^{\infty} A_i = \{s \in S | s \in A_i \ \ for \ \ some \ \ i\}.$$

The intersection consists of all simple outcomes s_j of S that are in all of the events.

$$\cap_{i=1}^{\infty} A_i = \{s \in S | s \in A_i \ \ for \ \ all \ \ i\}.$$

Example 7.1.1

(This and the next example are from Ross (2010, Chapter 2).) Let $S = (0, 1]$ and define

$$A_i = \left(\frac{i}{i+1}, 1\right], \quad i = 0,1,2,\ldots\ldots$$

The union and the intersection of these events are, respectively,

$$\cup_{i=0}^{\infty} A_i = \cup_{i=0}^{\infty} \left(\frac{i}{i+1}, 1\right] = (0,1] = S,$$

$$\cap_{i=0}^{\infty} A_i = \cap_{i=0}^{\infty} \left(\frac{i}{i+1}, 1\right] = \{1\}.$$

Example 7.1.2

Consider the collection of events $A_i = [i, i+1)$, $i = 0,1,2,\ldots\ldots\ldots$ Do these events make a partition of $S = [0,\infty)$? (See Definition 2.5.2 in Chapter 2 for definition of partition).

Check the "disjoint condition": $A_i \cap A_j = \varnothing$ *for all* $i, j, i \neq j$

$A_0 = [0,1);\ A_1 = [1,2);\quad A_2 = [2,3)\ \ldots$

Check the "union of events equal S condition": $\cup_{i=0}^{\infty} A_i = S = [0,\infty)$

Because the collection of events satisfies the two conditions, we say that this collection of events is a partition of the sample space S defined above.

7.2 Probability theory for a continuous random variable

Some physical phenomena such as the exact time that a train arrives at a specified stop, the lifetime of an atom, the stress of a beam, weight, height, the distance to the moon, and many other measurements are values in the real line. The random variable representing them is a continuous random variable. Probabilities for those random variables are areas under a function that we call the **density function.** We then define this new entity.

Definition 7.2.1

Let X be a random variable. Let a the smallest possible value of X (which could be $-\infty$) and b the largest (which could be $+\infty$). Any other real number in between is allowed. If there is a function $f(x)$ such that

- $f(x) \geq 0, \quad a \leq x \leq b$
- $\int_a^b f(x)dx = 1$

- for event $B = \{k < x < m\}$,

$$P(B) = \int_k^m f(x)dx,$$

then the function f is called the **probability density function** (**pdf**) of the random variable X, $a \leq X \leq b$, and we say that X is a continuous random variable. $P(a < X \leq b) = P(a \leq X < b) = P(a < X < b)$.

Definition 7.2.2

For the continuous random variable X in $[a, b]$ and density function $f(x)$,

$$\mu_x = E(X) = \int_a^b xf(x)dx.$$

$$E(X^k) = \int_a^b x^k f(x)dx,$$

and in general,

$$E(g(X)) = \int_a^b g(x)f(x)dx.$$

For example, the variance of the random variable X is

$$\sigma^2 = E[(X - \mu_X)^2] = \int_a^b (X - \mu)^2 f(x)dx = E(X)^2 - \mu^2.$$

The cumulative distribution function is

$$F(x) = P(X \leq x) = \int_a^x f(t)dt,$$

where F is the notation we use to denote the cumulative distribution function (**cdf**). We may use the cumulative distribution function to obtain the density function of X.

$$f(x) = \frac{dF(x)}{dx}$$

The cumulative distribution function can also be used to compute probabilities:

$$P(a \leq X \leq b) = F(b) - F(a)$$

Nothing much new has been said except that we now use integrals instead of the summation operator for calculating the same things we calculated for discrete random variables, and the entity containing the probabilities is an area under a density function.

The moment generating function of a continuous random variable is

$$M_X(t) = E(e^{tX}) = \int_a^b e^{tX} f(x)dx.$$

The reader does not need to delve deeper into how we made the transition from Σ to $\int$. But if there is interest, sources mentioned in section 3.1 of this book discuss in detail how we ended up here.

In this section, the reader will see a full example without any context to practice computing the usual things for a continuous random variable. We will use a Question and Answer approach.

Example 7.2.1

Let X be a continuous random variable which has density function

$$f(x) = 3x^2, \quad 0 \le X \le 1$$

This formula means that X has range in the real line interval [0,1]. The pdf of X is positive only on the interval [0,1]. Outside the domain (0,1) the $f(x)$ is 0. The f(x) is the function given.

We will illustrate next how we would compute all those relevant quantities in Definition 7.2.2 with this example. At the end of the computations you will be able to see a graphical display of some of the results.

Q: Is f(x) positive for all x and is the area under f(x) equal to 1? Are axioms of probability satisfied?

We can prove that $f(x)$ is a density function for the random variable X because we can show that the area under the curve in the range of X is 1.

$$\int_0^1 3x^2 \, dx = \frac{3x^3}{3} \Big|_0^1 = 1$$

Notice that we do not need to take the integral from $-\infty$ to ∞, because outside the interval (0,1), the $f(x)$ is 0, there is no area to account for.

Certainly, $f(x)$ is positive for all values of X in the range of X.

Thus, $f(x)$ is a legitimate density function for continuous random variable X. Axioms of probability are satisfied.

Q: How do I obtain the cumulative distribution function of continuous random variable X?

$$F(x) = P(X \leq x) = \int_0^x 3t^2 \, dt = \frac{3t^3}{3} = x^3, \quad 0 < x < 1$$

We can easily check that the derivative of $F(x)$ with respect to X is indeed the density function of X.

$$F'(x) = 3x^2 = f(x)$$

Box 7.1

Cumulative probability distribution of a continuous random variable

If X is a continuous random variable with density $f(x)$, and domain $a < X < b$, then the cumulative distribution function $F(x)$ is defined as

$$F(k) = P(X \leq k) = \int_a^k f(x)dx$$

This function has the following properties:

- F is monotonic increasing, $F(k) \leq F(r)$ when $k < r$
- $\lim_{x \to -\infty} F(x) = 0$ and $\lim_{x \to \infty} F(x) = 1$
- $\frac{dF(x)}{dx} = f(x)$

If we know the cumulative distribution function, then

$$P(m < X < n) = F(m) - F(n)$$

Box 7.2

Percentiles for a continuous random variable

We define the $100 \cdot q$th percentile of a random X defined in an interval [a,b] as the value c of the random variable such that

$$F(c) = P(X \leq c) = \int_a^c f(x)dx = q.$$

For example, the 90th percentile is the value c of X such that

$$F(c) = P(X \leq c) = \int_a^c f(x)dx = 0.9.$$

There are some percentiles that are particularly popular: the 50th (100*0.5)th percentile, the 25th percentile, and the 75th percentile. The 50th percentile is called the median, and the 25th and 75th percentiles are used to compute the interquartile range (IQR), which is defined as

$$IQR = 75th - 25th.$$

When a distribution is skewed, the median is a better measure of average than the expected value, and the interquartile range gives a better idea of the spread of the distribution. This is so because the expected value is too influenced by the tail values of the distribution, and therefore, the variance, which depends also on the expected value, is influenced as well.

Q: How do we compute probabilities?

We can compute the probability of a continuous random variable X in any interval contained in the range of X. For example,

$$P(0.5 < X < 0.8) = \int_{0.5}^{0.8} 3x^2 \, dx = \frac{3x^3}{3} \Big|_{0.5}^{0.8} = 0.387$$

There are advantages to knowing the cumulative distribution function. If we know it, then the probability that X is in any particular interval can be computed using the cumulative distribution formula. For example, we can show that the probability just calculated is also equal to

$$P(0.5 < X < 0.8) = F(0.8) - F(0.5) = 0.8^3 - 0.5^3 = 0.387$$

Q: How do I compute percentiles?

To compute percentiles you set the cumulative distribution function to the probability desired. For example, to find the 70th percentile, make

$$F(x) = 0.7.$$

Substituting for $F(x)$,

$$x^3 = 0.7,$$

and solving for x, we find that $x = 0.887904$.

Q: How do we compute expectations?

To find the expected value of the continuous random variables X we compute:

$$E(X) = \int_0^1 x(3x^2)dx = x\frac{3x^4}{4} \Big|_0^1 = 0.75$$

The expectation of any power of X can be found as follows:

$$E(X^k) = \int_0^1 x^k (3x^2)dx = \frac{3}{k+3}$$

To find the variance of X, we make use of the result:

$$Var(X) = E(X^2) - [(E(X))^2] = \frac{3}{5} - \left(\frac{3}{4}\right)^2 = 0.0375.$$

Q: How do we compute the moment generating function?

The moment generating function is also an expectation of a very peculiar function of the random variable. We can compute the moment generating function of X,

$M_X(t) = E(e^{tx}) = \int_0^1 e^{tx}(3x^2)dx$, which can be solved using integration by parts (left as an exercise).

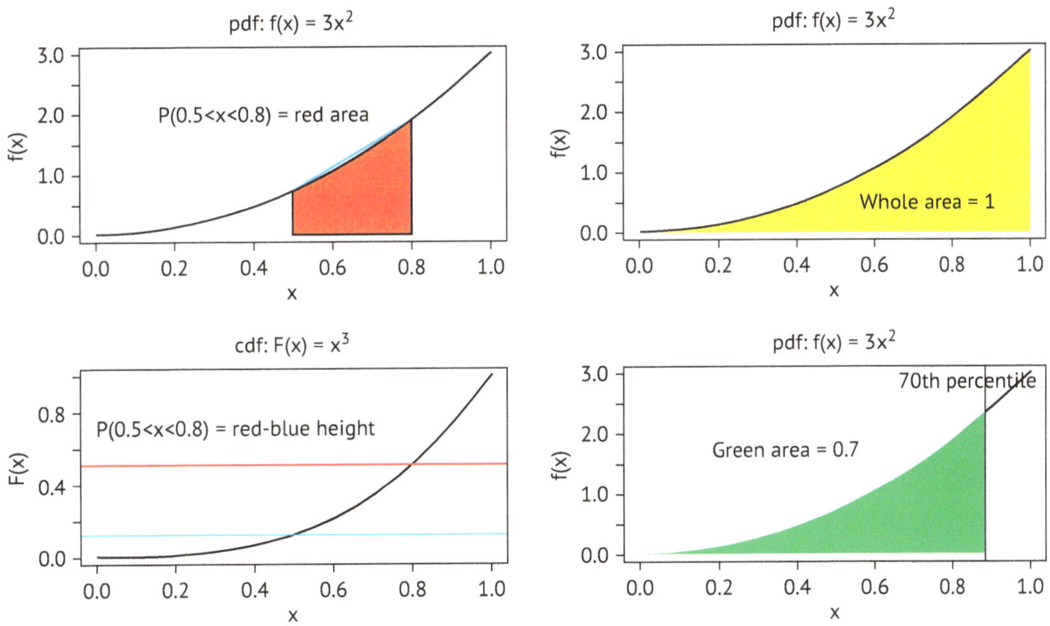

Figure 7.3 Figures correspond to Example 7.2.1. Probabilities are areas under the density function. A percentile is a value of X corresponding to some cumulative probability and cumulative probabilities viewed in a cdf plots are values on the vertical line of the plot.

7.2.1 Exercises

Exercise 1. (Based on Keeler and Steinhorst (2001).) An appliance repair firm has recorded the time it took to complete service calls during the past year. They found that time (Y) varies from a low of 0 hours (no repair required) to a high of 4 hours. The density function of Y is given in Figure 7.4.

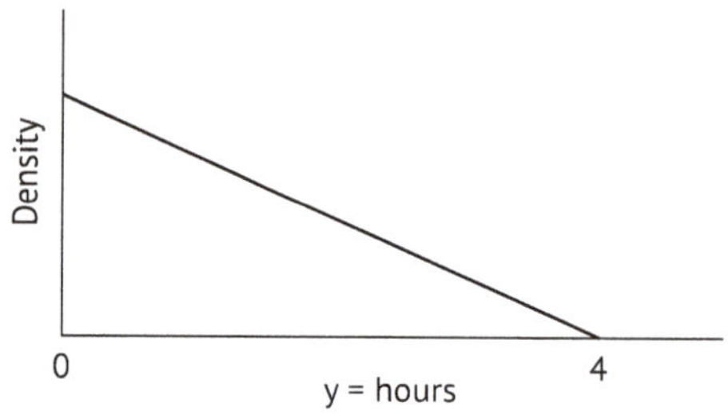

Figure 7.4

(i) Find the height of the triangle. That is, find the height where the triangle crosses the vertical axis.

(ii) If we consider a randomly selected call,

  (a) Find the probability that the service call will take at most an hour.

  (b) Find $P(Y > 2 \text{ hours})$.

(iii) Is μ less than 2 or greater than 2? Explain.

Exercise 2. Let X be the time that it takes to drive between point A and point B during the afternoon rush hour period on highway 4005. The density function of X is

$$f(x) = \frac{1}{2}x, \quad 0 \leq x \leq 2.$$

(i) Calculate the value of the 70th percentile. (ii) Calculate the interquartile range. (iii) Calculate $P(0.5 \leq X \leq 1.5)$. (iv) Find the median. (v) Find the moment-generating function of X.

Exercise 3. The time it takes a seasoned runner to complete the Mountain High half-marathon is a random variable X with cumulative distribution function

$$F(x) = \frac{1}{26}(2x^2 + x - 10), \quad 2 \leq x \leq 4.$$

Find the density function of X.

Exercise 4. If X is a continuous random variable with cumulative distribution function

$$F(x) = 1 - e^{-\frac{x}{4}}, \quad x \geq 0,$$

what is the probability that X is between two and five?

Exercise 5. A target is located at the point 0 on the horizontal axis. Let X be the landing point of a shot aimed at the target, a continuous variable with density function

$$f(x) = 1.5(1 - x^2), \quad 0 \le x \le 1.$$

(i) What is the expected landing point? (ii) What is the probability that the landing point is before 0.4? (iii) Find the cumulative distribution function of X. (iv) What is the standard deviation of X?

Exercise 6. The density function of a random variable X is given by

$$f(x) = a + bx^2.$$

If $E(X) = 3/5$, find a and b.

Exercise 7. The distribution of the amount of gravel (in tons) sold by a particular construction supply company in a given week is a continuous random variable X with density function

$$f(x) = 1.5\,(1 - x^2), \quad 0 \le x \le 1$$

(i)How often does this company sell less than 0.4 tons per week? (ii) What is the variability (in tons per week) of the amount sold each week? (iii) The company sells the gravel at a price of $10,000 per ton. But keeping the gravel in storage during the week costs $2000 per ton. On average, how much profit does the company make per week? (iv) The company keeps the gravel at a storage place each week. Gravel not sold that week is thrown away. Since the amount sold is a random variable, it doesn't pay for the company to keep one ton stored all the time. The company just keeps enough gravel to make sure that customers will have to leave without gravel only 20% of the time. How much does the company store each week? (v) Write down the cumulative distribution function of the random variable X and use it to compute $P(0.2 \le x \le 0.6)$. (vi) What is the probability that 2 out of 10 construction companies will sell more than 0.4 tons per week?

Exercise 8. (This problem is based on Scheaffer (1995, 150).) Daily total solar radiation for a certain location in Florida during the month of October has the following density function:

$$f(x) = \frac{3}{32}(x - 2)(6 - x), \quad 2 \le x \le 6,$$

where X is the solar radiation in hundreds of calories.
Find the expected daily solar radiation for October in that location.

Exercise 9. The lifespan in days of a certain bacterium is a random variable with a density function of the following form

$$f(x) = a^2 x\ e^{-ax}, \quad x \geq 0, \quad a = constant > 0.$$

Find the expected lifespan of the bacterium.

Exercise 10. (This exercise is based on Rice (2007).) Let X be the cosine of the angle at which electrons are emitted in muon decay. X is a random variable with the following density function:

$$f(x) = \frac{1 + \alpha x}{2}, \quad -1 \leq x \leq 1,$$

where α is a constant that can take values $-1 \leq \alpha \leq 1$. (i) Find the expected cosine of the angle at which electrons are emitted as a function of α. (ii) Find the variance.

7.3 Expectations of linear functions of a continuous random variable

We said in Section 7.2, and when we studied discrete random variables in Chapter 5, that, in addition to the functions g(X) considered so far, we can find the expectation of any other function of X; for example, if $f(x)$ is as in Example 7.2.1,

$g(X) = 20 + 2X^2$ has expected value

$$E[(g(X)] = E(20 + 2X^2) = \int_0^1 (20 + 2x^2)(3x^2)dx = 20 + 2\left(\frac{3}{5}\right) = 21.2.$$

The reader is advised to go to Chapters 5 and 6 to review the properties of expectations. The same properties apply to continuous random variables.

Example 7.3.1 (Example 7.2.1 continued)
Let X denote a random variable with density function

$$f(x) = 3x^2, \quad 0 \leq X \leq 1$$

And let $Y = a + bX$, where a and b are constants. What is the expected value and variance of Y? We use the definition of expectation and variance of a function of a random variable.

$$E(Y) = \int_0^1 (a + bx)3x^2\ dx = a\int_0^1 3x^2\ dx + b\int_0^1 (x)3x^2\ dx = a + bE(X).$$

We now compute the Variance of Y:

$$V(Y) = \int_0^1 (a+bx-(a+b\mu_x))^2 3x^2\ dx = b^2 \int_0^1 (x-\mu_x)^2 3x^2\ dx = b^2 Var(X)$$

How did we know earlier that we could compute the variance of X with the formula

$$Var(X) = E(X^2) - [(E(X))^2]?$$

By applying the definition of variance of a random variable,

$$\sigma_X^2 = V(X) = \int_0^1 (x-\mu)^2 f(x)\ dx = \int_0^1 (x-\mu)^2 3x^2\ dx$$

$$= \int_0^1 (x^2+\mu^2-2\mu x)3x^2\ dx$$

$$= \int_0^1 (x^2)3x^2\ dx + \mu^2\int_0^1 3x^2\ dx - 2\mu\int_0^1 x3x^2\ dx = E(X)^2 + \mu^2(1) - 2\mu^2$$

$$= E(X)^2 - \mu^2$$

7.3.1 Exercises

Exercise 1. Snowpack measurements on April 1st in a mountainous region have expected value 17.7 inches and standard deviation 3 inches, based on measurements made in gauging stations. In hydrology, there are lots of studies trying to measure the relation between spring river discharge and snowpack. A study found that this relation is

$$Y = 1.04 + 1.03\ X,$$

where X is snowpack depth on April 1st and Y is spring river discharge. Calculate the expected river discharge and the standard deviation of the discharge.

Exercise 2. The proportion of time X during a 40-hour work week that a health worker spends transporting blood samples from the clinic to the examination lab is a random variable with density function

$$f(x) = 2x, \quad 0 \le x \le 1.$$

The cost of transportation depends on the proportion of time used in transportation according to the following function:

$$Cost = 10 - 2X.$$

Find the expected value and standard deviation of the cost.

7.4 Sums of independent continuous random variables

According to the Bureau of Labor Statistics of the United States, the median annual wage for statisticians was $84,060 in May 2017. Overall employment of mathematicians and statisticians is projected to grow 33 percent from 2016 to 2026, much faster than the average for all occupations. Businesses will need these workers to analyze the increasing volume of digital and electronic data.

https://www.bls.gov/ooh/math/mathematicians-and-statisticians.htm

Many statisticians choose to become consultants. When billing customers, they take into account the time it takes to write the computer program they need to analyze data, the time it takes to interpret the results, the time it takes to write the report for the customer, and the time spent with the customer. That is a total of 4 different independent random variables, say X_1, X_2, X_3, X_4. Let's assume that these random variables are all equally distributed. What is the expected amount of time spent in a project by a typical statistical consultant?

We can calculate the expected sum of the four times and the variance of the four times as follows:

$$E(S_4) = E\left(\sum_{i=1}^{4} X_i\right) = \sum_{i=1}^{4} E(X_i) = \mu + \mu + \mu + \mu = 4\mu$$

$$Var(S_4) = Var\left(\sum_{i=1}^{4} X_i\right) = \sum_{i=1}^{4} Var(X_i) = \sigma^2 + \sigma^2 + \sigma^2 + \sigma^2 = 4\sigma^2$$

As we said in Section 5.5, sums of independent random variables are very important.

- Many credit cards offer rewards that consist of earning some amount of dollars for every $100 dollars spent. Credit card companies know the distribution of the usual expenses people make and can predict with a degree of certainty the expected total amount of money back that a customer's credit card will get.
- Research and Development (R&D) by corporations is an expensive multi-step process. When companies engage in the process of inventing new processes and products they must predict the cost at each step. The expected total cost will help the company decide whether the R&D project is worth pursuing.

Let's define the sum of *n* random independent continuous random variables as follows:

$$S_n = X_1 + X_2 + \ldots.. + X_n = \sum_{i=1}^{n} X_i,$$

where *S* is for sum and *n* for how many random variables we are adding. Then, without proving it, we claim that

$$E(S_n) = E\left(\sum_{i=1}^{n} X_i\right) = \sum_{i=1}^{n} E(X_i) = \mu + \mu + \ldots\ldots. + \mu = n\mu$$

$$Var(S_n) = Var\left(\sum_{i=1}^{n} X_i\right) = \sum_{i=1}^{n} Var(X_i) = \sigma^2 + \sigma^2 + \ldots\ldots. + \sigma^2 = n\sigma^2$$

Sums of independent and identically distributed random variables are of central importance in both Probability and Statistics.

The proof of this result was done for two discrete random variables in Chapter 6.

7.4.1 Exercises

Exercise 1. Consider the density function of exercise 10 in Section 7.2.1. Consider the cosine of the angle of 5 electrons. This is represented by 5 random variables, X_1, X_2, X_3, X_4, X_5. Of interest is comparing the expected value and variance (as functions of α), of the following functions of these random variables:

$$Y = 3\frac{\sum_{i=1}^{5} X_i}{5}$$

and

$$W = \frac{\sum_{i=1}^{5} X_i}{5}$$

Which one has the smallest variance?

7.5 Widely used continuous random variables, their expectations, variances, density functions, cumulative distribution functions, and moment-generating functions

Scientists that deal with data had to come up with models that allow them to make probabilistic predictions of random phenomena. Many phenomena have the same nature, and hence can be modeled the same way, albeit with different model parameters. The models described in this section have been found to be very useful in a wide range of applications. It is impossible to go over all the probability models for all continuous random variables. We emphasize in this section those that are widely adopted in many different areas of application. We can go ahead and repeat the computations in 7.2 for each of them, or we could just use what others have already proved and focus instead in the applications. Many web sites are dedicated to describe these distributions' properties. However, a word of caution: they may be using different parameterizations and notation, and that can be confusing for the beginner. For this reason, we illustrate in this chapter how to obtain all those quantities obtained in 7.2 for three examples. The reader can then obtain formulas for others not seen here in the references provided. The list of models seen here are:

- the uniform distribution on an interval in the real line
- the exponential distribution on the interval $[0, \infty)$
- the gamma distribution on the interval $[0, \infty)$
- the normal distribution
- the Weibull distribution
- the Beta distribution
- and many others that you will encounter in your work in the future

There exist applets online that allow you to calculate probabilities with these important distributions.

7.6 The Uniform Random Variable

▶ You call a friend to ask for a ride. Your friend declares that it will take between 10 and 30 minutes to arrive to pick you up. What is your uncertainty in this scenario? What uncertainty do we have regarding when the next earthquake will occur? How would you model the uncertainty in these two scenarios? What kind of density function would you use?

A continuous uniform random variable is used when the range of the random variable is known but we have no other knowledge about possible values that the random variable can have other than that the event may happen in any interval within the range. This random variable is often used to model the random phase of a sinusoidal signal, in that a uniform distribution of phase between 0 and 2π is frequently assumed. Also, in analog to digital conversion, this random variable is used to describe the errors due to rounding off in the conversion process.

Definition 7.6.1

A uniform random variable X on an interval (a,b) is a uniform random variable if its density $f(x)$ is constant on (a,b) and 0 elsewhere.

$$f(x) = \frac{1}{b-a}, \quad a \leq x \leq b$$

From this, we can derive all the relevant things that we have been focusing on:

$$F(c) = P(X \leq c) = \frac{c-a}{b-a},$$

$$a \leq x \leq b$$

The 100qth percentile will be found by setting

$$F(c) = \frac{c-a}{b-a} = q$$

and solving for c, for example, for the 90th percentile

$$F(c) = \frac{c-a}{b-a} = 0.9.$$

$$E(X) = \frac{a+b}{2}$$

$$Var(X) = \frac{(b-a)^2}{12}$$

$$M(t) = \frac{e^{tb} - e^{ta}}{t(b-a)}$$

The uniform distribution is also known as the "no knowledge" distribution.

Example 7.6.1

A friend is to appear in front of your door to pick you up sometime within the next 10 to 30 minutes. You will be ready if this friend shows up within 5 minutes of the 20 minutes since you called. Find the probability that you will be ready, assuming that your friend drops at a random time within that interval. Let X be the time your friend shows up.

Assume that

$$f(x) = \frac{1}{30-10}, \quad 10 < x < 30$$

The b = 30, a = 10.

$$P(15 < X < 25) = \frac{25-15}{30-10} = 0.5$$

The uniform distribution is important in the study of Poisson processes. Given a Poisson rate of appearance of an event in a unit interval, the way in which the event is scattered in the interval is uniformly distributed. What is the same, the location of the event is equally likely to be in any point in the interval.

The uniform distribution is also very useful in Bayesian statistics modeling. It is often used as an uninformative prior, when the researcher uses it to express the idea that there is no prior knowledge about an event.

Example 7.6.2

A teacher teaches a one-hour class that starts at 10 a.m. and usually asks a clicker question during the hour to assess whether students are following and are actively engaged in the lecture discussion. The time at which the teacher asks the question varies and is never the same. It depends on the lecture, the topic that the teacher thinks needs assessment throughout the lecture, how important that topic is, etc. A student would like to predict the probability that the teacher will ask the question between 10:20 and 10:40. What would this probability be?

Since we do not know when the teacher will ask the question, the uniform distribution seems appropriate as a model for the time when the teacher asks the question. Let X denote the time at which the teacher asks the question. X is uniform with domain 0 to 60 in minutes.

$$f(x) = \frac{1}{60-0}, \quad 0 \leq x \leq 60.$$

$$P(20 < x < 40) = \frac{40-20}{60} = \frac{1}{3}$$

Example 7.6.3

A commuter arrives at a train stop at 10:00 a.m., knowing that the train will arrive at some time uniformly distributed between 10:00 a.m. and 10:30 a.m. If at 10:15 the train has not yet arrived, what is the probability that the commuter will have to wait at least an additional 10 minutes?

This exercise is an illustration of the fact that the rules of probability learned in earlier chapters will continue to be used with random variables, since random variables are representing just events in the sample space.

Let X be the random variable representing the waiting time.

$$f(x) = \frac{1}{30-0}, \quad 0 \leq x \leq 30$$

The probability that the commuter will have to wait at least 10 additional minutes if the train arrives at 10:25 or later is

$$P(X > 25 \mid X > 15) = \frac{P((X > 25) \cap (X > 15))}{P(X > 15)} = \frac{P(X > 25)}{P(X > 15)} = \frac{\frac{5}{30}}{\frac{15}{30}} = \frac{1}{3}$$

7.6.1 Exercises

Exercise 1. Food trucks along a stretch of a mountain pass in Italy are prohibited. The views from that pass are beautiful and lots of people stop there. Consequently, food trucks try to stop at random locations not to be expected by police. They stop randomly along a two-kilometer part of the road that stretches across the best viewpoints and has as its midpoint the highest point of the mountain. What is the probability that a tourist will find the strawberry stand of a food truck within three meters of the top of the mountain?

Exercise 2. The failure of a circuit board interrupts work by a computer system until a new board is delivered. Delivery time, X, is uniformly distributed over the interval of from one to five days. The cost C of this failure and interruption consists of a fixed cost c_o for the new part and a cost that increases proportionally to X^2, so that

$$C = c_o + c_1 X^2$$

(i) What is the probability that the delivery time is two or more days? (ii) Find the expected cost of a single failure in terms of c_o and c_1.

Exercise 3. You arrive at a bus stop at 10 o'clock, knowing that the bus will arrive at some time uniformly distributed between 10:00 and 10:30. What is the probability that you will have to wait longer than 10 minutes?

7.7 Exponential random variable

- People wait in line to enter the Catalina Express ferry to Catalina Island. People entering the United States when returning or coming from a foreign country must wait in the passport checkpoint. Airplanes must wait in line to take off. Capacity building to service people in those lines requires that the capacity to service the people is in proper relation to the rate at which they arrive. If this service is too small, the queue will be disproportionately large and people and planes will wait an inordinate amount of time for service. On the other hand, if the capacity is too large, much of the service capacity will be underutilized or perhaps not utilized at all. How would you approach this problem?

Suppose an event happens at a constant rate λ. The random variable X measuring the time until the first event and the time between events is a random variable with range $[0, \infty)$ and density function

$$f(x) = \lambda e^{-\lambda x}, \quad x \geq 0$$

and cumulative distribution function

$$F(x) = 1 - e^{-\lambda x}$$

$$E(X) = \frac{1}{\lambda}$$

$$Var(X) = \frac{1}{\lambda^2}$$

$$M_x(t) = \frac{\lambda}{\lambda - t}$$

Figure 7.5 shows an exponential density function. The key property of the exponential random variable is that it is memoryless. For $s > 0$ and $t > 0$ the memoryless or Markov property states that

$$P(X > s + t | X > t) = P(X > s),$$

which implies that

$$P(X > s + t) = P(X > s)P(X > t).$$

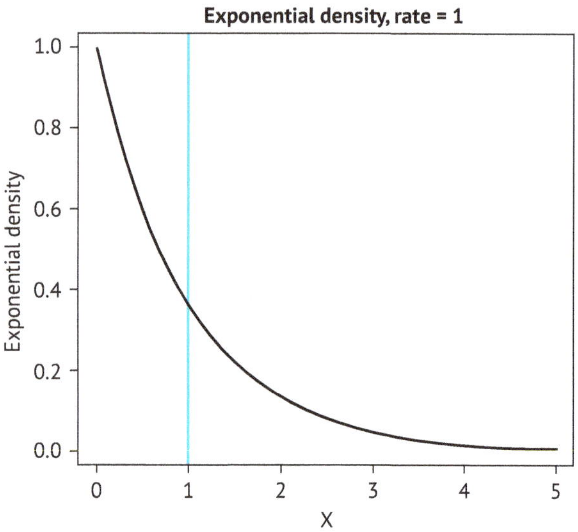

Figure 7.5 The density of an exponential with lambda = 1, the blue line marks the value of the expected value on the X axis.

The proof is left as an exercise.

Inanimate objects that do not wear out gradually, but stop functioning suddenly and unpredictably have random time to death well fit by an exponential density function.

Example 7.7.1

Traffic to an email server arrives in a random pattern (i.e., exponential arrival time) at a rate of 240 emails per minute. The server has a transmission rate of 800 characters per second. The message length distribution (including control characters) is approximately exponential with an average length of 176 characters. Assume a M/M/1 queueing system (i.e., exponential arrival times, exponential service time, and one server). What is the probability that 10 or more messages are waiting to be transmitted? See section 7.16.

Example 7.7.2

The lifetime of a light bulb, the length of a phone call, and radioactive decay are other examples of exponential random variables.

Example 7.7.3

Suppose that the amount of time it takes the bookstore to process the book purchase of a student at the beginning of the school year, in minutes, is an exponential random variable with parameter $\lambda = 1/10$.

Let X denote the length of processing the order in minutes. Then

$$f(x) = \frac{1}{10} e^{-\frac{1}{10}x}, \quad x \geq 0.$$

- If someone arrives immediately ahead of you at the bookstore, then the probability that you will have to wait more than 10 minutes is

$$\int_{10}^{\infty} f(x)dx = \int_{10}^{\infty} \frac{1}{10} e^{-\frac{1}{10}x} dx = -e^{-\frac{1}{10}x} \Big|_{10}^{\infty} = 0.368.$$

  We could realize that the probability computed is $1 - F(10)$.

- The probability that you will have to wait between 10 and 20 minutes can be computed using the cumulative distributions directly.

$$P(10 < X < 20) = P(X < 20) - P(X < 10) = F(20) - F(10) = e^{-2} - e^{-1} = 0.233$$

- The expected waiting time is

$$E(X) = \frac{1}{\lambda} = 10$$

minutes, give or take the standard deviation of 10 minutes,

$$SD(X) = \sqrt{\frac{1}{\lambda^2}} = 10.$$

- Each half hour that you waste in the bookstore line, you lose about \$10 from your work-study job at the statistics department. Compute the expected cost and the standard deviation of that cost.

$$C = \frac{1}{3}X.$$

Using the rules of expectations learned and used earlier,

$$E(C) = \frac{1}{3}E(X) = \frac{10}{3} = \$3.333\ldots$$

$$SD(C) = \sqrt{\left(\frac{1}{3}\right)^2 Var(X)} = \frac{1}{3}SD(X) = \$3.3333$$

7.7.1 Exercises

Exercise 1. Suppose that an experimenter studies the lifespan of members of a colony of bacteria. Let T be the lifespan of a randomly chosen member of the colony. Suppose the lifespan of the bacteria has an exponential distribution with expected value 100 minutes. What is the probability that a randomly chosen member of the colony lives more than 87 minutes?

Box 7.3

Integration by parts

To prove that the values of the expected values regarding an exponential random variable are what they are, you need to remember integration by parts.

You will need that in the exercises.

$$\int udv = uv| - \int vdu$$

Exercise 2. The time in hours required to repair a machine is an exponentially distributed random variable with expected value of two hours. (i) What is the probability that a repair time exceeds two hours? (ii) A company has four machines identical to those in (i) that need repairs. What is the probability that two of them require a repair time that exceeds two hours?

Exercise 3. The service times at a teller window in a bank were found to follow an exponential distribution, with a mean of five minutes. A customer arrives at a window at 2:00 p.m. (i) Find the probability that they will still be there at 2:06 p.m. Show work. (ii) Find the probability that the customer will still be there at 2:10 p.m., given that the customer was there at 2:06 p.m.

Exercise 4. Prove the Markov property of the exponential distribution using the definition of conditional probability of an event.

Exercise 5. Prove that the expected value of the exponential random variable with parameter λ is $1/\lambda$ and the variance is $\frac{1}{\lambda^2}$.

Exercise 6. The magnitude of earthquakes recorded in a region of North America can be modeled by an exponential distribution with a mean of 2.4 as measured on the Richter scale. Find the probabilities that the next earthquake to strike this region will have the following characteristics: (i) It will exceed 3.0 on the Richter scale. (ii) It will fall between 2 and 3 on the Richter scale. (iii) If all earthquakes happen independently of each other, what is the probability that the next three successive earthquakes will exceed 3 on the Richter scale?

Exercise 7. The time (in hours) required to repair a machine is exponentially distributed with parameter $\lambda = 1/2$. Calculate (i) the probability that a repair time exceeds 2 hours. (ii) The conditional probability that a repair takes at least 10 hours given that its duration exceeds 9 hours. (iii) the probability that the total repair time of 100 machines is greater than 180 hours.

Exercise 8. Prove that the moment-generating function of an exponential random variable with parameter λ is

$$M_x(t) = \frac{\lambda}{\lambda - t}$$

Exercise 9. Suppose that X is a random variable with p.d.f.:

$$f(x) = e^{-x}, \quad x \geq 0.$$

Find the third moment, i.e., the $E(X^3)$.

7.8 The gamma random variable

The gamma is a very important random variable. It applies to metrics that are nonnegative and have skewed distributions, but with less extreme exponential decay than the exponential. In fact, the exponential and the chi-square distributions are special cases of a gamma density.

You will research this random variable in Siegrist (1997). Go to http://www.randomservices.org/random/special/Gamma.html and visit the applets and simulators provided by the author. Try to do some of the exercises.

Example 7.8.1

The monthly salary of women that are in the labor force in a large town, Y, follows a gamma distribution with $\alpha = 2000$ and $\lambda = 4$. The expected salary of the women in this town is

$$E(Y) = \frac{\alpha}{\lambda} = \frac{2000}{4} = 500.$$

The variance is

$$Var(Y) = \frac{\alpha}{\lambda^2} = \frac{2000}{16} = 125.$$

7.8.1 Exercises

Exercise 1. Suppose that X is a gamma random variable with parameters α and λ. Define random variable $Y = \frac{X}{n}$, where n is an integer. Compute μ_y, σ_y and $E(Y^2)$.

Box 7.4

The gamma function

The gamma function, designated $\Gamma(s)$ is the value of the integral

$$\Gamma(s) = \int_0^{\infty} t^{s-1} e^{-t} dt = (s-1)!$$

And $\Gamma(1) = \Gamma(2) = 1$

Exercise 2. The response time at an online computer terminal follows, approximately, a gamma distribution, with expected value 4 seconds and variance of 8 seconds. Which of the following is the probability density function for the response times (all functions below have domain X from 0 to infinity?

$$f(x) = \frac{xe^{-(1/2)x}}{4}$$

$$f(x) = \frac{e^{-(1/2)x}}{2}$$

$$f(x) = \frac{e^{-(1/2)x}}{2} x$$

Exercise 3. A student proposes the following as a probability density function for a random variable Y. What would you tell this student?

$$f(y) = e^{-y} y^3, \quad y \geq 0.$$

7.9 Gaussian (aka normal) random variable

Carl Friedrich Gauss (1777–1855), whom many mathematical historians consider to have been the greatest mathematician of all time, was working as the royal surveyor for the King of Prussia. Surveyors measure distances. For instance, a survey crew may measure a distance to be 135.674m. To tell if that is the correct distance, they would check their work by measuring it again. The second time, they might get an answer of 135.677 m. So is it 135.674 m or 135.677 m.? They would have to measure it again. The next time, they might get an answer of 135.675 m. Which one is it? Each time they measured they got a different answer. Gauss would have them measure it about 15 times, and they would get, for example

135.674; 135.677; 135.675; 135.675; 135.676; 135.672; 135.675; 135.674; 135.676; 135.675; 135.676; 135.674; 135.675; 135.676; 135.675

We can interpret what Gauss did as an experiment. A trial of the experiment consists of one measurement of the distance. He requested 15 trials, i.e., measuring the same distance 15 times. Then he calculated how often the measurements were smaller than some number, for example less than 135.676. This experiment, and others like it, helped Gauss discover the famous normal or Gaussian distribution. We can see that the number that appears with higher frequency is 135.675. This led Gauss to conclude that 135.675 must be the value closest to the true distance. But other distances were recorded. If we do not know how frequently those other distances were recorded, we could not conclude anything about the true distance, i.e., we would not know that 135.675 is the most frequent distance. The other distance measurements recorded do not appear as often, indicating that they might be due to measurement error. This and other experiments left Gauss to conclude that many types of measurements are subject to error, but we can more or less know the frequency of those errors by experimentation. He came up with a mathematical formula for these frequencies, the Gaussian or normal model for errors of measurement.

The formula for the density function of a normal random variable X is

$$f(x) = \frac{1}{\sqrt{2\pi\sigma^2}} e^{-\frac{1}{2}(x-\mu)^2}, \quad -\infty < x < \infty$$

Errors in measurement arise for a myriad of reasons (the environment, the instrument, facts about the person making the measurement, the material being measured, to name a few). An awareness of measurement error, and having a distribution that allows the measurement of the probability of making certain error and has an expected value has proved invaluable in the history of science and in the maintenance of the quality of products and services. The awareness has made it possible to set government regulations and more reliable production processes and measurements, for example.

Example 7.9.1

Kinney (2002, 66), talks of a machine that fills soft drink cans advertised to contain 12 oz of soft drink but actually dispenses a volume (V) per can that is normally distributed with mean 11.8 oz and standard deviation 0.2 oz. Government regulations require that at least 99% of the cans advertised as containing 12 oz of soft drink actually contain 11.6 oz or more. Quality control statisticians are hired to use probability to design quality control experiments that constantly monitor the filling process to guarantee that this regulation is respected. The departure from 12 oz is due to chance. The quality control statisticians are not there to make it 12 oz, but to make sure that there are no systematic departures from it, such as always overfilling, that would indicate that the machine is broken.

Learn more about what statistical quality control statisticians do at:

https://www.britannica.com/topic/statistical-quality-control

Example 7.9.2

Kinney (2002, 67), also talks of a chemical process that requires a pH between 5.85 and 7.40. The pH of the process is a normal random variable with mean 6.0 and standard deviation 0.9. Assume that the process must be shut down if the pH falls outside the acceptable range of 5.85 to 7.40. By constantly monitoring that the pH falls within that range, the quality control statistician must make sure that the process will not be shut down. Working with the engineers, the statistician can figure out how to adjust the mean pH of the process to maximize the probability that the process will not be shut down.

Box 7.5

Awareness of measurement error helps you be a better manager

W. Edwards Deming was a famous statistician and management expert. To help managers, he would make them aware of the fact that half of the workers would be making measurements below the expected value and half above the expected value, just by chance (due to numerous factors beyond the control of the worker such as the nature of the equipment used, the information received, etc.). Talking to workers to warn them because they are below or above the expected value is likely to do more harm than good, Deming told managers.

7.9.1 Which things other than measurement errors have a normal density?

The normal or Gaussian distribution was first observed by DeMoivre, an eighteenth-century mathematician. Later on, Laplace discovered that the normal or Gaussian model is also followed by sums of random variables (discovering the famous Central Limit Theorem that we study in chapter 9). The model was later used for measurements of people by Quetelet. When used for measurements other than error, the normal density is just demonstrating the variability of many measurements due to many factors affecting the value observed. This variability arises for different reasons, for example, in the case of the height of people, many factors (genetic, environmental) determine height. Height varies because people vary, not because there is any error in measurement (well, perhaps the instrument is not accurate and then there is both, but you cannot say that one person measuring 150 cm and another measuring 200 cm is because there is an error; we look at the two people and they are certainly one taller than the other).

The reader should be aware of the fact that sometimes people carry over ancient interpretations of things. Names such as standard deviation, for example, should be called standard distance. The fact that a person is above average in height means that this person is at some distance from the expected value for the population the person is coming from.

Similarly, the normal model arises naturally in nature when there are many unrelated factors affecting the metric we are interested in. The following website gives a very interesting example in physics: dust particles moving in water. Why is the distance traveled by dust particles in water normally distributed? The author of this applet simplifies the explanation of Brownian Motion for you.

http://webphysics.davidson.edu/Applets/Galton/BallDrop.html

Example 7.9.3

The annual rainfall in inches in a certain region is normally distributed with the parameters $\mu = 20$, and $\sigma = 4$. What is the probability that, starting with the current year, it will take over 10 years before the occurrence of a rainfall over 28 inches?

Use the normal model to find probability q = P(rainfall > 28) = 0.02. Let p = 1 − q. To interpret the question asked, think that seeing rainfall over 28 inches for the first time after 10 years implies that during the 10 years all we have seen is rainfall at most 28 inches. Let Y be a random variable representing the number of years with rainfall ≤28. Then

$$P(Y = 10 \mid p = 0.98, n) = 0.817.$$

> **Box 7.6**
>
> **Maxwell's law of velocities**
>
> The velocity of a molecule (with mass M) in a gas at absolute temperature T, according to Maxwell's law of velocities, obeys a normal probability law with parameters $\mu = 0$ and $\sigma^2 = \frac{M}{kT}$, where k is the physical constant called Boltzmann's constant. (Parzen (1960), 237.)

Example 7.9.4

(This example is from Samuels (2016).) When red blood cells are counted using a certain electronic counter, the standard deviation of repeated counts of the same blood specimen is about 0.8% of the true value, and the distribution of repeated counts is approximately normal. For example, this means that if the true value is 5,000,000 cells/mm³, then the standard deviation is 40,000.

(i) If the true value of the red blood count for a certain specimen is 5,000,000 cells/mm³, what is the probability that the counter would give a reading between 4,900,000, and 5,100,000?
(ii) A hospital lab performs counts of many specimens every day. For what percentage of these specimens does the reported blood count differ from the correct value by 2% or more?

7.9.2 Working with the normal random variable

The normal density function is difficult to integrate. Thus, finding probabilities for a normal random variable is usually done by approximating the integral, with software or tables. To do this approximation, here are the steps to follow.

- Convert your problem on X to a problem on $Z = \frac{X - \mu_X}{\sigma_X}$, which is the expression of the normal random variable X in standard units. The $E(Z) = 0$, $Var(Z) = 1$ and Z is normally distributed as well. The distribution of Z is called the standard normal density.
- Finding $P(a < X < b) = P\left(\frac{a - \mu_X}{\sigma_X} < Z < \frac{b - \mu_X}{\sigma_X}\right)$. Thus we use the distribution of Z to approximate the probabilities for X. The probabilities for the normal can not be found by the usual means of integration, thus we recur to software, applets, or, as they did many years ago, a table which contains all cumulative probabilities.

- Look at a normal table
- Alternatively, all the steps above can be done using an applet or some normal curve calculator, of which the reader will find a lot on the internet and for which you do not need to convert to the Z at all. For how to do it with R, follow these directions: You must provide the x, mu, and sigma values

```
pnorm(x, mean= mu, sd=sigma) # Calculates P(X<x)
qnorm(quantile, mean=mu, sd=sigma)
```

To calculate with an applet, see, for example, the Allan Rossman and Beth Chance applet. http://www.rossmanchance.com/applets/NormCalc.html. For example, if X is a normal random variable with mean 50 and standard deviation 3, you may enter 50 for mean and 3 for standard deviation and variable X. Do not check mark the third line. Then click on "Scale to Fit" and you will see the normal curve drawn. Below check mark the first line and enter the X, the Z, or the probability, and hit return to obtain the others. Do not check mark the second line. You can change the $>$ or $<$ as well.

Example 7.9.5

A retailer is contemplating opening a bike shop in a residential area. This retailer does not plan to sell over the internet. To know what kind of bikes to stock, the retailer needs to know the age of the population where the store will be. If there are many kids, the retailer would order bikes for kids. If there are many elderly people, the retailer would like to be able to offer them recreational bikes for leisure rides. The retailer manages to get the distribution of ages in the population of this residential area. The distribution is displayed in Figure 7.6 along with the corresponding values of the Z variable. The age of residents has mean 48 and standard deviation 10.

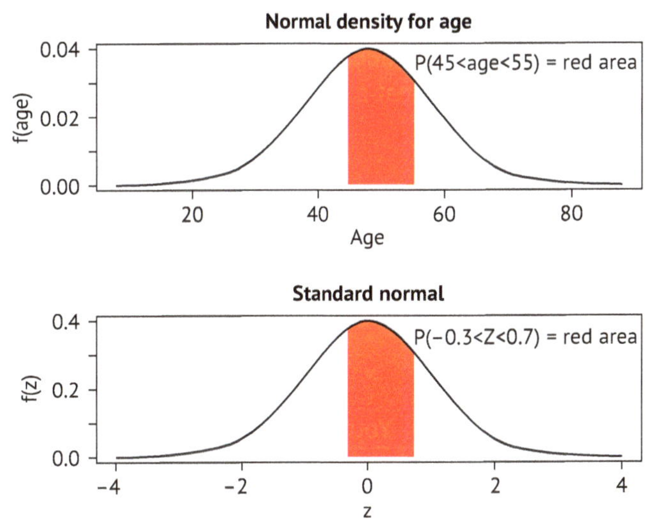

Figure 7.6 Normal distribution of ages in a residential neighborhood and the standard normal. Probability of an age interval and corresponding probability under the standard normal curve.

In the top of Figure 7.6 we see the distribution of ages. In the bottom, we see the standard normal distribution. The points of the distribution of ages from 70 to 80 years correspond to which Z values? The shaded area represents the probability. Thus, the probability that an age lies between 70 and 80 equals the area under the standard normal curve for which Z values?

Example 7.9.6

If a random variable X is normally distributed with mean 100 and standard deviation 50, find the probability that X is less than 200.

$$P(X < 200) = P\left(Z < \frac{200-100}{50}\right) = P(Z < 2) = 0.9772$$

Example 7.9.7

Scores in an exam, which we will denote by X, are normally distributed with expected value $\mu_X = 70$ and standard deviation $\sigma_X = 6$. What is the probability that a randomly chosen student will score higher than 76 on the exam?

$$P(X > 76) = P\left(Z > \frac{76-70}{6}\right) = P(Z > 1) = 1 - P(Z < 1) = 1 - 0.84.$$

Example 7.9.8

The time it takes a driver to react to the brake light on a decelerated vehicle is critical in avoiding rear-end collisions. Someone suggests that reaction time for an in-traffic response to a brake signal from standard brake lights can be modeled with a normal distribution with expected value 1.25 seconds and standard deviation 0.46 seconds. What is the probability that a driver's reaction time will be between 1 and 1.75 seconds?

$$P(1 \le X \le 1.75) = P(-0.54 \le Z \le 1.086) = 0.5653.$$

To find percentiles for the normal random variable we may also use the standard normal random variable or use an applet or normal curve calculator. The steps are as follows:

- Find the value c corresponding to the probability $P(Z \le c) = q$
- Convert the Z value to an X vale using the inverse formula $X = \mu + Z\sigma$

You may compute percentiles with the Rossman/Chance applet by typing the probability and leaving the $<$ sign.

Another applet that is widely used is that of David Lane, which can be found at http://onlinestatbook.com/2/calculators/normal_dist.html

Example 7.9.9

What is the 67th percentile in the standard normal random variable Z?

We first find the 0.67 area within the body of the table. We then identify that with a Z value of 0.4399132. In other words,

$$P(z \le 0.4399132) \approx 0.67$$

Example 7.9.10

The amount of distilled water dispensed by a certain machine has a normal distribution with $\mu = 64$ ounces and $\sigma = 0.78$ ounces. What container size c will ensure that overflow occurs only 0.5% of the time? Let X denote the amount of water dispensed.

We are looking for the 99.5th percentile, so find the value of Z under the standard normal curve that leaves 99.5 percent of the total area to the left of it. That is $Z = 2.58$.

$$2.58 = \frac{c - 64}{0.78}$$

Solving for c we get that the 99.5th percentile is 66.0124 ounces.

7.9.3 Linear functions of normal random variables are normal

One very important property of the normal random variable is that any linear function of a normal random variable is also normally distributed. That is, if X is normally distributed with mean μ and variance σ^2, and if $Y = aX + b$, with a, b constants, then Y is also normally distributed with $\mu_Y = E(Y) = a + b\mu_X$ and $\sigma_Y^2 = Var(Y) = a^2\ \sigma_X^2$

Example 7.9.11

If X is N ($\mu = 2, \sigma^2 = 0.2$), then $Y = 4X - 2$ is N($\mu = 6, \sigma^2 = 3.2$). We use the linear property of the expectation operator and the rules of expectations we know to prove this.

Box 7.7

Importance of the Gaussian distribution in statistics

- Statisticians and data scientists study data, and the normal distribution is a good model for many measurements that are affected by a myriad of factors.
- The probability distributions of sums of random variables and of averages are asymptotically Gaussian.
- The Binomial distribution approaches the Normal, which is so important in the sampling of populations.

7.9.4 Exercises

Exercise 1. Wires manufactured for a certain computer system are specified to have a resistance of between 0.10 and 0.17 ohms. The actual measured resistances of the wires produced by company A have a normal probability density distribution, with expected value 0.13 ohms and standard deviation 0.005 ohms. If three independent such wires are used in a single system and all are selected from company A, what is the probability that they all will meet the specifications?

Exercise 2. The temperatures in June in Los Angeles are distributed normally with mean 77° Fahrenheit and standard deviation 5° Fahrenheit. (i) What is the probability that the temperature on a randomly chosen June day next year in Los Angeles is at least 84° Farenheit? (ii) In order to convert degrees Fahrenheit to degrees Celsius, we subtract 32 from the degrees Fahrenheit and multiply the result by $\frac{5}{9}$. What is the density model for the temperature in degrees Celsius? (iii) Compute the probability that a randomly chosen day in June in Los Angeles next year will have a temperature of 28° degrees Celsius or higher? (iii) How cold are the coldest 10% of June days in Los Angeles, in degrees Celsius?

Exercise 3. Among first-year students at a certain university, scores on the math SAT followed the normal curve, with an average of 500 and a standard deviation of 100. At what percentile was a student who scored 350?

Exercise 4. Prove that if X has normal density with mean μ_X and variance σ_X^2 then $Z = \frac{X - \mu_X}{\sigma_X}$ has expected value 0 and standard deviation 1.

Exercise 5. If X is a normal random variable with parameters $\mu = 3$ and $\sigma^2 = 9$, find

$$P(|X - 3| > 6).$$

Exercise 6. Statisticians use the theory we learn in probability about the normal density function to determine whether data that they observe might follow the normal model. This is how they operate. They use data tools to calculate the percentiles of the data, the mean of the data, and the standard deviation of the data. Then they assume a theoretical normal model with the same mean and the same standard deviation. The idea then is to compare the percentiles of the model with the percentiles of the data set.

You are going to apply this technique to SAT scores. Research and find the average verbal SAT score in 2017 in the United States. Find also the standard deviation and a few percentiles. Compute the same percentiles for the normal model. Are they the same? Would you conclude that the normal model is a good model for these data?

Exercise 7. How many standard deviations above and below the expected value do the quartiles of any normal distribution lie?

Exercise 8. Federal Services advertises that its average delivery time is 30 hours. Federal Services' standard deviation of its delivery time is 5 hours. What is the probability that a document arrives in less than 36 hours?

Exercise 9. Family branding occurs when a firm applies one brand name to its entire product line, such as Levi's. Individual branding occurs when a firm uses individual brand names for its products, for example, Procter & Gamble's Pringles, Crisco, and Tide. GSP Inc. is trying family branding for a new toothpaste in 20 test cities. The mean and standard deviation of units sold per week are 2,250 and 250 respectively. GSP is also test marketing the toothpaste using individual branding in 20 similar cities. The mean and standard deviation in units sold per week are 2,250 and 500. GSP will select the strategy that maximizes its chance of selling at least 2,350 units per week. This will ensure that it meets its return on the project's investment goal. Which marketing approach—family or individual branding—should GSP select? Assume sales are normally distributed.

Exercise 10. The number of chocolate chips in an 18-ounce bag of Nabisco's Chips Ahoy! chocolate chip cookies was found to be normally distributed with mean 1,261 and standard deviation 117.6, based on analysis of many bags way back in the late 1990s. This was found in response to Nabisco's "Chips Ahoy! 1000 Chips Challenge (Warner and Rutledge 1999). Nabisco asked for confirmation that there are at least 1,000 chips in every 18-ounce bag and some content participants used the normal model to find the answer. Other approaches to answer Nabisco's question were tried. Use the normal density assumption to find the probability that there are at least 1000 chips in a randomly chosen 18-ounce bag.

Exercise 11. The weight of anodized reciprocating pistons produced by a company follows a Gaussian distribution with $\mu = 10$ lb and standard deviation 0.2 lb. A sampling inspection scheme designed by the quality control engineers calls for rejecting the heaviest 2.5% of the pistons. What weight, in pounds, determines the overweight classification?

7.9.5 Normal approximation to the binomial distribution

If the parameter **n** is very large and p small, such that $np > 10$ and $n(1-p) > 10$ then a binomial random variable can be approximated in distribution by a normal random variable with $\mu = np$ and $\sigma^2 = np(1-p)$.

Example 7.9.1

It is known that 45% of home improvement loan applications are approved. If 500 applications are chosen at random, what is the probability that less than 200 are approved?

$$X = \#\ of\ applications\ out\ of\ 500\ approved.\ X \sim Bin(n = 500, p = 0.45)$$
$$E(X) = np = 500(0.45) = 225$$
$$Var(X) = np(1-p) = 500((0.45)(0.55) = 123.75$$

The standard deviation of X is 11.124

We see that np > 10, n(1-p) > 10. Therefore we may use the normal approximation to binomial:

$$P(X < 200) = P(\ Z < (200 - 225) / 11.124) = P(\ Z < -2.247) = 0.01253$$

If we had wanted to calculate this probability exactly, we would have calculated

$$P(X < 200) = 1 - \sum_{k=201}^{500} \binom{500}{k} (0.45^k)\left((1-0.45)^{500-k}\right).$$

Example 7.9.2

A poll conducted by ABC News (https://abcnews.go.com/Technology/Traffic/story?id=485098&page=1) reported that 17% of the 1,204 sampled drivers run a stop sign or light. If this proportion were true for the population of drivers at large, what would be the probability that more than 250 drivers in a sample of 1,204 run a stop sign or red light?

$$X = \text{\# of drivers out of 1204 that run stop sign or red light. } X \sim \text{Bin}(n = 1204, p = 0.17)$$

$$E(X) = np = 1204\,(0.17) = 204.68$$

$$Var(X) = np\,(1-p) = 1204((0.17)\,(0.83) = 169.8844$$

The standard deviation of X is 13.03397.

np > 10, n(1-p) > 10. Therefore we may use the normal approximation to the binomial:

$$P(X > 250) = P(\,Z > (250 - 204.68) / 13.03397) = P(\,Z > 3.477068) \approx 0.$$

It is unlikely that there would be so many drivers running a stop sign or red light in a sample of 1204 if 17% of drivers in the population do.

7.9.6 Exercises

Exercise 1. According to the Census Bureau, 15.3% of Californians live below poverty level. A random sample of 1,000 Californians is taken. What is the probability that less than 450 people in the sample live below the poverty level?

Exercise 2. According to government data, 30% of married Americans marry after age 30. A study of married people chooses a random sample of 400 married Americans and asks each person in the sample their age at marriage. What is the probability that more than 200 people in the sample married after age 30?

Exercise 3. Approximately 1.07% of the population in the United States has chronic hepatitis C, according to the Center for Disease Control. If you randomly select individuals to test, how many should be selected for the normal approximation to hold?

Exercise 4. In 1,000 flips of a fair coin, heads came up 560 times and tails 440 times. Are these results consistent with a fair coin?

Exercise 5. Inspired by Mansfield (1994). Department stores in the United States offer store credit cards that come with some perks. Customers with a store credit card can get a 20% greater discount than customers without a store credit card. Department store Mysees knows from past experience that on days when there is a substantial storewide sale and Mysees offers an additional 20% off if the customer opens an account, 25% of the customers without an account will open one. If 1,000 customers without the credit card visit Mysees on a particular day, what is the probability that more than 300 open a credit card account?

Exercise 6. Vralp supermarket must stock items on a daily basis to make sure that only a small percentage of the customers that want a product leave unsatisfied because the product is not in stock. Cash registers at supermarkets are automated, which allows supermarkets to keep track of what products are bought and with what frequency on a daily basis. It is known, for example, that a popular brand of canned soup is bought by 20% of customers. Looking ahead at tomorrow, when 1,000 customers are expected, the store asks: What is the minimum number of cans of the popular soup that they must have in stock if the probability is to be at most 3% that they will run out?

Exercise 7. This exercise is adapted from Mansfield (1994, 206). Telephone companies use probability to solve many kinds of engineering problems. This problem is about landlines, the static phones servicing many homes. But it is applicable to cell phones.

A telephone exchange at A was to serve 2,000 telephones in a nearby exchange at B. It is too expensive to install 2,000 trunk lines from A to B. Instead, the telephone company decided to install trunk lines so that only 1 out of 100 calls would fail to find an unutilized trunk line immediate at its disposal. Under typical conditions, the probability that one of the 2,000 telephone subscribers will require a trunk line to B is 1/30 (this probability will be different if there are natural disasters, or other hazardous events). The telephone company wants to determine how many trunk lines it should install so that when 1 out of the 2,000 subscribers puts through a call requiring a trunk line to B during the busiest hour of the day, he or she would find an unutilized trunk line to B immediately at the subscriber's disposal in 99 out of 100 cases.

(See also the version of this problem in Parzen (1960, 246).)

7.10 The lognormal distribution

Under a normal distribution assumption, the negative realizations of a random variable of interest are possible. Meanwhile, most of the environmental indices are positive by nature because they represent some positive measures. Their approximation using the normal distribution can lead to erroneous conclusions.

If the log of a random variable is a normal curve with parameter μ and σ, then that variable is lognormal with mean μ and σ. The lognormal model is used for the description of attributes of biological objects (organisms or colonies). Concentration of the main air pollutants, for example, tend to have lognormal distribution. Constituents of ambient air such as concentration of CO, ozone and NO_2 at a certain location are distributed as lognormal distribution. Ott (1995) suggested a simple model of successive random dilution for explanation of origins of the lognormal distribution for these variables.

7.11 The Weibull random variable

Waloddi Weibull (1887–1979) was a Swedish physicist. He was an inventor, engineer, and professor. Weibull was interested in the strength properties of brittle material and conducted experiments to measure such strength. He was well aware that no physical measurement is exact. Thus he conducted experiments that consisted on measuring the strength of a material under a given stress repeatedly. His measurements lead him to discover that material strength under the same experimental conditions varied, as he expected, very much like Gauss discovered that distances were different depending who measured them. But the normal or Gaussian distribution was not a good mathematical model for Weibull's measurements. To have a reference that engineers could use, he came up with a mathematical probability model that came to be known as the Weibull distribution (Weibull (1951)). With this model, engineers could determine the proportion of times that measured strength would be found to be less than a given value.

It turned out that Weibull's model gained wide applicability in engineering as a model for many types of strength of different materials. But the health sciences also found the model useful to measure the strength of people after a medical surgery or some traumatic event. The field survival analysis relies to a large extent on the Weibull model to measure survival, or time until death after a surgery.

Whereas the Gaussian distribution is pretty symmetric, the Weibull distribution tends to be skewed right.

The formula that Weibull found for his distribution was

$$f(x)=\frac{\alpha}{\beta}\left(\frac{x}{\beta}\right)^{\alpha-1}e^{-\left(\frac{x}{\beta}\right)^{\alpha}}, \quad x\geq 0,$$

where α, β are positive shape and scale parameters, respectively. Changing β stretches or compresses the x scale without changing the shape of the distribution. As in any density function, the area under the curve represents probability.

Example 7.10.1

An engineer found that the Weibull distribution model that fits fracture strength, X, of silicon nitride braze points had parameters $\alpha=5$ and $\beta=125$. Find the quartiles of this distribution and the value of the interquartile range.

The model is

$$f(x)=\frac{5}{125}\left(\frac{x}{125}\right)^{5-1}e^{-\left(\frac{x}{125}\right)^{5}}, \quad x\geq 0.$$

To have a reference for future problems that you may do with this distribution, it pays to do the computations first with generic parameters and then plug in the values of the parameters given.

To find the first quartile, we compute

$$\int_0^c \frac{\alpha}{\beta}\left(\frac{x}{\beta}\right)^{\alpha-1} e^{-\left(\frac{x}{\beta}\right)^{\alpha}}\, dx = 0.25$$

$$-e^{-\left(\frac{x}{\beta}\right)^{\alpha}}\Bigg|_0^c = 0.25$$

which reduces to

$$1 - e^{-\left(\frac{c}{\beta}\right)^{\alpha}} = 0.25$$

Simplifying further,

$$e^{\left(-\frac{c}{\beta}\right)^{\alpha}} = 0.75$$

or taking natural logs

$$\left(-\left(\frac{c}{\beta}\right)^{\alpha}\right) = log\,(0.75)$$

$$\left(\left(\frac{c}{\beta}\right)^{\alpha}\right) = -log\,(0.75)$$

Taking logs on both sides,

$$\alpha log(c) - \alpha log(\beta) = log(-log(0.75))$$

so

$$log(c) = \frac{\alpha log(\beta) + log(-log(0.75))}{\alpha}$$

which gives

$$c = e^{\frac{\alpha log(\beta)+log(-log(0.75))}{\alpha}}$$

Substituting the values of the parameters given in the problem, we find that $c = 97.42997$. We leave it as an exercise for the reader to find the 75th percentile and the interquartile range.

7.11.1 Exercises

Exercise 1. Consider the problem solved in Example 7.10.1. Complete the exercise by finding the 75th percentile and then computing the interquartile range.

7.12 The beta random variable

The beta distribution models random variables that take values between 0 and 1. A proportion is that type of variable because a proportion is a number between 0 and 1. Thus, this distribution is widely used in Bayesian modeling as a prior distribution of the parameter p of a binomial distribution. But the beta has many other uses.

The beta is another density function that you will research in the random project of Siegrist (1997), at this site http://www.randomservices.org/random/special/Beta.html. This author has applets for you to see the shapes of the density. Study that section of the random project for this section on the beta random variable. Do some of the exercises.

7.13 The Pareto random variable

Have you heard the saying that 80% of wealth is in the hands of 20% of the population? This has been known in many circles as the Pareto principle, after Vilfredo Pareto (1848–1923). Twenty percent of the causes is responsible for 80% of the outcomes. The Pareto density is widely used to measure income distributions, and income inequality indexes have been built around it.

In this section, you will research the Pareto density function with the applets and simulators found in Siegrist (1997). Visit http://www.randomservices.org/random/special/Pareto.html to acquaint yourself with this density. Do some of the exercises.

7.14 Skills that will serve you in more advanced studies

When doing proofs and discovering new formulas, it helps to remember the formulas for probability density functions and the property that the area under continuous density functions is 1.

Example 7.15.1

Consider the following expression:

$$\frac{1}{\sqrt{2\pi}}\int_{-\infty}^{\infty} e^{(t\sigma)^2/2}e^{t\mu}e^{-\frac{(z-t)^2}{2}+\frac{t^2}{2}}dz.$$

Can it be simplified? It can, if you notice that there is this term embedded in the formula:

$$\int_{-\infty}^{\infty}\frac{1}{\sqrt{2\pi}}e^{-\frac{(z-t)^2}{2}}dz,$$

that is the area under a normal curve with mean t and variance 1, and therefore this integral is 1.

Recognizing that, you are left with the expression:

$$e^{(t\sigma)^2/2}e^{t\mu}e^{\frac{t^2}{2}}\int_{-\infty}^{\infty}\frac{1}{\sqrt{2\pi}}e^{-\frac{(z-t)^2}{2}}dz = e^{\frac{(1+\sigma^2)t^2}{2}+t\mu}$$

The moment generating function of the normal random variable is of the form:

$$e^{\frac{(Variance)t^2}{2}+t(Mean)}$$

In the placeholder for the variance we have $(1+\sigma^2)$. So the expression that we started this example with is the moment generating function of a normal random variable with mean μ and variance $(1+\sigma^2)$.

7.15 Mini quiz

Question 1. In a large lecture course, the scores on the final examination followed the normal curve closely. The average score was 60 points and three-fourths of the class got between 50 and 70 points. The SD of the scores was

a. larger than 10 points
b. smaller than 10 points
c. 0
d. equal to 10 points

Question 2. What is the constant k that makes the following function a valid density?

$$f(x) = kx^9(1-x)^2, \quad 0 \le x \le 1.$$

a. 0.00151
b. 2
c. 660
d. 210

Question 3. Let X be the time that it takes to drive between point A and point B during the afternoon rush hour period in highway 4005. The density function of X is

$$f(x) = \frac{1}{2}x, \quad 0 < x < 2.$$

Find the value of the 50th percentile.

a. 1.414
b. 1.0
c. 5/16
d. 0.034

Question 4. The cumulative distribution function of a random variable X is

$$F(x) = \frac{x-2}{5}, \quad 2 \le x \le 7$$

What is the density function of this random variable?

a. Standard normal, $N(\mu = 0, \sigma^2 = 1)$
b. Uniform $(a = 2, b = 2/5)$
c. Exponential with parameter $\lambda = \frac{2}{5}$
d. Uniform $(a = 2, b = 7)$

Question 5. Systolic blood pressure in normal healthy individuals is normally distributed with $\mu = 120$ and $\sigma = 10$ mm Hg. A systolic blood pressure of 136.45 mm Hg is at what percentile?

a. 95th percentile
b. 25th percentile
c. 88th percentile
d. 69th percentile

Question 6. Suppose X is a normal random variable with mean μ and standard deviation σ. Under what circumstances is $\frac{X}{\sigma}$ the standard normal random variable?

a. μ_X is 0
b. $\sigma_x = 1$
c. μ_X is 1
d. $\sigma_x = 0$

Question 7. Let X be the change (in dollars per share) next year in the stock price of Apple. A financial analyst found that X is normally distributed with $\mu = 2$ and standard deviation $\sigma = 3$. The probability of a price change larger than 2.5 would be?

a. 0.1131
b. 0.4338
c. 0.0001
d. 0.1364

Question 8. The probability density function of a continuous random variable X is such that the following applies: $P(X < 16) = 0.75$, $P(X < 10) = 0.25$, $P(X < 12) = 0.5$, $P(X < 19) = 0.9$, $P(X < 5) = 0.10$. Which of the following is true?

a. The Interquartile range is 6, and the distribution is symmetric
b. The range is 4, and the distribution is skewed
c. The interquartile range is 7, and the distribution is skewed.
d. The interquartile range is 6 and the distribution is skewed

Question 9. The distribution of X, the time it takes women between 50 and 55 to run a 10k race, is such that the event A (X between 40 and 60 minutes) has $P(40 < X < 60) = 0.8$; the event B, taking between 50 and 90 minutes has $P(50 < X < 90) = 0.2$; and the event C, taking between 50 and 60 minutes, has $P(50 < X < 60) = 0.1$. For a runner woman chosen at random from the population in this age group, compute the probability that she is in at least one of A or B.

a. 0.3
b. 0.9
c. 0.8
d. 0.2

Question 10. Consider a set of 20 independent and identically distributed uniform random variables in the interval (0,1). What is the expected value and variance of the sum S of these random variables?

a. $E(S) = 10$, $Var(S) = 1.66667$
b. $E(S) = 200$, $Var(S) = 16.6667$
c. $E(S) = 100$, $Var(S) = 200$
d. $E(S) = 10$, $Var(S) = 12$

7.16 R code

This simulation is about queues and waiting and service times. Think of yourself waiting in the ATM line. You are the customer, the ATM is the service center, which takes more or less time with each customer. Your total time in the system is the service time for your transaction plus the time you had to wait in the queue, if any. Then there is a queue of other customers. We will figure out the outcome of interactions in a complex system like this. It is very important that you read line by line and do line by line, except where indicated, to avoid errors. Also keep notes of what is going on. Answer the questions asked.

Queueing theory is a very important modeling technique that represents a computer system as a network of service centers, each of which is treated as a queueing system. That is, each service center has an associated queue or waiting line where customers who cannot be served immediately queue (wait) for service. The customers are, of course, part of the queueing network. "Customer" is a generic word used to describe workload requests such as CPU service, I/O service requests, requests for main memory, etc. All these arrive at random to the service facility. Queueing theory models are often used to determine the effects of changes in the configuration of a computer system. But queuing is an issue not only in computers, but also in banks, or any service industry where satisfactory resolution of the queue is necessary for the system to work properly, like for example, Starbucks, or the textbook store.

The simplest queueing theory model is the {M/M/1} model (which stands for Memoryless, Memoryless, and 1 server, and assumes: (a) that customers arrive in accordance with a Poisson process with average rate δ and thus the inter arrival times are exponentially distributed with mean $\frac{1}{\delta}$. Service time by the server is assumed exponential with parameter μ. Expected service time is then $\frac{1}{\mu}$. (c) The customers are served one at a time by a single server. If the server is busy upon the customer`s arrival, then the customer waits in the queue. Another simple model is one where the service time is uniform, instead of exponential. That is the model we will start with.

The set of random variables involved in either case is summarized in the diagram displayed in Figure 7.7, which is an adaptation of a diagram in Allen (1990, 251).

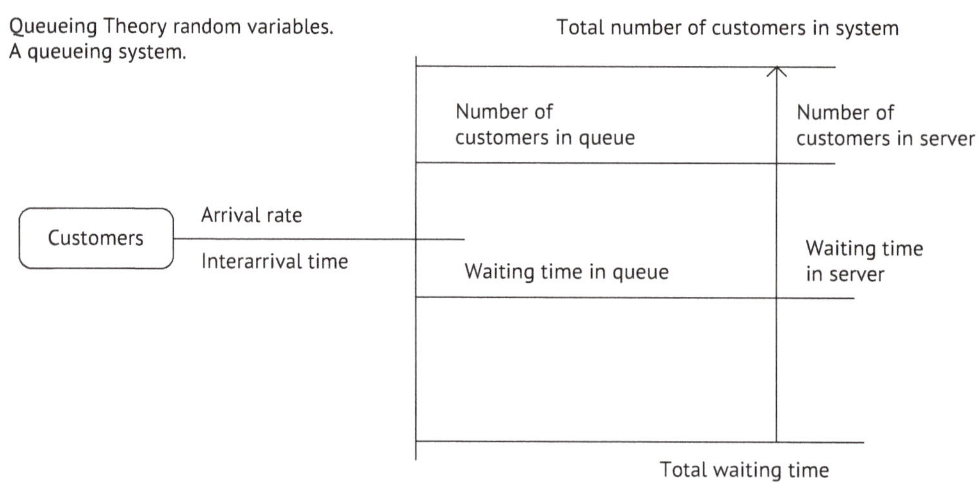

Figure 7.7 Queueing system.

For these models, we are usually interested in determining, among other things, the average number of customers in the system (or in the queue) and the average amount of time a customer spends in the system, among other things.

We simulate together (I give you the complete code for it) the queueing system where the service time is uniform. We will leave it as an exercise to simulate the M/M/1 system. That will require minor changes. Issues that may arise in these systems are: What values of the parameters of the generating distributions guarantee that there is a probability distribution towards which we converge? What values of parameters lead to queues that grow and grow without limits? Which parameters make the queue grow faster?

7.16.1 Simulating an M/Uniform/1 system together

It is very important that you read line by line and do line by line, except where indicated, to avoid errors. We will follow the simulation steps we learned.

Suppose that customers arrive at a single server according to a Poisson process with arrival rate $\lambda = 4$ per hour. This implies that the time between two customers` arrivals, (inter arrival time called **it** is exponential with parameter $\lambda = 1/15$.

Assume that the service time (called **s**) of a customer (in minutes) is uniformly distributed on the interval (5, 15).

One trial consists of generating a random number from the exponential distribution and a random number from the U(5,15) for each customer. That is, we generate inter arrivals times and service times.

Repetition of trials is done many times then.

Type this code and run it line by line.

```
#### Generating the time between customer arrivals ####
it =c(rep(0,1000)) # create space to put the trials
lambda =1/15 # fix the parameter of exponential
it =rexp(1000,lambda) # draw random numbers from exponential
```

To interpret what the numbers mean, type

```
head(it) # to see the first 6 numbers of the 1000 you created.
```

We want to see some of the numbers you got for inter arrival times. So in R, after the above commands, type

```
head(it)
```

To summarize the output in **it**, you can do a histogram

```
hist(it, freq=F, main="Exp(1/15) Inter arrival times" )
```

Describe the shape of the histogram. Is it what you expected? Why?

Now generate the service time (**s**) in minutes, which we are assuming is uniform. Generate now a sequence of uniform random numbers in the interval (5,15).

```
s= c(rep(0,1000))
s=runif(1000,5,15)
head(s) # View first numbers in s
```

You may see the distribution of service times generated by doing a histogram:

```
hist(s, freq=F, main = "Histogram of service times")
```

Now extract what you need from the random numbers generated.

Using the **it** numbers, now create a new variable "arrival times" **(at)**, that contains the cumulated interarrival times of all customers. This will give the time of arrival of each customer. That is, if, for example, the first four values of **it** are 15, 16, 17, and 8, the first four values of the variable **at** should be: 15, 31, 56, and 63. That is, customer 1 in this example arrived at minute 15, customer 2 arrived at minute 31, and so on.

```
at= c(rep(0,1000)) # create space to put the cumulative numbers.
at=cumsum(it) # Cumulative sum of the numbers in it
head(at) # View the first few numbers created
```

Compare the first six numbers you obtained for **it** earlier in this handout with the numbers just obtained for **at**. Is the computer doing the right job for your numbers? I.e., is it giving you the right arrival times for your customers? Show by showing the computation the computer is doing.

The information obtained so far about arrival times, **at**, and service time, **s**, will help us simulate the operation of the queueing system as follows:

To start with, create a variable for the queueing time of each customer, **qt**, and another for the exit time of each customer, **exit**. These are the time the customer waits in line after arriving to the system, and the time at which the customer is done and leaves the system.

```
qt = c(rep(0,1000)) # open space to put the numbers
exit = c(rep(0,1000)) # open space to put the numbers
```

Now determine the queueing time and the exit time for each customer. The following code does the following: if the next customer $i + 1$ arrives before the arrival time **at** plus the service time **s** plus the queuing time **qt** of last customer i is over, then the new customer $i + 1$ has to wait in the queue (**qt** > 0). Otherwise, it does not have to queue (**qt** $= 0$). The program calculates the queuing time. You run this bunch of code all as one: copy and paste the whole thing into R after you have typed it in your script file.

```
for(i in 1:999) {
if(at[i] +s[i] + qt[i] <= at[i+1]){
qt[i+1]=0
exit[i]= at[i]+s[i]+qt[i]
```

```
}
else if(at[i] + s[i]+qt[i] > at[i+1]){
exit[i]=at[i]+s[i]+qt[i]
qt[i+1] = at[i]+s[i]+qt[i] -at[i+1]
}
}
exit[1000]=at[1000]+qt[1000]+s[1000]
```

We know the service time, the queueing time and the exit time for each customer. So we can now figure out the total time spent by each customer in the system. Let's call it **tts**.

```
tts = c(rep(0,1000))
tts = exit - at # total time= exit time-arrival time.
```

We can do a histogram to see the distribution of the total time in the system as follows

```
hist(tts, freq=F, main = "Histogram of total time in the system")
```

Then we can find the sample mean, standard deviation and other summaries of the generated data:

```
mean(tts)
sd(tts)
summary(tts)
```

> Enter here the mean, median, first quartile, third quartile, minimum, maximum, and standard deviation of the total time spent in the system by a customer.
> Describe the shape of the histogram of total time in the system.

We would also like to know the average number of customers in the system. For that, we first need to put the operation of the system in real time. The following program puts the arrival times and exit times one after the other for each customer. Then in a separate column, if arrival, a number 1 is entered(customer in), if exit, then a -1 is entered. After that, the times of arrival and exit are sorted from lowest to highest. The objective is to be able to count how many number 1's we have in an interval of time. Let us follow the code more in detail.

```
at=matrix(at) # convert at to matrix object
exit = matrix(exit) #convert exit to matrix object
atexit = cbind(at,exit) # combine the two in one matrix
tatexit = t(atexit) # transpose the matrix.
realtime=matrix(tatexit,2000,1,byrow=T) # create the real time.
head(realtime)
```

Odd row numbers of this last matrix give us the arrival time, and even numbers give us the exit time. So we can add one column to the matrix, which tells us that with every arrival there is a new customer, and with every departure one customer less.

```
customer =matrix(rep(1,2000),2000,1)
realtime=cbind(realtime,customer)
# The next three lines must be run as a block together.
for(i in 1:1000) {
realtime[2*i,2]=-1
}
```

The following two commands order the times of arrivals and exits and then helps us see at each point in time whether there was a customer.

```
oo =order(realtime[,1])
trace =realtime[oo,]
```

We can help the computer see this with the command head(trace):

```
head(trace)
```

The first time in column 1 (row 1) is the entry time of customer 1, which was the first line of realtime. The second time in column 1 is the entry time of customer 3, which was the third line of realtime. The third time in column 1 is the exit time of customer 1, which was line 2 of realtime. The fourth time in column 1 is the entry time of customer 3, which was line 5 in realtime. The fifth time in column 1 is the departure time of customer 2, which was line 4 in realtime. Finally, the last time in column 1 is the departure time of customer 3, which was line 6 in realtime.

Now we would like to focus on the number of customers. So we create a variable that tells us how many customers are in the system at each real time of entry or exit.

```
tracecum= cumsum(trace[,2])
```

We can see the result of cumsum looking at the first 6 we looked at earlier:

```
head(tracecum)
```

We can find the average number and other summaries of the number of customers at each of the minutes in which action is taking place by typing in R:

```
mean(tracecum)
sd(tracecum)
summary(tracecum)
```

Of interest is whether the system converges. That is, are the service times and arrival times such that the number of customers in the system explodes? That would happen maybe if the service time is too slow relative to the arrival rate. So playing with the parameter values of the service time and the inter arrival times, you can generate all sorts of behavior. Let us see which behavior we generated.

```
plot( tracecum,type="l")
```

Does it seem like the system is stable? That is, does it look like customers come and go and do not accumulate too much in the system? Sketch the type of plot you get.

Exercise to do on your own:

To see whether the behavior of the random variables analyzed change when we change the assumption about service time, repeat the analysis done in this activity, but this time, in the probability model, the service time is exponential with parameter $\delta = \frac{1}{15}$ instead of uniform (5,15). The rest of the code should not change. Then answer all the questions again. In particular, explain well what happens at the end by running all the code several times and comparing the output you get. Is this new system more likely to converge? Is the final picture different from the uniform service time case? Experiment with the randomness of the situation by doing the simulation several times. That way, you will appreciate how each time you run it the data is different (since it is random).

Notice: you will have to touch only this command.

```
s=rexp(1000, lambda)
```

The rest of the code should be as for the uniform service time.
Note: This simulation has also benefitted from Goodman(1988).

7.17 Chapter Exercises

Exercise 1. The distribution of hemoglobin in g/dl of blood is approximately normal with expected value 14 and standard deviation 1. Find the IQR.

Exercise 2. After completing a study, a company in Kansas City concluded the time its employees spend commuting to work each day is normally distributed with a mean equal to 15 minutes and a standard deviation equal to 3.5 minutes. One employee has indicated that she commutes 22 minutes or more per day. (i) What is the probability that a randomly chosen employee commutes that much? (ii) An employee commuting exactly 22 minutes per day is at what percentile? (iii) The company will pay each employee the cost of commuting by giving them 10 cents per minute. How much should the company budget per employee? What is the variability of the cost of commuting per employee?

Exercise 3. Four fuses were shipped to a customer, before being tested, on a guarantee basis. That is, if the customer finds some of the fuses defective, s/he may return the shipment for repair. The number of defectives in a shipment of four fuses is a random variable Y with $E(Y) = 0.4$ and $Var(Y) = 0.36$. The cost of repairing the defective fuses is given by

$$C = 3Y^2.$$

What is the expected repair cost?

Exercise 4. Let X be uniformly distributed over (0,1). Draw the density curve and calculate $E(X^3)$.

Exercise 5. Suppose that one is told that the time one has to wait for a bus on a certain bus stop is a continuous random variable with a probability density function given by

$$f(x) = 4x - 2x^2 - 1, \quad 0 \le x \le 2$$

Is $f(x)$ a density function?

Exercise 6. (This exercise is based on William F. Stout (1999).) A certain insect species has a mean length of 1.2 centimeters and a standard deviation of 0.1 centimeters. If there are estimated to be 10,000 of these insects in a terrarium, how many of them would be expected to be less than 0.8 cm in length. Assume that length is normally distributed.

Exercise 7. Prove that if X is a random variable with mean $\mu = 3$ and standard deviation $\sigma = 0.5$ then

$$\frac{X-\mu}{\sigma} \text{ has mean 0 and standard deviation 1.}$$

Exercise 8. The Weibull distribution has the following density function

$$f(x) = \frac{\alpha}{\beta^\alpha} x^{\alpha-1} e^{-\left(\frac{x}{\beta}\right)^\alpha}, \quad x > 0$$

Under what conditions will this density be equal to the exponential density?

Exercise 9. Prove that the variance of a uniform random variable with range (a,b) is

$$Var(X) = \frac{(b-a)^2}{12}$$

Exercise 10. If X is a gamma random variable, what is the following equal to?

$$\int_x x^{\alpha-1} e^{-\frac{x}{\beta}} dx$$

Exercise 11. A six-sided die is tossed 200 times. The number 1 or 2 or 3 came up 125 times. Is this die fair?

Exercise 12. The length of human pregnancies from conception to birth varies according to a normal distribution, with a mean $\mu = 266$ days and standard deviation $\sigma = 16$ days. Calculate the median length of pregnancies.

Exercise 13. For a group of the US population in 1992, the average income was \$35,000 and the standard deviation of income was \$23,000. Only 1/10 of 1% had incomes above \$150,000. Was the percentage with incomes larger than \$35,000 larger or smaller than 50%?

Exercise 14. Consider the cumulative distribution of the proportion of income in a country. That is a distribution that has proportion of income on the horizontal axis and proportion of persons that have that proportion of income or less on the vertical axis. With these criteria, draw the cumulative distribution of income when there is complete equality, i.e., everybody gets the same amount of income. In the same graph, draw the cumulative distribution of income when there is inequality of income distribution. Explain why you drew your graph that way.

Exercise 15. The cumulative distribution function of a random variable X is

$$F(x) = 1 - e^{-\left(\frac{x}{\beta}\right)^{\alpha}}$$

where α, β are parameters. Find the probability density function of X.

Exercise 16. The coffee chain Starbucks created an app that supports mobile ordering at 7,400 of its stores in the United States, giving users the opportunity to order and pay for their drinks before they even arrive at their local Starbucks. Starbucks estimates the typical wait time given in the app will average around 3–5 minutes at most stores, with variability depending on the number of mobile orders in the queue and what a call order entails. After the transaction is completed in the app, users can skip the line and instead head to the pick-up counter where they can ask a barista for their order. Suppose that at one of the stores the waiting time in seconds has moment generating function given by

$$M_X(t) = (1 - 200t)^{-1}$$

If you enter your order immediately after another customer, what is the probability that your order will be ready in 300 seconds? (ii) If 300 seconds have passed and you arrive at the counter and your coffee is not ready, what is the probability that you will have to wait an additional 50 seconds?

Exercise 17. The length of life, X, of a fuse has the following probability density function:

$$f(x) = \frac{1}{\theta} e^{-\frac{x}{\theta}}, \quad x \geq 0, \theta > 0.$$

Three such fuses operate independently. Find the joint density of their lengths of life.

Exercise 18. Consider the following random variables:

(a) the volume of acorns
(b) the number of people in a town that think that marijuana should remain illegal
(c) survival time after diagnosis with fatal disease

Which probability density function model would be appropriate to consider for each of these random variables?

Exercise 19. Suppose the distribution of math scores on the SAT test has a roughly unimodal and symmetric distribution with mean equal to 500 and standard deviation equal to 100. You happened to earn a 600 on the math part of the SAT. Where do you stand among all students who took this math portion of the SAT? (i) Draw the distribution of scores and label the points that are one SD, two SDs, and three SDs away from the mean. (ii) What percentage of students scored below 400? (iii) What percentage of students scored above 700? (iv) What percentage of students scored between 500 and 600? (v) What percentage of students scored above 620? (vi) What score did you have to receive to be above the 90th percentile? What about the 10th percentile?

Exercise 20. In the United States, many universities require applicants to submit scores on standardized tests, such as the SAT tests. The college your friend wants to apply to says that while there is no minimum score required, the middle 50% of their students have SAT scores between 1020 and 1220. You would feel confident if you knew her score was in the top 25%, but unfortunately she took the ACT test, an alternative standardized test. How high must her score be on the ACT to be comparable to the top quarter of equivalent SAT scores? (Note: The mean SAT score for all college-bound seniors is about 1000 and the standard deviation is about 200 points. For the same group, the ACT average is 20.8 with a standard deviation of 4.8. You can assume both score distributions are nearly normal.)

Exercise 21. (This exercise is from Allen(1990, 113, 162).) The interactive computer system at Gnu Glue has 20 communication lines to the central computer system. The lines operate independently and the probability that any particular line is in use is 0.6. (i) What is the probability that 10 or more lines are in use? Compare the exact binomial solution and the normal approximation solution to this problem. (ii) Do the assumptions needed for the normal approximation make sense?

Exercise 22. (This exercise is based on Degroot (1974, 247).) Suppose that F is a continuous cumulative distribution function on the real line, and let a and b be numbers such that $F(a) = 0.3$ and $F(b) = 0.8$. If 25 observations are selected at random from the distribution for which the distribution function is F, what is the probability that six of the observed values will be less than a, ten of the observed values will be between a and b, and nine of the observed values will be greater than b?

7.18 Chapter References

Allen, Arnold O. 1990. *Probability, Statistics and Queueing Theory with Computer Science Applications*. Academic Press, Inc.

Denny, Mark and Steven Gaines. 2000. *Chance in Biology*. Princeton University Press.

Goodman, R. 1988. *Introduction to Stochastic Models*. Benjamin/Cummings Publishing Co., Inc.

Hamming, Richard W. 1991. *The Art of Probability*. Addison Wesley-Publishing Company, Inc.

Harris, Frank E. 2014. *Mathematics for Physical Sciences and Engineering*. Elsevier.

Keeler, Carolyn and Steinhorst. 2001. "A New Approach to Learning Probability in the First Statistics Course." *Journal of Statistics Education* 9, no. 3.

Kinney, John J. 2002. *Statistics for Science and Engineering*. Addison-Wesley.

Mansfield, Edwin. 1994. *Statistics for Business and Economics: Methods and Applications*. Fifth Edition. W.W. Norton & Company.

Ott, Wayne R. 1995. *Environmental Statistics and Data Analysis*. Lewis Publishers.

Parzen, Emanuel. 1960. *Modern Probability Theory and Its Applications*. New York. John Wiley and Sons, Inc.

Pitman, Jim. 1993. *Probability*. Springer Verlag.

Rice, John A. 2007. *Mathematical Statistics and Data Analysis*. Thomson Brooks/Cole

Ross, Sheldon. 2010. *A First Course in Probability*. 8th Edition. Pearson Prentice Hall.

Samuels, Myra L., Jeffrey A. Witmer, and Andrew A. Schaffner. 2016. *Statistics for the Life Sciences*. Fifth Edition. Pearson.

Scheaffer, Richard L. 1995. *Introduction to Probability and Its Applications*, Second edition. Duxbury Press.

Siegrist, Kyle. 1997. The Random Project. http://www.randomservices.org/random/

Stout, William F. 1999. *Statistics: Making Sense of Data*. 3rd Edition. Mobius Communications Ltd.

Warner, Brad, and Jim Rutledge. 1999. "Checking the Chips Ahoy! Guarantee." *Chance*, 12, no. 1: 10–14.

Weibull, Waloddi. 1951. A Statistical Distribution Function of Wide Applicability. *Journal of Applied Mechanics*, 18:293–7.

Chapter 8

Models for More Than One Continuous Random Variable

> A particular fast-food outlet is interested in the joint behavior of the random variables X, defined as the total time between a customer's arrival at the store and the customer's leaving the service window, and Y, the time that the customer waits in line before reaching the service window. Because X includes the time a customer waits in line, we must have $X > Y$. How do you propose to find the probability that the time at the service window is larger than 1 minute?

8.1 Bivariate joint probability density functions

Suppose we are interested in the joint behavior of two continuous random variables, X and Y, where X and Y might represent, for example,

- the mass of a body and the time of descent of this body from a given height to the earth's surface (keeping other things such as height, air density and initial velocity of the body constant)
- methane concentration in a sample of the earth's atmosphere and the sample's carbon dioxide concentration
- the diameter and length of logs on the deck of a sawmill

The fact that each of the pairs of random variables are obtained from the same object is the motivation for wanting to study them together. The probability of occurrence of the pair of random variables is controlled by the rules governing the probabilities of multiple events.

The notion of a pdf of a continuous random variable X can be extended to the notion of the pdf of two or more random variables, a surface in three or more dimensions. We will focus on a pair of random variable, X and Y.

Definition 8.1.1

The joint density function of a pair of continuous random variables, $f(X, Y)$, is a function of two random variables X, Y asociated with a sample space such that:

a. $f(X, Y)$ always nonnegative,
b. If $A = \{[a,b]\}$ and $B = \{[c,d]\}$ are sets of real numbers, then the joint probability of X and Y being in those sets is

$$P(X \in A, Y \in B) = P(a \le x \le b, c \le y \le d) = \int_a^b \int_c^d f(x,y)dydx.$$

Thus, the probability of an event involving the two random variables is a volume under the joint bivariate density function.

c. The volume under the density function throughout the whole domain of the function is 1. More formally, we say that

$$\int_{-\infty}^{\infty} \int_{-\infty}^{\infty} f(x,y)dxdy = 1,$$

where it is understood that the integration will be done throughout the range in which $f(x, y)$ is not 0.

The reader can visualize a bivariate continuous surface at Figure 8.1.

Solving probability problems involving two random variables require multivariable differential and integral calculus.

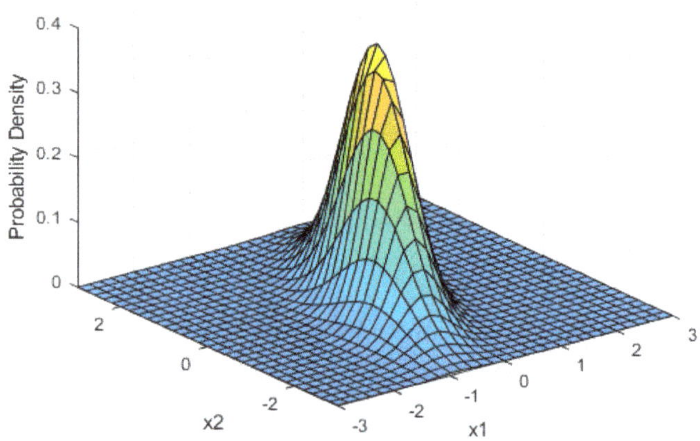

Figure 8.1 A bivariate probability density function.

Example 8.1.1.

Let

$$f(x,y) = 6x^2y, \quad 0 \le x \le 1, \quad 0 \le y \le 1.$$

and let the event $A = \{(x,y): 0 < x < \frac{3}{4}, \frac{1}{3} < y < 1\}$. Then

$$P(A) = P\left(0 < x < \frac{3}{4}, \frac{1}{3} < y < 1\right) = \int_{\frac{1}{3}}^{1} \int_{0}^{\frac{3}{4}} 6x^2ydxdy = 3/8.$$

Example 8.1.2

The two indices BOD (biochemical oxygen demand) and DO (dissolved oxygen) are among the parameters that determine the quality of water in a river. BOD is a relative measure of the biologically degradable organic matter present in the water. The higher the DO level, the better the self-cleaning ability of the water. For a "healthy" river, the BOD has to be low and DO has to be high.

As a rule, there is a strong correlation between the BOD and DO; the higher the BOD, the lower the DO, and vice versa. BOD is measured in mg/L and goes from 1.5 to 5.0 and DO in mg/L goes from 5 to 10. One problem of interest is the probability that the water in the river is unpolluted and healthy provided that it has high ability for self-cleaning.

In problems relating to this subject, we might also want, for example, $P(DO > 7.5 \text{ and } BOD < 3.2)$ (Khilyuk et al 2005, 68).

8.1.1 Exercises

Exercise 1. An institution that prepares applicants for the sequence of actuarial exams yields applicants that have two main characteristics. Let X denote the proportion of applicants who feel very confident about their passing the first exam, and let Y denote the proportion of applicants who feel confident about passing all of the exams. The joint pdf of X and Y can be modeled by

$$f(x,y) = 2(1-x), \quad 0 \le x \le 1, \quad 0 \le y \le 1.$$

Find $P(x \le 0.4, 0.3 \le y \le 0.6)$.

8.2 Marginal probability density functions

When we talked about the marginal or total probability of a particular value of a discrete random variable in Chapter 6 we added the joint probabilities that involved that value of the random variable in question. The equivalent operation in the continuous case is to add the infinite number of joint probabilities that involve that value of the random variable. The mathematical operation to do that in the continuous case is integration.

Consider two random variables X and Y. Then to compute probabilities for X we will add over Y, and viceversa if we want to compute probabilities for Y

$$f(x) = \int_{-\infty}^{\infty} f(x,y)dy$$

$$f(y) = \int_{-\infty}^{\infty} f(x,y)dx,$$

where again, the integration will actually be done over the range of X and Y for which $f(x, y)$ is not 0.

With marginal density functions, we can compute total or marginal probabilities, marginal expectations and variances, cumulative distribution functions for X and for Y, marginal percentiles, and any other thing we can compute with a univariate density function. The reader is now instructed to go back to review univariate random variable concepts in chapter 7, for example:

$$E(x) = \int_{-\infty}^{\infty} xf(x)dx,$$

$$E(y) = \int_{-\infty}^{\infty} yf(y)dy.$$

Notice that we may not, in general, obtain the joint pdf from the marginal pdfs. See Section 8.3 for an exception to this statement.

Example 8.2.1

Let $f(x,y) = 6x^2y, \quad 0 \le x \le 1, \quad 0 \le y \le 1$. Then

$$f(x) = \int_0^1 6x^2ydy = 3x^2, \quad 0 \le x \le 1$$

$$f(y) = \int_0^1 6x^2ydx = 2y, \quad 0 \le y \le 1$$

$$E(X) = \int_0^1 xf(x)dx = \int_0^1 x3x^2dx = \frac{3}{4}$$

$$E(Y) = \int_0^1 yf(Y)dy = \int_0^1 y2ydx = \frac{2}{3}$$

$$E(X^2) = \int_0^1 x^2f(x)dx = \int_0^1 x^2 3x^2dx = \frac{3}{5}$$

$$E(Y^2) = \int_0^1 y^2f(y)dy = \int_0^1 y^2 2ydx = \frac{1}{2}$$

$$\sigma_x^2 = E(X^2) - \mu^2 = \frac{3}{5} - \left(\frac{3}{4}\right)^2 = \frac{3}{80}$$

$$\sigma_y^2 = E(Y^2) - \mu^2 = \frac{1}{2} - \left(\frac{2}{3}\right)^2 = \frac{1}{18}$$

We can use the marginal pdfs to compute probabilities as well. For example,

$$P(X > 0.3) = \int_{0.3}^{1} 3x^2 \; dx = 0.973.$$

Example 8.2.2

(This problem is based on Scheaffer (1995, 231).) Gasoline is to be stocked in a bulk tank once each week and then sold to customers. Let X denote the proportion of the tank that is stocked in a particular week, and let Y denote the proportion of the tank that is sold in the same week. Due to limited supplies, X is not fixed in advance but varies from week to week. Suppose that a study of many weeks shows the joint relative frequency behavior of X and Y to be such that the following joint density function provides an adequate model:

$$f(x,y) = 3x, \quad 0 \le y \le x \le 1.$$

We can see that

$$f(y) = \int_y^1 3x \, dx = \frac{3}{2}(1 - y^2), \quad 0 \le y \le 1.$$

The probability that more than 50% of the tank is sold in a week is

$$P(Y > 0.5) = \int_{0.5}^1 \frac{3}{2}(1 - y^2) dy = \frac{5}{16}$$

On the other hand,

$$f(x) = \int_0^x 3x \, dy = 3x^2, \quad 0 \le x \le 1.$$

And the expected amount of gasoline stocked for a week is

$$E(X) = \int_0^1 x \, 3x^2 dx = \frac{3}{4}.$$

Example 8.2.3

Let

$$f(x,y) = 15e^{-3x-5y}, \quad x \ge 0, y \ge 0$$

$$f(x) = 3e^{-3x}, \quad x \ge 0$$

$$f(y) = 5e^{-5y}, \quad y \ge 0$$

8.2.1 Exercises

Exercise 1. Suppose X is the number of hours it takes for a new pair of blue jeans mail ordered from a store to arrive to customers after ordering, and Y is the time it takes between ordering and the customer trying the new pair of jeans. The joint density of these two random variables is

$$f(x,y) = \frac{1}{125000}, \quad 0 \le x \le y \le 500.$$

What is the probability that the delivery takes longer than 250 hours?

Exercise 2. The joint pdf of the random variables X and Y is:

$$f(x,y) = 25e^{-5y}, \quad 0 \le x \le 0.2;\ y \ge 0$$

Calculate the expected value of X. Calculate the expected value of Y.

Exercise 3. Consider the joint density in Section 8.1.1, Exercise 1. Find (i) the marginal density functions of X and Y; (ii) the expected value and variance of X, and the expected value and variance of Y; and (iii) the cumulative distribution function of X and the cumulative distribution function of Y.

Exercise 4. Consider Example 8.2.3. Find the cumulative distribution function of X and of Y.

Exercise 5. Consider the joint density function in Example 8.2.2. (i) Calculate

$$P(X < 0.3, Y > 0.1).$$

(ii) Find the cumulative distribution function of X.

8.3 Independence

▶ Do you remember what independence implied about two events A and B? About two discrete random variables X, Y? Do you remember how independence was defined in those contexts? List all the things you remember about these. Then check Chapters 3 and Chapter 6 to see how much you remembered correctly.

Independence of two continuous random variables is also very important in Statistics applications. Two continuous random variables X and Y are independent if their joint probability density function, $f(x, y)$, equals the product of the marginal density functions. Finding that two continuous random variables are independent is a good thing for us because a lot of conclusions about the two random variables can be deduced from that fact.

Definition 8.3.1

Let X and Y be two continuous random variables with joint pdf $f(x,y)$. Then X and Y are independent if $f(x,y) = f(x)\,f(y)$.

It must be pointed out that if X and Y are independent and $g(X)$ and $h(Y)$ are functions of X and Y respectively, then $g(X)$ and $h(Y)$ are also independent.

Example 8.3.1

If, as in Example 8.2.1,

$$f(x,y) = 6x^2y,$$

then

$$f(x)f(y) = 3x^2(2y) = 6x^2y,$$

and therefore X and Y are independent random variables.

Example 8.3.2

For the following joint density function, the two random variables are not independent.

$$f(x,y)=3x, \quad 0\le y\le x\le 1.$$

As we saw in Example 8.2.2,

$$f(y)=\frac{3}{2}(1-y^2), \quad 0\le y\le 1, \quad f(x)=3x^2, \quad 0\le x\le 1.$$

We can see that $f(x)f(y)=\frac{3}{2}(1-y^2)\,3x^2$ is not equal to $f(x,y)=3x$. Therefore, X and Y are not independent random variables.

8.3.1 Exercises

Exercise 1. Consider the random variables in Exercise 1, Section 8.1.1. Are X and Y independent random variables?

Exercise 2. Consider the random variables in Example 8.2.3. Are X and Y independent random variables?

Exercise 3. Prove that if two random variables X and Y are independent, then

$$P(X>x, Y>y)=P(X>x)P(Y>y).$$

8.4 Conditional density functions

A person will arrive at work between 9 and 10 o'clock in the morning. Sometime before 10 o'clock an important call must be placed. Assume that the time of arrival is uniformly distributed between 9:00 and 10:00, and assume that the time that the call is placed is uniformly distributed between the time of arrival and 10:00. What is the joint distribution of the time of arrival and the time the call is placed?

Let X denote the time of arrival and Y the time at which the call is placed. If we measure time in hours starting at 9:00, then X is uniformly distributed on the interval [0,1].

$$f(x)=1, \quad 0\le x\le 1.$$

But the information on Y is conditional on X. Given the information on X, Y is uniformly distributed on the interval [x,1].

(Page (1989))

$$f(y|x)=\frac{1}{1-x}, \quad x<y<1.$$

Definition 8.4.1

The conditional density functions of X and Y can be defined as follows:

$$f(x|y)=\frac{f(x,y)}{f(y)}$$

$$f(y|x)=\frac{f(x,y)}{f(x)}$$

The domain of these functions will depend on the relation between the domain of each other.

It follows from the definition of the conditional densities that

$$f(x,y)=f(x|y)f(y)$$

and

$$f(x,y)=f(y|x)f(x)$$

Thus, we may obtain the joint density from the conditional densities.

Therefore, the joint density, f(x,y) is

$$f(x,y) = f(y|x)f(x) = \frac{1}{1-x}, \quad 0 \le x \le y \le 1.$$

If two variables are independent, then, by definition, it is not hard to see that $f(x \mid y) = f(x)$ and $f(y \mid x) = f(y)$.

Conditional density functions allow us to make more interesting statements than we can make with just the marginal densities. That is, we can now specialize the probabilities to specific groups instead of offering generalized conclusions.

Example 8.4.1

$$f(x,y) = 2, \quad 0 \le x \le y \le 1.$$

$$f(x) = \int_x^1 2dy = 2(1-x), \quad 0 \le x \le 1.$$

$$f(y) = \int_0^y 2dx = 2y, \quad 0 \le y \le 1.$$

We can see that $f(x,y) \neq f(x)f(y)$ and therefore the two random variables are not independent. The conditional density functions are:

$$f(y|x) = \frac{f(x,y)}{f(x)} = \frac{2}{2(1-x)} = \frac{1}{(1-x)}, \quad x \le y \le 1,$$

$$f(x|y) = \frac{f(x,y)}{f(y)} = \frac{2}{2y} = \frac{1}{y}, \quad 0 \le x \le y.$$

There are an infinite number of continuous conditional densities $f(x \mid y)$, one for each value of Y in the real line. Similarly, there are an infinite amount of continuous conditional densities $f(y \mid x)$, one for each value of X in the real line.

Once we specialize a conditional density to the value of the other random variable, we can compute conditional expectations, conditional variances, conditional probabilities, conditional percentiles and so on.

A conditional density function is a univariate distribution.

Example 8.4.2

Let's see what we can do with the conditional densities of Example 8.4.1.

First, we specialize to a value of X. Suppose we are interested in the situation when $X = 0.5$.

$$f(y|X=0.5) = \frac{1}{(1-0.5)} = 2, \quad 0.5 \le y \le 1.$$

This is a univariate density function now. The computations that we are about to do are similar to those we did for univariate densities in chapter 7 and marginal densities in this chapter 8. But the names we will give them will be different.

The conditional expectation of Y when $X = 0.5$ is calculated using the conditional density of Y when $X = 0.5$:

$$E(Y|X = 0.5) = \int_{0.5}^{1} y2dy = 3/4,$$

$$E(Y^2|X = 0.5) = \int_{0.5}^{1} y^2 2dy = 7/12.$$

So the conditional variance of Y when $X = 0.5$ is

$$\sigma^2_{Y|X=0.5} = Var(Y|X = 0.5) = E(Y^2|X = 0.5) - (E(Y|X = 0.5))^2 = \frac{7}{12} - \left(\frac{3}{4}\right)^2 = 0.0208333.$$

We may also compute conditional probabilities. For example, the conditional probability that Y is larger than 0.6 given that X is 0.5 is

$$P(Y > 0.6\ |X = 0.5) = \int_{0.6}^{1} 2dy = 0.8.$$

8.4.1 Conditional densities when the variables are independent

Obviously, if two random variables are independent,

$$f(x|y) = \frac{f(x,y)}{f(y)} = \frac{f(x)f(y)}{f(y)} = f(x),$$

$$f(y|x) = \frac{f(x,y)}{f(x)} = \frac{f(x)f(y)}{f(x)} = f(y).$$

Example 8.4.3

If two random variables X and Y have joint density

$$f(x,y) = 15e^{-3x-5y}, \quad x \geq 0, y \geq 0$$

then

$$f(x|y) = f(x) = 3e^{-3x}, \quad x \geq 0$$

$$f(y|x) = f(y) = 5e^{-5y}, \quad y \geq 0$$

Thus

$$P(X < 2|Y = 4) = P(X < 2) = F_x(2) = 1 - e^{-6}$$

8.4.2 Exercises

Exercise 1. Teachers get courses assigned to teach each semester. For each instructor, there are the courses that the instructor can teach based on the skill set of the instructor, and there are courses that the teacher would rather teach all the time, closer to their specialization.

To be able to teach in any department, a teacher must be able to teach more than the favorite courses. Let X denote the proportion of teachers who teach the whole spectrum of courses taught in a department, and Y the proportion of teachers who teach the courses they specialize in. Let X and Y have the joint density function

$$f(x,y) = 2(x+y), \quad 0 < y < x < 1.$$

(i) Given that 10% of the teachers teach the whole spectrum of courses, what is the probability that fewer than 5% teach their favorite courses? (ii) What is the expected percentage of teachers teaching their favorite courses when the proportion teaching the whole spectrum is 0.7?

Exercise 2. In Section 8.4, we constructed the joint density of two random variables using the conditional and marginal density. For that problem, (i) what is the probability distribution of the time at which the call is placed? (ii) When is the expected time for the call to be placed?

8.5 Expectations of functions of two random variables

Most of the time, as it happened with univariate continuous random variables, we are not interested in the random variables per se, but on functions of them. For example, suppose we take two strength measurements on the same section of a cable and we are interested in the difference between the measurements. This may allow us to compare the two instruments used.

Definition 8.5.1

Let $g(X, Y)$ be a function of two continuous random variables. We define the expectation of this function as follows:

$$E(g(X,Y)) = \int_{-\infty}^{\infty}\int_{-\infty}^{\infty} g(x,y)f(x,y)dxdy,$$

where, as usual, the domain of integration will be just where the density is not 0. Similarly, the variance is defined as follows:

$$Var(g(X,Y)) = \int_{-\infty}^{\infty}\int_{-\infty}^{\infty} (g(x,y) - E(g(x,y)))^2 f(x,y)dxdy = E(g(x,y))^2 - E(g(x,y))^2$$

Example 8.5.1

Let's consider again the case of the gasoline of Example 8.2.2. A quantity of interest to a gas station is the difference between the amount stocked and the amount sold, because this allows the gas station to predict shortages or excess inventory. Let $D = X - Y$. Then,

$$E(D)=\int_0^1\int_0^x (x-y)3xdydx=\int_0^1\int_0^x x3xdydx-\int_0^1\int_0^x y3xdydx$$

$$=\int_0^1 x\int_0^x 3xdydx-\int_0^1 y\int_y^1 3xdxdy=\int_0^1 xf(x)dx-\int_0^1 yf(y)dy=E(X)-E(Y)=\frac{3}{4}-\frac{3}{8}=3/8.$$

Similarly, we can find the variance of D as follows:

$$Var(D)=\int_0^1\int_0^x (x-y-(\mu_x-\mu_y))^2 3xdydx$$

$$=\int_0^1\int_0^x [(x-\mu_x)^2+(y-\mu_y)^2-2(x-\mu_x)(y-\mu_y)]3xdydx$$

$$=Var(X)+Var(Y)-2Cov(X,Y)=0.0593.$$

Example 8.5.2

A very special function of two random variables that is very helpful in computing the correlation later on in the chapter is $g(X, Y) = XY$ (the product of the two random variables). In the case of example 8.2.2,

$$E(XY)=\int_0^1\int_0^x xy3xdydx=\frac{3}{10}$$

8.6 Covariance and correlation between two continuous random variables

As in the discrete bivariate case studied in Chapter 6,

$$Cov(X,Y)=E[(X-\mu_X)(Y-\mu_Y)]=E(XY)-E(X)E(Y),$$

and the correlation between X and Y is defined as

$$\rho_{(X,Y)}=\frac{Cov(X,Y)}{\sigma_X\sigma_Y},\quad -1\le\rho\le 1.$$

Example 8.6.1

Continuing with Examples 8.2.2, 8.5.1 and 8.5.2, we know that $E(X)=3/4$, $E(Y)=3/8$, $E(XY)=3/10$, $Var(X)=0.0375$, $Var(Y)=0.0594$. Thus

$$\rho=\frac{Cov(X,Y)}{\sigma_X\sigma_Y}=\frac{E(XY)-E(X)E(Y)}{\sigma_X\sigma_Y}=\frac{\left(\frac{3}{10}\right)-\left(\frac{3}{4}\right)\left(\frac{3}{8}\right)}{(\sqrt{0.0375})(\sqrt{0.0594})}=0.3972$$

The correlation between variables that we expect to be positively correlated sometimes is not perfectly one for a good reason. For example, a better score in calculus is highly correlated with a better score in Introduction to Probability. But performance in Probability depends on other factors such as attendance to class, motivation, hours of study, knowing what is going on in the class, and capability of modeling the context studied using mathematics, to name a few factors.

8.6.1 Properties of covariance

To study the properties of covariance we will make extensive use of the expectation operator. Consider for example the following linear functions of X and Y and of M and N:

$$W = a\ X + bY, \quad T = cM + d\ N.$$

By definition of covariance,

$$Cov(W,T) = E[(W - E(W))(T - E(T))].$$

All we have to do is first of all substitute, expand, simplify, put terms that make sense to put together inside the brackets.

Step 1.

$$\begin{aligned} Cov(W,T) &= E[(aX + bY - E(aX + bY))(cM + dN - E(cM + dN))] \\ &= E[(aX + bY - (aE(X) + bE(Y)))(cM + dN - (cE(M) + dE(N)))] \\ &= E[(aX - aE(X) + bY - bE(Y))(cM - cE(M) + dN - dE(N))] \\ &= E[(a(X - E(X)) + b(Y - E(Y)))(c(M - E(M) + d(N - E(N)))]. \end{aligned}$$

Now that we have simplified, we cross multiply the terms.

Step 2.

$$\begin{aligned} Cov(W,T) = E[&(a(X - E(X))c(M - E(M) + (a(X - E(X)d(N - E(N)) \\ &+ (b(Y - E(Y))c(M - E(M) + (b(Y - E(Y)d(N - E(N))]. \end{aligned}$$

And now we bring the expectation operator inside the brackets, term by term, and simplify a little bit further.

Step 3.

$$\begin{aligned} Cov(W,T) = acE[&(X - E(X)(M - E(M)] + adE[(X - E(X))(N - E(N))] \\ &+ bcE[(Y - E(X))(M - E(M))] + bdE[(Y - E(Y))(N - E(N))]. \end{aligned}$$

Step 4.
Now we recognize definitions.

$$Cov(W,T) = acCov(X,M) + adCov(X,N) + bcCov(Y,M) + bdCov(Y,N)$$

8.6.2 Exercises

Exercise 1. Let X and Y have joint density function

$$f(x,y) = kxy, \quad 0 \leq x \leq 1, \quad 0 \leq y \leq 1.$$

(i) Determine the value of k. (ii) Find the correlation between X and Y.

Exercise 2. What would be the covariance of the following two functions of the random variables X and Y?

$$W = a\ X + bY; \quad T = cX + d\ Y.$$

8.7 Expectation and variance of linear combinations of two continuous random variables

- If X is the time your friend arrives to a party and Y is the time you arrive, what would be the expected gap of time between your arrivals to a future party?

8.7.1 When the variables are not independent

As we did in Chapter 6, now that we know the concept of covariance of two continuous random variables, we can study a new important result concerning the variance of a linear combination of two continuous random variables. Consider the following function of two random variables:

$$g(X,Y) = aX + bY.$$

The expected value of this sum is

$$E(g(X,Y) = aE(X) + bE(Y).$$

Similarly, we can prove that

$$Var(aX + bY) = a^2Var(X) + b^2Var(Y) + 2abCov(X,Y).$$

8.7.2 When the variables are independent

When X and Y are independent, covariance is 0, and, therefore,

$$Var(aX + bY) = a^2Var(X) + b^2Var(Y).$$

Example 8.7.1

Let's revisit the sum of the rolls of two dice. What would be the expected sum and the variance of the sum obtained when rolling two fair six-sided dice?

$$E(X + Y) = E(X) + E(Y) = 3.5 + 3.5 = 7$$

$$Var(X + Y) = Var(X) + Var(Y) = 2.92 + 2.92 = 5.84$$

Thus, if two continuous random variables are independent, then the expected value of the sum is the sum of the expectations, as in the non-independent case, and the variance of the sum is the sum of the variances. Let X and Y be two independent continuous random variables. Then

$$E(X+Y)=\int_y\int_x (x+y)f(x)f(y)dxdy$$

$$=\int_y\int_x (x)f(x)f(y)dxdy+\int_y\int_x (y)f(x)f(y)dxdy$$

$$=\int_y f(y)\int_x (x)f(x)dxdy+\int_x f(x)\int_y (y)f(y)dydx$$

$$=\int_y f(y)\mu_x dy+\int_x f(x)\mu_y dx$$

$$=\mu_x\int_y f(y)dy+\mu_y\int_x f(x)dx$$

$$=\mu_x+\mu_y$$

Similar calculations allow us to show that the variance of the sum is the sum of the variances.

$$Var(X+Y)=\int_y\int_x (x+y-(\mu_x+\mu_y))^2 f(x)f(y)dxdy$$

$$=\int_y\int_x (x-\mu_x)^2 f(x)f(y)dxdy+\int_y\int_x (y-\mu_y)^2 f(x)f(y)dxdy+0$$

$$=\int_y f(y)\int_x (x-\mu_x)^2 f(x)dxdy+\int_x f(x)\int_y (y-\mu_y)^2 f(y)dydx$$

$$=\int_y f(y)\sigma_x^2 dy+\int_x f(x)\sigma_y^2 dx$$

$$=\sigma_x^2\int_y f(y)dy+\sigma_y^2\int_x f(x)dx$$

$$=\sigma_x^2+\sigma_y^2$$

8.7.3 Exercises

Exercise 1. What is $\int_y\int_x (x-\mu_x)(y-\mu_y)f(x)f(y)dxdy$ equal to?

Exercise 2. Let X denote the amount of gasoline stocked in a bulk tank at the beginning of a week, and let Y denote the amount sold during the week:

$$f(x,y)=3x, \quad 0\leq y\leq x\leq 1$$

The random variable $D=X-Y$ represents the amount left over at the end of the week. Find the mean and variance of D.

Exercise 3. Prove that when X and Y are two random variables that are not independent. Prove that

$$Var(aX+bY)=a^2Var(X)+b^2Var(Y)+2abCov(X,Y).$$

8.8 Joint distributions of independent continuous random variables: Applications in mathematical statistics

If X and Y are two independent continuous random variables, we have seen in Definition 8.3.1 that $f(x,y)=f(x)f(y)$. This condition generalizes to more than two random variables. If we have n independent random variables, $X_1,X_2,\ldots\ldots,X_n$, each of them with density $f(x_i)$, then the joint density of all these random variables is

$$f(x_1,x_2,\ldots\ldots,x_n)=f(x_1)f(x_2)\ldots\ldots f(x_3)$$

Example 8.8.1

The lifetime of a system component is a random variable X with density function

$$f(x)=\frac{1}{3}e^{-\frac{1}{3}x}, \quad x\geq 0.$$

A system contains 6 of these independent components in series. The density function of the joint lifetime of all components is

$$f(x_1,x_2,\ldots\ldots,x_n)=f(x_1)f(x_2)\ldots\ldots f(x_6)=\left(\frac{1}{3}\right)^6 e^{-\frac{1}{3}\sum_1^6 x_i}$$

Continuing with the example in the context of probability, i.e., the joint density and known parameter, we could compute the probability that the system works longer than 10 years. That will happen if all components last longer than 10 years.

$$P(X_1>10,\ldots\ldots,X_6>10)=\int_{10}^{\infty}\int_{10}^{\infty}\ldots\ldots\int_{10}^{\infty}\left(\frac{1}{3}\right)^6 e^{-\frac{1}{3}\sum_1^6 x_i}dx_1dx_2,\ldots\ldots dx_6$$

Box 8.1

Joint distributions and Maximum Likelihood in Statistics

This operation that we just did in Example 8.8.1 is the typical setup of problems of maximum likelihood estimation in Statistics. A problem of maximum likelihood consists in assuming that the n independent random variables that represent a random sample taken from the population have a joint density that can be obtained by multiplying their individual densities. The parameter of the individual densities is not known, though. Because of that, the expression for the joint density is called the *likelihood function*. Statisticians do maximum likelihood statistical inference by maximizing the likelihood function with respect to the unknown parameter. Mathematical statistics studies this method.

For example, a statistician would not know that the parameter of the exponential random variable in Example 8.8.1 is 1/3. Using the joint probability for $n = 6$ random variables, for example,

$$likelihood\ \ function = f(x_1, x_2, \ldots\ldots, x_n) = f(x_1)f(x_2)\ldots\ldots f(x_6) = (\theta)^6 e^{-\theta\sum_1^6 x_i}$$

the statistician would maximize the log likelihood

$$\log likelihood\ \ function = 6\log\theta - \theta\sum_1^6 x_i$$

with respect to θ to obtain an estimator for θ, namely

$$\hat{\theta} = \frac{6}{\sum_1^6 x_i}$$

Using the data values for the X's the statistician gets an estimate. On the other hand, the mathematical statistician uses properties of expectations to determine whether the expected value of that estimator in general, for any n, for any random sample, has good properties, i.e., gives good estimates of θ.

Because of independence, all those integrals factor out, and the result is just

$$P(X_1 > 10, \ldots\ldots, X_n > 10) = \left(\int_{10}^{\infty} \left(\frac{1}{3}\right)^6 e^{-\frac{1}{3}x} \right)^6$$

8.8.1 Exercises

Exercise 1. The length of life of a fuse, X, has density

$$f(x) = \frac{e^{-\frac{x}{\theta}}}{\theta}, \quad x > 0, \theta > 0.$$

Three such fuses operate independently. Find the joint density of their lengths of life, simplifying it as much as possible.

Exercise 2. If two random variables X and Y have joint density

$$f(x,y) = 15e^{-3x-3y}, \quad x \geq 0, y \geq 0,$$

is $P(X > 3, Y > 5) = F_x(3)F_y(5)$? Why? Why not?

8.9 The bivariate normal distribution

The bivariate normal density function, and in general the multivariate normal, are perhaps the most widely used distributions in statistics. In this section, we are going to describe the form of the joint, marginal and conditional distributions. Operating with these distributions is not different than operating with the generic ones we have been talking about in this chapter. But because conditionals and marginal are normal densities, you would have to use the same procedures we used with the normal density introduced in chapter 7.

The joint density of two bivariate normal random variables X and Y is

$$f(x,y) = \frac{1}{2\pi\sigma_x\sigma_y\sqrt{1-\rho^2}} exp\left\{-\frac{1}{2(1-\rho^2)}\left[\left(\frac{x-\mu_x}{\sigma_x}\right)^2 + \left(\frac{y-\mu_y}{\sigma_y}\right)^2 - 2\rho\left(\frac{x-\mu_x}{\sigma_x}\right)\left(\frac{y-\mu_y}{\sigma_y}\right)\right]\right\}$$

Notice that this density function has five parameters that we have been concerned about throughout our discussion in this chapter, namely: mean of X, mean of Y, standard deviation of X, standard deviation of Y, and correlation between X and Y. The difference between the normal and the other densities seen so far in this chapter is that all the parameters of the bivariate normal appear in the joint density formula, but that was not the case in the other examples we have seen.

The marginal densities of X and Y are both normally distributed

$$f(x) \sim N(\mu_x, \sigma_x), \quad f(y) \sim N(\mu_y, \sigma_y).$$

The conditional densities of X and Y are a little bit more complicated. They are normal density functions as well, but their expectations and variances consist of formulas that depend on the parameters:

$$f(x \mid y) \sim N\left(\mu_{x|y} = \mu_x + \rho\frac{\sigma_x}{\sigma_y}(y-\mu_y),\ \sigma_{x|y} = \sqrt{(1-\rho^2)\sigma_x^2}\right)$$

$$f(y \mid x) \sim N\left(\mu_{y|x} = \mu_y + \rho\frac{\sigma_y}{\sigma_x}(x-\mu_x),\ \sigma_{y|x} = \sqrt{(1-\rho^2)\sigma_y^2}\right)$$

Example 8.9.1

At a certain university, the joint probability density function of X and Y, the grade point averages of students in the first and last year at school, respectively, is bivariate normal. From the grades of past years, it is known that

$$\mu_x = 3, \quad \mu_y = 2.5, \quad \sigma_x = 0.5, \quad \sigma_y = 0.4, \quad \rho = 0.4$$

What is the probability that a student with grade point average equal to 3.5 in the first year earns a grade point average of at least 3.2 in the last year?

We want $P(Y \geq 3.2 \mid X = 3.5)$. This will require finding the z score for the corresponding normal random variable. We must first find what this normal random variable is. Following the formulas just given for the conditional densities,

$$f(y \mid x) \sim N\left(\mu_{y|x=3.5} = 2.5 + 0.4\frac{0.4}{0.5}(3.5-3) = 2.66,\ \sigma_{y|x} = \sqrt{(1-0.4^2)0.4^2} = 0.36666\right)$$

Thus

$$P(Y \geq 3.2 \mid X = 3.5) = P\left(Z > \frac{3.2-2.66}{0.36666}\right) = P(Z > 1.4727) = 0.0704.$$

8.9.1 Exercises

Exercise 1. A portfolio has two assets. The return in the two assets is bivariate normal. The return R_A on asset A has Expected value \$20 and standard deviation \$3. The return on asset B, R_B has expected value of \$15 and standard deviation of \$2. The correlation between the return of the two assets is 0.7. Let a = 0.3 denote the share of wealth invested in asset A and b = 0.7 the share of wealth invested in asset B. Then the portfolio return is $T = aR_A + bR_B$. (i) Find the expected T; (ii) Find the portfolio risk (standard deviation). (iii) Find the expected return for asset A, when asset B has return equal to 14; (iv) Find the probability that the return for asset A is larger than 16 when asset B has return equal to 14.

Exercise 2. Consider students at a university. Let X be their math SAT scores and Y their verbal SAT scores. Suppose a study reveals that

$$\mu_x = 600, \quad \sigma_x^2 = 64, \quad \rho = 0.6$$
$$\mu_y = 500, \quad \sigma_y^2 = 60.$$

What is the probability that a student with a verbal SAT score of 450 has a math SAT score larger than 650?

Exercise 3. Scientists studied the relationship between the length of the body of a bullfrog and how far it can jump. Mean body length is 149.64 mm and the standard deviation is 14.47 mm. The mean maximum jump is 103.99 cm and the standard deviation is 17.94 cm. The correlation between body length and maximum jump is 0.28. The two random variables follow a bivariate normal distribution. (i) What jump size should be expected in a bullfrog that is 140 mm long? (ii) What is the probability that jump size is larger than 100 in a randomly chosen bullfrog?

Exercise 4. Students in a school were asked to participate in a study of the effects of a new teaching method on reading skills of 10th graders. To determine the effectiveness of the new method, a reading test was given to each student before applying the new method (pre-test).

Another test was given to the same students after applying the new method (post-test). Experts consider that a good measure of improvement is given by the following formula:

$$\text{Improvement} = 20(\text{Post-test score}) - 10(\text{Pre-test score}).$$

An improvement > 300 is considered good. It is known that the average score in the pre-test is 40, average score in the post-test is 40, standard deviation in the post-test is 6, standard deviation in the pre-test is 5, and correlation between the scores in the two tests is 0.5. (i) What is the probability that a student with a pretest score of 38 has good improvement? (ii) What is the standard deviation of the improvement?

Exercise 5. A certain type of cable car has a maximum weight capacity X, with mean and standard deviation of 5000 and 300 pounds, respectively. In a touristic site at high elevation, the cable car loading Y has mean and standard deviation 4000 and 400 pounds, respectively. For any given time that the cable car is in use, find the probability that it will be overloaded, assuming that X and Y are independent and normally distributed. How does this problem differ from the other problems done in this Chapter? What is common to other problems done in this chapter?

Exercise 6. We have said in Chapter 6 and in this Chapter that if two random variables are independent, their covariance and therefore their correlation is 0. We have also said that the converse is not generally true. An exception is the multivariate normal density function. If two random variables are bivariate normal and their covariance is 0, then the random variables are independent. Prove this result. Hint: you may want to start with the formula we gave for the joint density function in Section 8.9.

Exercise 7. Let X denote the carapace length of common shrimp and let Y denote the postpinuou position of the carapace. It is known that $\mu_X = 249,\ \mu_Y = 177.53,\ \sigma_X = 6.7,\ \sigma_Y = 5.18,\ \rho = 0.83.$ (i) Compute the probability that carapace length is larger than 260 for shrimp with postpinuou position equal to 170. (ii) What is the probability that a randomly chosen shrimp has postpinuou position smaller than 180? (iii) What is the expected carapace length for shrimp with postpinuou position equal to 170?

8.10 Mini quiz

Question 1. When we talk about the joint density function of two random variables X, Y, ($f(x, y)$), for constants a and b,

$$P(X < a, Y > b)$$

is

a. An area
b. A volume
c. Always 1
d. 0

Question 2. Let X and Y be two random variables that are not independent. Var(3X–4Y) is equal to:

a. $3Var(X)+4Var(Y)-8Cov(3X,-4Y)$
b. $9Var(X)+16Var(Y)-24Cov(3X,-4Y)$
c. $9Var(X)+16Var(Y)-24Cov(X,Y)$
d. $3Var(X)+4Var(Y)-24Cov(X,Y)$

Question 3. The future lifetimes (in months) of two components of a machine have the following joint density function:

$$f(x,y)=\frac{6}{125000}(50-x-y), \qquad 0<x<50-y<50.$$

What is the probability that both components are still functioning 20 months from now?

a. $\int_0^{20}\frac{6}{125000}\int_0^{20}(50-x-y)\,dy\,dx$
b. $\int_{20}^{30}\frac{6}{125000}\int_{20}^{50-x-y}(50-x-y)\,dy\,dx$
c. $\int_{20}^{30}\frac{6}{125000}\int_{20}^{50-x}(50-x-y)\,dy\,dx$
d. $\int_{20}^{50}\frac{6}{125000}\int_{20}^{50-x-y}(50-x-y)\,dy\,dx$

Question 4. $Cov(2X+3Y-Z,5M)$ where X, Y, Z, M are random variables, is equal to

a. $Cov(X,M)+10\,Cov(Y,Z)-15\,Cov(M,Y)$
b. $Cov(X,M)+Cov(Z,M)+Cov(Y,X)$
c. $10\,Cov(X,M)+15\,Cov(Y,M)-5Cov(Z,M)$
d. $2Cov(X,M)+3\,Cov(Y,M)-Cov(Z,M)$

Question 5. If $g(X,Y)=(X-\mu_X)(Y-\mu_Y)$ and X and Y have joint density $f(x, y)$, what will the following calculation give us?

$$\int\int(x-\mu_X)(y-\mu_Y)f(x,y)dxdy$$

a. $Cov(g(X,Y))$
b. $Var(XY)$
c. $Cov(X,Y)$
d. $Var(X)Var(Y)$

Question 6. Let

$$f(x,y)=2, \quad 0 \le y \le x \le 1.$$

Which of the following is the marginal probability density function of X?

a. $2x, 0 < x < 1$
b. $2xy, 0 < x < 1$
c. $2-x, 0 < x < 1$
d. $2x^2, 0 < x < 1$

Question 7. Consider the joint density function

$$f(x,y)=6x^2y, \quad 0 \le x \le 1, 0 \le y \le 1$$

$E(XY)$ equals

a. 3/12
b. 4/9
c. 1/2
d. 7/8

Question 8. Consider the joint density function

$$f(x,y)=6x^2y, \quad 0 \le x \le 1, 0 \le y \le 1$$

The correlation between X and Y is

a. 1
b. 0.7
c. 0.6
d. 0

Question 9. Consider the following conditional density function extracted from a joint bivariate normal density function. A student thinks that there is a mistake, that the two variables were uncorrelated. If that is the case, and everything else is correct, what is the new $\mu_{y|x=3.5}$ and $\sigma_{y|x}$?

$$f(y \mid x) \sim N\left(\mu_{y|x=3.5} = 2.5 + 0.4\frac{0.4}{0.5}(3.5-3) = 2.66, \quad \sigma_{y|x} = \sqrt{(1-0.4^2)0.4^2} = 0.36666\right)$$

a. A univariate normal density with mean 0 and standard deviation 1
b. A univariate normal density with mean 2.5 and standard deviation 0.4
c. A bivariate normal density with mean of X equal 3.5 and mean of Y equal 0.4
d. A bivariate normal density with variance 0.4 and mean 0

Question 10. (This question is from Devore (2004, 208).) A bank operates both a drive-up facility and a walk-up window. On a randomly selected day, let X = the proportion of time that the drive-up facility is in use (at least one customer is being served or waiting to be served) and Y = the proportion of time that the walk-up window is in use. Suppose the joint probability density function of X and Y is given by

$$f(x,y) = \frac{6}{5}(x + y^2), \quad 0 \leq x \leq 1, \quad 0 \leq y \leq 1.$$

The probability that neither facility is busy more than one-quarter of the time is

a. 0.67
b. 0.0109
c. 0.0004
d. 0.11

8.11 R code

Gibbs sampling is a form of Markov Chain Monte Carlo simulation that consists of drawing random numbers from conditional distributions repeatedly in order to obtain the marginal distributions. It is usually used in Bayesian statistics to do inference about unknown parameters, but it can be used with any random variable.

Below you will find the code to generate random draws from two conditional density functions. At the end you will see the marginal empirical distribution of each of the variables.

```
####### Gibbs sampling #############
theta1=rep(0,5000)
theta2=rep(0,5000)
theta2[1]=2.5
theta1[1]= rnorm(1,0.8*theta2[1], sqrt(0.36))
for(i in 2:5000){
 theta2[i]  = rnorm(1,0.8*theta1[i-1], sqrt(0.36))
 theta1[i]=rnorm(1,0.8*theta2[i],sqrt(0.36))
 }
######## Summarize the marginals #######
#### after removing first 500 draws ########

 ###### marginal for theta1 ########
hist(theta1[500:5000])
summary(theta1[500:5000])
sd(theta1[500:5000])
######## marginal for theta2 ########
```

```
hist(theta2[500:5000])
summary(theta2[500:5000])
sd(theta2[500:5000])
####### plot the traces of the markov process #####
plot(theta1[500:5000], type="l", lty=1)
lines(theta2[500:5000],type="l",lty=2)
######## joint posterior for theta1 theta2 ######
plot(theta1[500:5000],theta2[500:5000],type="l") # line plot to see
the traces
plot(theta1[500:5000],theta2[500:5000], type="pt") # point plot
#######################################
```

8.12 Chapter Exercises

Exercise 1. Stores A and B, which belong to the same owner, are located in two different towns. If the probability density function of the weekly profit of each store, in thousands of dollars, is given by

$$f(x) = \frac{x}{4}, \quad 1 \le x \le 3,$$

$$f(y) = \frac{y}{4}, \quad 1 \le y \le 3,$$

and the profit of one store is independent of the other, what is the probability that next week one store makes at least \$500 more than the other store?

Exercise 2. Mary and Antonio plan an eight-hour hike in the mountains. Let X and Y denote the time it takes, respectively, for Mary and Antonio to arrive at the top of the mountain. Assume that Mary always arrives first, and that X and Y are uniform on the region $0 \le X \le Y \le 8$. The joint density is:

$$f(x,y) = \frac{1}{32}, \quad 0 \le x \le y \le 8.$$

What is the expected value of $Y - X$, which is the time interval between Mary and Antonio arriving to the top?

Exercise 3. Consider the following joint density

$$f(x,y) = \lambda^2 e^{-\lambda y}, \quad 0 \le x \le y \le \infty.$$

(i) Find the marginal density function of X and determine what family it is from, if one of the important densities. (ii) Find the conditional density of Y when $X = 5$.

Exercise 4. Prove, using integration, that if X and Y are two random variables,

$$Var(aX+bY)=a^2Var(X)+b^2Var(Y)+2abCov(X,Y).$$

Exercise 5. Prove, using integration, that if X and Y are two random variables that are independent,

$$E(XY)=E(X)E(Y).$$

Exercise 6. Prove, using integration, that if X and Y are any two random variables for which the expectation exists,

$$E(aX+bY)=aE(X)+bE(Y).$$

Exercise 7. The length (X) of the leaf of a special kind of flower and the width (Y) of the leaf are bivariate normally distributed with

$$\mu_x=3\ cm,\quad \sigma_x^2=0.1,\quad \rho=0.6$$
$$\mu_y=1\ ccm,\quad \sigma_y^2=0.3,$$

(i) What is the $E(Y\mid X=1/2)$ equal to? (ii) Complete what is missing in the parenthesis containing the letters A, B, and C in the formula for the conditional density of width given length $=1/2$

$$f(y|x=0.5)=\frac{1}{\sqrt{2\pi}(A)}exp\left\{-\frac{[y-(B)]^2}{2(C)}\right\}$$

Exercise 8. Suppose X_1, X_2, and X_3 are three independent and identically distributed normal random variables. Write the joint density of these random variables, simplifying it as much as possible.

Exercise 9. (This problem is from Degroot (1975, 178).) Suppose that X and Y are random variables such that $Var(X)=9$, $Var(Y)=4$, and correlation between X and Y is $-1/6$. (i) Determine $Var(X+Y)$; (ii) determine $Var(X-3Y+4)$.

Exercise 10. Suppose X and Y are jointly uniform distributed on the square delimited by $0\le x\le 1, 0\le y\le 1$. What is the probability of the event $X>Y$?

Exercise 11. (This exercise is from Page (1989, 134).) In a certain device, X represents the voltage drop across a component and Y represents the current in another part of the device. The two are related probabilistically, however. Let's suppose that X and Y have a joint density function

$$f(x,y)=c$$

in the part of the first quadrant bounded by the curve $y = 4 - x^2$ and the x axis, with f(x, y) = 0 outside this region.

(i) Find the value c must have to make this a joint density
(ii) Find the marginal densities of X and Y.
(iii) What is the probability that 3X is greater than Y?

Exercise 12. (This exercise is from Pitman (1993, 351).) Suppose X and Y are jointly uniformly distributed in the region $0 < x < y < 1$. (i) Find the joint density of X and Y. (ii) Find the marginal densities of X and Y. (iii) Are X and Y independent?

Exercise 13. (This exercise is from Grami (2016, chapter 4).) Suppose that two random variables X and Y are independent and we have $g(X, Y) = f(X)h(Y)$. Show that $E(g(X, Y)) = E(f(x))E(h(y))$.

8.13 Chapter References

Degroot, Morris H. 1975. *Probability and Statistics*. Addison-Wesley Publishing Company.

Devore, Jay L. 2004. *Probability and Statistics for Engineering and the Sciences*. Sixth Edition. Thomson Brooks/Cole.

Grami, Ali. 2016. *Introduction to Digital Communications*. Elsevier.

Khilyuk, Leonid F., George V. Chillingar, and Herman H. Rieke. 2005. *Probability in Petroleum and Environmental Engineering*. Elsevier.

Page, Lavon B. 1989. *Probability for Engineering with Applications to Reliability*. Computer Science Press.

Pitman, Jim. 1993. *Probability*. Springer Texts in Statistics.

Scheaffer, Richard L. 1995. *Introduction to Probability and Its Applications*. Duxbury Press.

Society of Actuaries/Casualty Actuarial Society. 2007. *Exam P Probability*. Exam P Sample Questions. P-09-07. https://www.soa.org/education/exam-req/edu-exam-p-detail.aspx

Chapter 9

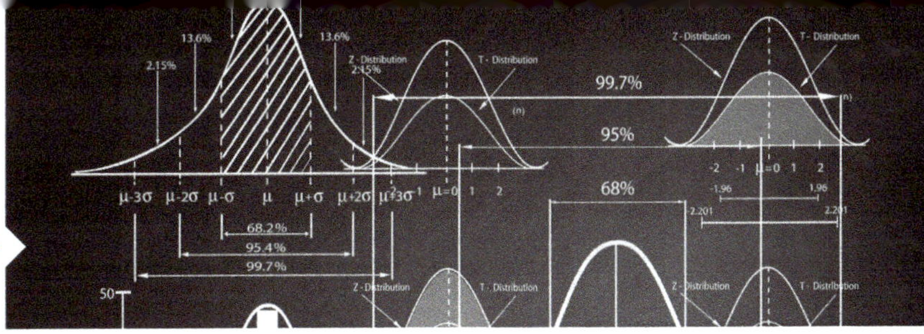

Some Theorems of Probability and Their Application in Statistics

- How do insurance providers determine their rates? Why do insurance rates vary from one individual to another?
- How do statisticians determine whether their tests of hypotheses about populations of interest provide reliable evidence to support their claims?

9.1 Bounds for probability when only μ is known. Markov bounds

Until now, probabilities have been arrived at by assuming that the random variables followed a particular distribution, or else the probabilities were given to us. But how could we find probabilities if none of these conditions are satisfied? There are theorems that allow us to find approximations to the probabilities for a single random variable. We study these theorems here not only because they allow us to approximate probabilities but also because they are very valuable when doing proofs involving several random variables.

Markov's inequality shows how to obtain probability bounds when the only thing we know about the random variable is its expected value, and nothing else.

Example 9.1.1

The average number of fallen trees in a week at the Green Tree National Forest is two. What is the probability that three or more trees will fall next week?

Theorem 9.1.1 Markov's inequality

Let X be a nonnegative random variable with expected value $\mu = E(X)$ and let a be a constant. Then

$$P(X \geq a) \leq \frac{E(X)}{a}$$

gives the tail bounds of a nonnegative random variable when all we know is its expectation. The larger the variance, the more accurate this bound is. For the complement event, the theorem implies that

$$P(X < a) > 1 - \frac{E(X)}{a}$$

By its very nature, Markov's inequality is conservative. A way to check this claim is to compare what you obtain with the bound and what you would obtain if you knew the true distribution of the random variable. Of course, you would not want to use Markov's bound if you know the probability distribution of the random variable; we are using this suggestion to convince ourselves that Markov's inequality works.

As the reader can see, we know only that the average is two. Let X be the number of trees falling per week. According to Markov's theorem,

$$P(X \geq 3) \leq \frac{2}{3} = 0.66666$$

Thus the probability that more than three trees will fall per week is at most 0.66666.

Suppose now that the actual distribution of X is Poisson with mean 2. Then, according to the Poisson,

$$P(X \geq 3) = 1 - P(X < 3) = 1 - \left[\frac{2^0 e^{-2}}{0!} + \frac{2^1 e^{-2}}{1!} + \frac{3e^{-2}}{3!}\right] = 0.323 < 0.66666$$

We can see that the probability bound proposed by Markov's theorem is twice as large as that obtained exactly using the Poisson, but Markov is not wrong: 0.323 is indeed smaller than 2/3 (at most 0.66666 means less than 0.66666). Markov's does not tell us exactly how much. It just gives us a point of reference.

9.1.1 Exercises

Exercise 1. The time it takes a pedestrian to react to the change of a traffic light from red to green is expected to be 10 seconds. What is the probability that the next pedestrian waiting for the green light will take more than 12 seconds? Calculate your answer first by assuming that you only know the expected value. Then calculate what the probability would be if it is known that the reaction time is exponentially distributed with the same expected value of 10 seconds. Compare your answers.

Exercise 2. The average income per capita in Madagascar is about \$400 per year. Find an upper bound for the percentage of families with incomes over \$1,000.

Exercise 3. A retailer advertises that its average delivery time is 10 days. Estimate the probability that the response time is larger than 20 days. What would be that probability if delivery times were exponentially distributed, with the same expected value?

9.2 Chebyshev's theorem and its applications. Bounds for probability when μ and σ known

Chebyshev's inequality shows how to obtain probability bounds when the only things we know about the random variable are its expected value and the standard deviation, and nothing else. The interval produced by Chebyshev's inequality is a generalization of Markov's. Like Markov's it is a conservative bound.

Theorem 9.2.1

Let X be any random variable with expected value $\mu = E(X)$ and let $k>0$ be a constant. Then

$$P\{|X-\mu|<k\sigma\} = P(\mu-k\sigma < X < \mu+k\sigma) \geq 1-\frac{1}{k^2}$$

Which implies, using the complement rule, that

$$P\{|X-\mu|\geq k\sigma\} = P(X<\mu-k\sigma \text{ or } X>\mu+k\sigma) \leq \frac{1}{k^2}$$

Another version of the theorem is

$$P\{|X-\mu|<k\} = P(\mu-k < X < \mu+k) \geq 1-\frac{\sigma^2}{k^2}$$

Which implies, using the complement rule, that

$$P\{|X-\mu|\geq k\} = P(X<\mu-k \text{ or } X>\mu+k) \leq \frac{\sigma^2}{k^2}$$

Proof

The proof of the theorem uses Markov's inequality, by making a in Theorem 9.1.1 equal to k and our random variable is now $(X-\mu)^2$.

$$P\{(X-\mu)^2 \geq k^2\} \leq \frac{E(X-\mu)^2}{k^2}$$

But $(X-\mu)^2 \geq k^2$ if $|X-\mu| \geq k$. So

$$P\{|X-\mu|\geq k\} \leq \frac{\sigma^2}{k^2}$$

Which implies that

$$P\{|X-\mu|<k\}\leq 1-\frac{\sigma^2}{k^2}$$

And similarly, for the other version,

$$P\{|X-\mu|\geq \sigma k\}\leq \frac{1}{k^2}$$

And

$$P\{|X-\mu|<\sigma k\}\leq 1-\frac{1}{k^2}$$

Example 9.2.1

The daily consumption of carbohydrates by a healthy individual eating a healthy diet in a given community averages 225 grams, with a standard deviation of 10 grams. (i) What can be said about the fraction of individuals for which the carbohydrate intake falls between 205 and 245 grams?. (ii) Find the shortest interval about the mean certain to contain at least 90% of the individual daily carbohydrate intakes.

(i) The interval from 205 to 245 represents $\mu-2\sigma$ to $\mu+2\sigma$ with $\mu=225$ and $\sigma=10$. Thus k = 2 and $1-\frac{1}{k^2}=1-\frac{1}{4}=3/4$. We conclude that at least 75% of all individuals have a carbohydrate intake between 205 and 245 grams. So at most, 25% fall outside that interval.

(ii) To find k, we must set $1-\frac{1}{k^2}=0.9$. Then $\frac{1}{k^2}=0.1$; $k^2=10$. Which implies that

$$k=\sqrt{10}=3.16$$

The interval is $\mu-3.16\sigma$ to $\mu+3.16\sigma$ or $225-3.16(10)$ to $225+3.16(10)$ or (193.4, 256.6). We say that at least 90% of individuals in this community eat between 193.4 and 256.4 grams of carbs. Which implies that at most 10% eat beyond that bound.

9.2.1 Exercises

Exercise 1. The average income per capita in Madagascar is about \$400 and the standard deviation of incomes is \$400. Find the shortest interval about the mean certain to contain at least 90% of the individual incomes.

Exercise 2. The number of machines used in a gym by the gym members during the peak hour is closely monitored by the gym management, since it is critical to the efficient operation of the gym. The number of machines in use averages 20 during peak hour, with a standard deviation or 2. (i) Find an interval that includes at least 90% of the peak-hour figures for the number of machines in use. (ii) In the advertisement of the gym, the management promises that there will always be at least two machines available for a member in any peak hour. Is the director safe in making this claim?

Exercise 3. Average monthly precipitation in Mumbai, India, is 241 mm and the standard deviation is 280 mm. Find an interval that will contain 75% of the monthly rain values in Mumbai.

9.3 The weak law of large numbers and its applications

- When Gallup Polls (https://news.gallup.com) tell us that 40% of the population approve of a particular candidate for the presidency of the United States, how accurate is that? After all, they ask only a small group of people, usually no more than 5,000 people.

Let's go back to Venn's quote on cows in chapter 1, section 1. What does that quote have to do with insurance?

When a person buys car insurance, insurance companies do not know how risky the person is. There is no way for insurance companies to know whether people are accident prone or not. But they have past data on drivers with the age of the insured, where they live, doing similar things they do and with similar automobile. They know that for that large group the probability of accident is, say, p. According to that, they determine the insurance premium for that person. However, insuring a single person or a small number of individuals like that person only would not guarantee anything to the insurance company. In order to predict how much the company will lose or gain, the insurance company needs to have a lot of policy holders. The rate p will be true for many, but not for a few only. If that p is stable, the company can predict its expected losses or gains. With a few policy holders the proportion that have an accident is hard to predict, as if fluctuates depending on the number of policy holders.

The insurance company is using the law of large numbers twice: (i) to set p and decide what the individual policyholder's premium will be (using information on many individuals in the population) and (ii) to reduce risk exposure for the company (by issuing many automobile policies). However, the law of large numbers is rendered less effective when risk-bearing policyholders are dependent of one another. This is most easily seen in the health and fire insurance industries, because diseases and fire can spread from one policy holder to another if not properly contained. This problem is known as contagion. To read more about this issue, go to Behind the Law of Large Numbers in the Insurance Industry | Investopedia https://www.investopedia.com/articles/personal-finance/081616/behind-law-large-numbers-insurance-industry.asp#ixzz5N275G4zj

As in insurance, the laws of probability are basic to the understanding of software engineering and management.

The problem of insurance companies is no different than wanting to find out whether a die is fair or not. If it is fair, repeated rolling of the die a zillion times should give us 1/6 of the tosses or close with the number 5 in them. We call that an empirical probability, because it varies depending on the number of tosses we make. But as the number of tosses increases, the empirical probability gets more stable. It is that number towards which the relative frequency tends to go that we consider to be the true probability of getting a 5 in the roll of the die. That number would tell us whether the die follows the model of a fair die that we have in our minds.

The law of large numbers says that the process just described is very likely to give us the true probability that way. In words, the law says that, given an event A, the probability that a relative frequency of event A occurring converges to the true probability of (A) goes to 1 as the number of trials increases.

Box 9.1

The relevance of the law of large numbers

Statisticians and other data scientists have talked about the law of large numbers as follows:

> *The fundamental empirical fact upon which are based all applications of the theory of probability. (Parzen 1960)*

> *It is a striking fact that we can start with a random experiment about which little can be predicted and, by taking averages, obtain an experiment in which the outcome can be predicted with high degree of certainty. (Grinstead and Snell 1997)*

Theorem 9.3.1

Let n be the number of trials of an experiment. Let p be the true probability of an event in each trial and let $\hat{p}_n$ be the empirical frequency of the event up to trial n. Then the law says that for any $\varepsilon > 0$

$$P(|\hat{p}_n - p| > \varepsilon) \underset{n\to\infty}{\longrightarrow} 0$$

or,

$$P(|\hat{p}_n - p| < \varepsilon) \underset{n\to\infty}{\longrightarrow} 1$$

To prove this law, we will use Chebyshev's theorem.

$$P(|\hat{p}_n - p| > \varepsilon) \leq \frac{Var(\hat{p}_n)}{\varepsilon^2} = \frac{p(1-p)}{n\varepsilon^2} \underset{n\to\infty}{\longrightarrow} 0$$

This theorem is saying that $\hat{p}_n$ converges in probability to the true p.

The theorem applies to any random variable and is more generally stated for any average quantity as follows:

Theorem 9.3.2

Let $X_1, X_2, \ldots, X_n$ be a sequence of independent random variables with finite expected value $\mu = E(X_j)$ and finite variance $\sigma^2 = V(X_j)$. Let $S_n = X_1 + X_2 + \ldots + X_n$. Then for any $\varepsilon > 0$,

$$P\left(\left|\frac{S_n}{n} - \mu\right| > \varepsilon\right) \underset{n\to\infty}{\longrightarrow} 0$$

Put another way,

$$P\left(\left|\frac{S_n}{n} - \mu\right| < \varepsilon\right) \underset{n\to\infty}{\longrightarrow} 1$$

And, again, Chebyshev's theorem can be used to prove this version of the theorem. We will leave the proof as an exercise.

It is the version of the law of large numbers in Theorem 9.3.2 that makes people sometimes call the law of large numbers the "Law of Averages."

Example 9.3.1

Let's toss a fair coin n times and let S_n be the number of heads in the n tosses. Then $\hat{p} = \frac{S_n}{n}$ represents the fraction of times heads appear in the n tosses. The law of large numbers predicts that the outcome for $\hat{p}$ will be near 1/2 for large n.

Example 9.3.2

Let's now consider n rolls of a fair six-sided die and let X_j denote the outcome of the jth roll. The $E(X_j) = 7/2$. Let $S_n = X_1 + X_2 + \ldots + X_n$ be the sum of the first n rolls. Then for any $\varepsilon > 0$,

$$P\left(\left|\frac{S_n}{n} - \frac{7}{2}\right| \geq \varepsilon\right) \underset{n\to\infty}{\longrightarrow} 0$$

Put another way,

$$P\left(\left|\frac{S_n}{n} - \frac{7}{2}\right| < \varepsilon\right) \underset{n\to\infty}{\longrightarrow} 1$$

9.3.1 Monte Carlo integration

Suppose we want to do the following integral of a mathematical function $g(x)$:

$$I(f) = \int_0^1 g(x)dx$$

A Monte Carlo approach to doing this mathematical integral consists of the following steps:

- Draw n random numbers from a uniform distribution defined on the interval [0,1]. Denote these numbers by $x_1, x_2, \ldots, x_n$
- Compute $\widehat{I(f)} = \frac{1}{n}\sum_{i=1}^{n} g(x_i) \approx E[g(X)]$ *by law of large numbers when n is large*
- Realize that $E[g(X)] = \int_0^1 g(x)f(x)\,dx = \int_0^1 g(x)\,1\,dx = \int_0^1 g(x)\,1\,dx$

Example 9.3.3

Evaluate the following integral

$$\int_0^1 \cos(2\pi x^2)dx.$$

using Monte Carlo integration. Compare with the exact answer. We used R to do the integration. See the R section to see how as we increase n, the solution approaches 0.244127. You can check yourself by going to Wolfram Alpha (www.wolframalpha.com) and typing: *integrate cos(2pix^2), from 0 to 1*

9.3.2 Exercises

Exercise 1. Prove the following result using Chebyshev's theorem?

$$P(|\bar{X} - \mu| > \varepsilon) \underset{n\to\infty}{\rightarrow} 0$$

Exercise 2. Jaron Lanier, author of the book "Ten Arguments for Deleting your Social Media Accounts Right Now," writes the following:

> *Behavior modification, especially the modern kind implemented with gadgets like smartphones, is a statistical effect, meaning it's real but not comprehensively reliable; over a population, the effect is more or less predictable, but for each individual it's impossible to say. To a degree, you are an animal in a behaviorist's experimental cage. But the fact that something is fuzzy or approximate does not make it unreal.* (Lanier 2018, p.11)

What does the author mean by statistical effect? What does this comment have in common with the observations we made about insurance companies at the beginning of Section 9.3?

Exercise 3. As pointed out in Grinstead and Snell (1997, 308), although Chebyshev's inequality proves the law of large numbers, it is a crude inequality for the probabilities involved. This problem is based on a problem in their book. Let $X_1, X_2, \ldots, X_n$ be a set of independent and identically distributed uniform random variables defined in the interval [0,1]. Assume $\varepsilon = 0.1$. How large must n be for the $P(|\bar{X} - \mu| > \varepsilon)$ to approach 0?

Exercise 4. The law of large numbers is a probability statement, it tells us that we can be more and more certain of being close to the value of $E(X)$ if we compute the average or proportion of many observations. Where there is probability statements there is a distribution. What distribution is the law talking about and what happens to that distribution as n increases?

Exercise 5. There are two options: (a) You roll a die 100 times, if the number is less than four, you win one dollar, if the number is larger than three, you lose one dollar.

(b) You draw 100 times at random with replacement from a box containing a ticket worth \$1 and another ticket worth $-\$1$.

Which option is better? Or are they the same?

Exercise 6. (This example is from Freedman, Pisani and Purves (1998).) Basketball players who make several baskets in succession are described as having a "hot had." Fans and players have long believed in the hot hand phenomenon, which refutes the assumption that each shot is independent of the next. What other explanation is possible for the hot hand?

9.4 Sums of many random variables

As we have mentioned several times throughout the book, applications of probability often call for the use of a random variable that is itself the sum or a linear combination of other random variables. For example,

- The study of downtimes of computer systems might require knowledge of the sum of the random downtimes each hour of the day. The downtime of each hour is a random variable and the sum of the independent downtimes of each of the 24 hours of the day is a sum of random variables giving us the total downtime per day.
- The random total cost of a building project can be studied as the sum of the random costs for the major independent components of the project.
- The random size of an animal population can be modeled as the sum of the random sizes of the independent colonies within the population.
- At the end of the summer the total weight of seeds accumulated by a nest of seed-gathering ants will vary from nest to nest. We may be interested in the sum of the total weights of seeds of all nests.
- The total weight of people riding an elevator is important to know to prevent overloading the particular elevator.
- An insurance company may want to know the total yearly claim by all the automobile policy holders.

We have talked about sums of two discrete random variables in Chapter 6, and sums of two continuous random variables in Chapter 8. In those chapters, we proved several results regarding the expected value and the variance of the sum of two random variables. We showed in those chapters that

$$E(X+Y)=E(X)+E(Y).$$

For variables that are not statistically independent random variables X, Y,

$$Var(aX + bY) = a^2Var(X) + b^2Var(Y) + 2abCov(X,Y).$$

When X and Y are independent,

$$Var(aX + bY) = a^2Var(X) + b^2Var(Y).$$

In particular, if $a = b = 1$, and X and Y are independent,

$$E(X + Y) = E(X) + E(Y),$$
$$Var(X + Y) = Var(X) + Var(Y).$$

We also mentioned in Chapters 6 and 8 that those results extend to the sum of more than two random variables although we did not prove it. The proof is beyond the scope of this book, as it requires using joint distributions of more than two variables. Let's denote the sum of n random independent random variables by

$$S_n = X_1 + X_2 + \ldots.. + X_n = \sum_{i=1}^{n} X_i,$$

where S is for sum and n for how many random variables we are adding. Then, without proving it, we claim that

$$E(S_n) = E\left(\sum_{i=1}^{n} X_i\right) = \sum_{i=1}^{n} E(X_i) = \mu + \mu + \ldots\ldots.. + \mu = n\mu$$

$$Var(S_n) = Var\left(\sum_{i=1}^{n} X_i\right) = \sum_{i=1}^{n} Var(X_i) = \sigma^2 + \sigma^2 + \ldots\ldots.. + \sigma^2 = n\sigma^2$$

Sums of independent and identically distributed random variables is of central importance in both Probability and Statistics.

Example 9.4.1

A fair six-sided die is rolled 100 times. Calculate the expected sum of the 100 rolls and the variance of the sum of the 100 rolls.

$$E(S_{100}) = E\left(\sum_{i=1}^{100} X_i\right) = \sum_{i=1}^{100} E(X_i) = \mu + \mu + \ldots\ldots.. + \mu = n\mu = 100\left(\frac{7}{2}\right) = 350,$$

$$Var(S_{100}) = Var\left(\sum_{i=1}^{100} X_i\right) = \sum_{i=1}^{100} Var(X_i) = \sigma^2 + \sigma^2 + \ldots\ldots.. + \sigma^2 = 100(2.916667) = 291.6667.$$

Example 9.4.2

Three flour mills receive raw corn in bulk. The amount of corn that one mill can process in one day can be modeled as having an exponential distribution with a mean of 4 tons for each of the three mills. If the three mills are independent, what is the expected value and variance of the total amount of corn processed by the three mills together, respectively?

$$E(S_3) = E\left(\sum_{i=1}^{3} X_i\right) = \sum_{i=1}^{3} E(X_i) = \mu + \mu + \mu = 3\mu = 3(4) = 12$$

$$Var(S_3) = Var\left(\sum_{i=1}^{3} X_i\right) = \sum_{i=1}^{3} Var(X_i) = \sigma^2 + \sigma^2 + \sigma^2 = 3(16) = 48$$

Example 9.4.3

Random variable X denotes the number of classes that a typical student at College Bliss has in a given Monday. The probability mass function of X is given below:

x	0	1	2
$P(X = x)$	¼	½	¼

We find the expected value of X and the variance of X to be

$$\mu_X = 1, \quad \sigma^2 = 0.5.$$

List all possible values of the sum $X_1 + X_2$, where the two random variables have the probability mass function given above.

9.4.1 Exercises

Exercise 1. A machine has n identical components each of which has an exponentially distributed lifetime T with expected value 10. What is the Expected value and variance of the total lifetime of all n machines?

Exercise 2. The number of patients entering a randomly chosen hospital in a city is a Poisson random variable with expected value 10 per day. If there are 25 similar hospitals in town, what is the expected value and variance of the total number of patients entering hospitals in a given day (assuming all hospitals are independent)?

9.5 Central Limit Theorem: The densify function of a sum of many independent random variables

This section goes beyond the results seen in section 9.4 and introduces the Central Limit Theorem (CLT) which concerns the density function of the sum of several independent random variables.

Box 9.2

CLT and its implications

In the limit, as n goes to infinity, the sum S_n of the random behavior of many individuals, regardless of the distribution of the individual behavior, approaches the Gaussian distribution. By Chapter 7, functions of Gaussian random variables are Gaussian. Thus, in the limit, $\bar{X} = \frac{S_n}{n}$ is Gaussian, $aS_n + b$, where a, b are constants is Gaussian, $S_n + S_m$ is Gaussian.

The CLT is at the core of most inference methods that statisticians apply to their data to learn about populations, and is another indication of why the Gaussian distribution plays such a prominent role in Statistics and data science.

Theorem 9.5.1

New in this section is the result that if a random variable S_n is itself the sum of a large number of independent and identically distributed random variables which may individually have distributions of finite variance that are quite different from the normal distribution, i.e. $S_n = X_1 + X_2 + \ldots.. + X_n = \sum_{i=1}^{n} X_i$, then for each fixed value of z, as n tends to infinity,

$$P\left(\frac{S_n - n\mu}{\sigma\sqrt{n}} > z\right)$$

approaches the probability that the standard normal random variable Z exceeds z. This result is known as Central Limit Theorem (CLT). In practical terms, this means that if n is large we may use the standard Gaussian density to approximate the answer to probability questions about sums of independent and identically distributed random variables even if we do not know what the distribution of X is.

Example 9.5.1

Suppose that the number of automobiles per single family housing unit, X, can be modeled by a Poisson probability mass function with expected value $E(X) = \lambda = 3$. If there are $n = 100$ independent single family housing units in a town, the number of automobiles in each of these units is a set of identically distributed Poisson random variables, $X_1, X_2, \ldots\ldots\ldots X_n$ with expected value $E(X_i) = \lambda = 3$, $i = 1, \ldots., n$. If we were asked what is the probability that the total number of automobiles in town is larger than 400, we could use the normal curve (by a consequence of the central limit theorem) to find that probability.

We first find the expected value and the variance of the sum.

$$E(S_n) = E\left(\sum_{i=1}^{n} X_i\right) = \sum_{i=1}^{n} E(X_i) = \mu + \mu + \ldots\ldots. + \mu = n\mu = 100(3) = 300,$$

$$Var(S_n) = Var\left(\sum_{i=1}^{n} X_i\right) = \sum_{i=1}^{n} Var(X_i) = \sigma^2 + \sigma^2 + \dots\dots + \sigma^2 = n\sigma^2 = 100(3) = 300.$$

By the CLT,

$$P(S_n > 400) = P\left(\sum_{i=1}^{n} X_i > 400\right) = P\left(Z > \frac{400 - 300}{\sqrt{300}}\right) = P(Z > 5.773) \approx 0$$

Using moment generating functions, we could prove that the exact distribution of the sum of the 100 Poisson random variables in this problem is Poisson with expected value 300 and

$$P(S_{100} > 400) = 1 - \sum_{S_{100}=0}^{400} \frac{300^{S_{100}} e^{-300}}{S_{100}!} \approx 0.$$

Example 9.5.2

An insurance company has 10,000 automobile policyholders. The expected yearly claim per policy holder is \$240 with a standard deviation of \$800. What is the probability that the total yearly claims exceed \$2.7 million?

$$E(S_n) = E\left(\sum_{i=1}^{10000} X_i\right) = \sum_{i=1}^{10000} E(X_i) = \mu + \mu + \dots\dots + \mu = 10000\mu = 10000(240) = \$2400000$$

$$Var(S_n) = Var\left(\sum_{i=1}^{10000} X_i\right) = \sum_{i=1}^{10000} Var(X_i) = \sigma^2 + \sigma^2 + \dots\dots + \sigma^2 = 10000\sigma^2 = 10000(800^2)$$

By the CLT,

$$P(S_n > 2700000) = P\left(\sum_{i=1}^{n} X_i > 2700000\right) = P\left(Z > \frac{2700000 - 2400000}{\sqrt{10000(800^2)}}\right) = P\left(Z > \frac{30}{8}\right)$$
$$= P(Z > 3.75) \approx 0$$

Example 9.5.3

Resistors of a certain type have resistances that are exponentially distributed, with parameter $\lambda = 0.04$ Ohms. 50 such independent resistors are connected in series, which causes total resistance in the circuit to be the sum of individual resistances. We will calculate the probability that the total resistance is larger than 1245. First,

$$E(S_n) = E\left(\sum_{i=1}^{50} X_i\right) = \sum_{i=1}^{50} E(X_i) = \mu + \mu + \dots\dots + \mu = 50\mu = 50\left(\frac{1}{0.04}\right) = 1250,$$

$$Var(S_n) = Var\left(\sum_{i=1}^{50} X_i\right) = \sum_{i=1}^{50} Var(X_i) = \sigma^2 + \sigma^2 + \dots\dots + \sigma^2 = 50\sigma^2 = 50\left(\frac{1}{0.04^2}\right) = 31250.$$

By the CLT,

$$P(S_n > 1245) = P\left(\sum_{i=1}^{50} X_i > 1245\right) = P\left(Z > \frac{1245 - 1250}{\sqrt{31250}}\right) = P(Z > -0.02828) = 0.5112823$$

Example 9.5.4. The CLT and a random walk in one dimension

Pitman (1993, 197) presents the following example.

Physicists use *random walks* to model the process of diffusion, or random motion of particles. The position S_n of a particle at time n can be thought of as a sum of displacements $X_1, X_2, \ldots, X_n$. Assuming the displacements are independent and identically distributed, the central limit theorem applies. Here is an illustration with a problem.

Suppose at each step a particle moving on sites labeled by integers is equally likely to move one step to the right, one step to the left, or stay where it is. Find approximately the probability that after 10,000 steps the particle ends up more than 100 sites to the right of its starting point.

Let *X* represent a single step. The probability mass function for *X* is

x	−1	0	1
P(X = x)	1/3	1/3	1/3

Then $E(X) = 0$.

$$\text{Var}(X) = E(X^2) - 0^2 = \frac{-1^2}{3} + \frac{0^2}{3} + \frac{1^2}{3} = \frac{2}{3}$$

and

$$\text{SD(X)} = \sqrt{\frac{2}{3}} = 0.8165$$

The problem is to find

$$P(S_{10000} > 100),$$

where

$$E(S_{10000}) = 10000E(X) = 0$$

$$SD(S_{10000}) = \sqrt{10000}\ SD(X) = 100(0.8165) = 81.65$$

The normal approximation gives

$$P(S_{10000} > 100) = P\left(\frac{S_{10000} - 0}{81.65} > \frac{100 - 0}{81.65}\right) = 1 - P(Z < 1.22474) \approx 0.11$$

9.5.1 Implications of the central limit theorem

We stated in chapter 7 that functions of normal random variables are also normal. Thus, as n goes to infinity, the CLT also implies that a random variable of high interest to statisticians, namely, the so called sample mean, denoted by $\overline{X}$, a function of the sum, defined as

$$\overline{X}_n = \frac{S_n}{n'}$$

and expressed alternatively as $\overline{X}_n = \frac{X_1}{n} + \frac{X_2}{n} + \ldots.. + \frac{X_n}{n}$, is itself a sum of random variables, and therefore follows theorem 9.5.1. We just need to put the right expected value and variance in the formula. That is,

$$P\left(\frac{\overline{X}_n - \mu}{\frac{\sigma}{\sqrt{n}}} > z\right) = P\left(\frac{S_n - n\mu}{\sigma\sqrt{n}} > z\right)$$

approaches the probability that the standard normal random variable Z exceeds z. This is so because the factor of n in the $\overline{X}_n$ does not affect the standardized variables.

Example 9.5.5

Let

$$\overline{X}_n = \frac{S_n}{n} = \frac{\sum_{i=1}^{n} X_i}{n}$$

where

$$X_i, i = 1,\ldots,n$$

is a Poisson random variable with expected value λ. The expected value and variance of $\overline{X}_n$ are:

$$E(\overline{X}_n) = \lambda$$

$$Var(\overline{X}_n) = \frac{\lambda}{n}$$

Example 9.5.6

The average hospital inpatient length of stay (the number of days that on average a person stayed in the hospital) in the United States was 4.5 days in 2017 with a standard deviation of approximately 7. What is the probability that a random sample of 30 patients will have an average stay longer than 6 days next year if the information about 2017 still holds for next year?

$$P\left(\frac{\sum_{i=1}^{30} X_i}{30} > 6\right) = P\left(Z > \frac{6-4.5}{\frac{7}{\sqrt{30}}}\right) = P(Z > 1.173691) = 1 - P(Z < 1.173691) = 0.121.$$

Example 9.5.7

Consider the number of months since a patient had the last medical examination. This is a random variable that varies across patients. At a given point in time, this distribution can be assumed to be uniform between 4 and 20 months. Consider 150 patients randomly chosen. What is the probability that the average number of months since the last examination is 12 or larger?

Let X denote the number of months per patient.

$$P\left(\frac{\sum_{i=1}^{150} X_i}{150} > 12\right) = P\left(Z > \frac{12-12}{\frac{4.618802}{\sqrt{150}}}\right) = P(Z > 0) = 0.5.$$

9.5.2 The CLT and the Gaussian approximation to the binomial

The Central Limit theorem is what was at work when we talked about the normal approximation to the binomial. We said there that the Binomial random variable converges to a normal distribution when n is large but we did not prove it. Consider now the Binomial random variable divided by the n. That is a linear combination of a normal when n is large. Thus, it is also normal. That quantity, we said, is what statisticians call $\hat{p}$, also known as the sample proportion.

Example 9.5.8

Approximately 16.2 percent of Americans purchase private individual health plans in the United States. If we take a random sample of 200 Americans what is the probability that there will be more than 14 percent with private individual health plan?

$$P(\hat{p} > 0.14) = P\left(Z > \frac{0.14-0.162}{\sqrt{\frac{0.162(1-0.162)}{200}}}\right) = P(Z > -0.844193) = 1 - P(Z < -0.844193) = 0.8.$$

9.5.3 How to determine whether n is large enough for the CLT to hold in practice?

The CLT gives just an approximation to the distribution of sums of iid random variables. The approximation will be as good as our satisfying the assumptions for it to hold is. There is a way to double-check whether your approximation is good. Do you recall the study of the normal density function in Chapter 7? There, we said that a normal density has almost all of the probability within 3 standard deviations from the expected value. Checking what values of S_n we get beyond three standard deviations from the expected value of S_n and determining whether those make sense would be one possible approach to determine if n is large enough.

Example 9.5.9 The sum of the roll of two dice cannot be negative or larger than 12

Recall the roll of two fair six-sided dice experiment. Table 9.1 shows how it produces a triangular distribution (which we have rotated) for the sum of the two dice. One might be tempted to claim that it looks like a normal density function. But is it normal?

Table 9.1

S_2	Outcomes in the corresponding event	Probability
2	{1,1}	1/36
3	{1,2}{2,1}	2/36
4	{1,3}{3,1}{2,2}	3/36
5	{4,1}{1,4}{2,3}{3,2}	4/36
6	{5,1}{1,5}{3,3}{4,2}{2,4}	5/36
7	{3,4}{4,3}{5,2}{2,5}{6,1}{1,6}	6/36
8	{4,4}{3,5}{5,3}{6,2}{2,6}	5/36
9	{3,6}{6,3}{2,7}{7,2}	4/36
10	{5,5}{6,4}{4,6}	3/36
11	{5,6}{6,5}	2/36
12	{6,6}	1/36

The average of the two rolls is 7 and the standard deviation is 2.415229.

$$\mu - 3\sigma = 7 - 3 * 2.415229 = -0.2456,$$
$$\mu + 3\sigma = 7 + 3 * 2.415229 = 14.24569.$$

We can see that, by going three standard deviations to the left of the expected value, we are putting ourselves in negative values, which are impossible, and by going three standard deviations to the right of the expected value we are putting ourselves at higher values of the random variable than are possible. These two results are an indication that the normal approximation is not very good yet. We need to add many more dice for the normal curve to be a good approximation to the distribution of the sum.

Similarly, if the problem involves the sample average $\bar{X}$ we can check whether going beyond three standard deviations from the expected value puts us in forbidden values of $\bar{X}$.

Example 9.5.10 The average length of hospital stays cannot be negative.

In Example 9.5.6, the approximate distribution of $\sum_{i=1}^{30} \frac{X_i}{30}$ has expected value 4.5 and standard deviation $\frac{7}{\sqrt{30}} = 1.278$. Three standard deviations to the left of 4.5 is 0.666. So it is reasonable to assume that the normal is reasonable approximation despite the skewness of the distribution of X. However, had the standard deviation been higher (which is not unheard of in this type of random variables), for example, 12, then the standard deviation of $\sum_{i=1}^{30} \frac{X_i}{30}$

would be 2.19 approximately. Three standard deviations to the left of 4.5 is -2.07, implying that $\sum_{i=1}^{30} \frac{X_i}{30}$ is negative. But length of stay cannot be negative. This result would be an indication that the normal approximation has not been reached under the given conditions. Similar result could have been obtained if the n is smaller than 30. For example, less than 15. Do you see why?

Example 9.5.11 How many Poisson random variables does it take for the CLT to hold?

Figure 9.1 shows how the distribution of the sum of independent and identically distributed $X_1, X_2, \ldots, X_n$ Poisson random variables with parameter $\lambda = 1$ approaches the Gaussian density.

As a general rule, the more symmetric the distribution, and the thinner the tails, the faster the approach to normality as n increases (Dinov, Christou and Sanchez 2008).

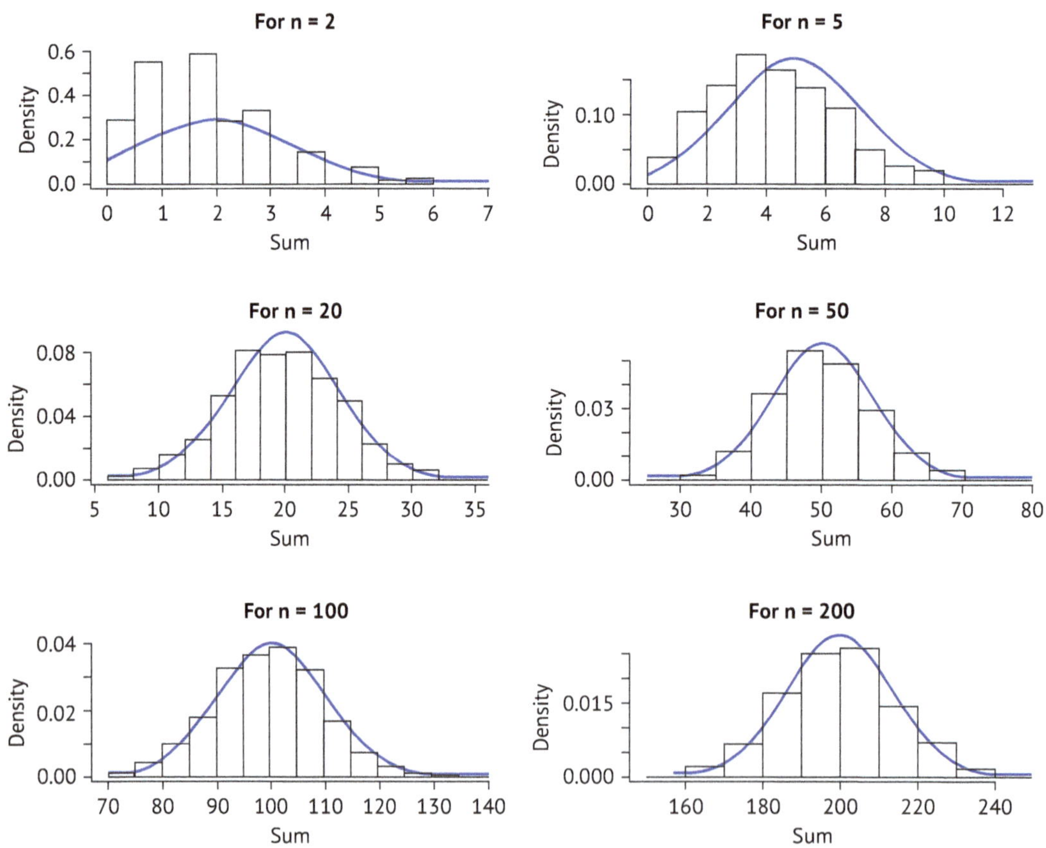

Figure 9.1 The Figure shows that as we increase n, the distribution of the sum of n Poisson random variables with expected value 1 approaches the Gausisian distribution. The Gaussian is the continuous curve in blue.

An alternative approach to determining whether n is large enough is to use the skewness coefficient. The Gaussian density has value of the skewness coefficient equal to 0. The farther the skewness coefficient is from 0 the larger is the deviation from the Gaussian density curve.

9.5.4 Combining the central limit theorem with other results seen earlier

Sometimes, we may be interested in comparing two sets of iid random variables. In that case, it is helpful to remember what we reviewed at the beginning of Section 9.5.1 about Gaussian random variables. The following example illustrates what we mean.

Example 9.5.12

Iron deficiency among infants is a major problem. The table below contains the average blood hemoglobin levels at 12 months of infants following different feeding regimens: breastfed infants, and baby formula without any iron supplements. Here are summary results. (Note: none of the babies take both feeding regimens).

Group	μ	σ
Breast-fed	13.3	1.7
Formula	12.4	1.8

Let $X_1, X_2, \ldots, X_{100}$ represent the blood hemoglobin of 100 unrelated breast-fed babies and let $Y_1, Y_2, \ldots, Y_{100}$ the blood hemoglobin of 100 unrelated formula babies. What is the probability that the difference in average hemoglobin levels at 12 months is bigger than 2?

We first identify the random variable needed. We can see that in this problem, we are being asked to compare two random variables: the $\bar{X}_n - \bar{Y}_n$.

We can use what we learned in this chapter to show that

$$\mu_{\bar{X}_n - \bar{Y}_n} = E(\bar{X}_n - \bar{Y}_n) = \mu_X - \mu_Y = 13.3 - 12.4 = 0.9.$$

We can also use what we learned in this chapter to show that

$$\sigma_{\bar{X}_n - \bar{Y}_n} = \sqrt{Var(\bar{X}_n - \bar{Y}_n)} = \sqrt{\frac{1.7^2}{100} + \frac{1.8^2}{100}} = 0.24758$$

$$P(\bar{X}_n - \bar{Y}_n > 2) = P\left(Z > \frac{2 - 0.9}{0.24758}\right) = P(Z > 4.443) \approx 0.$$

9.5.5 Applications of the central limit theorem in statistics. Back to random sampling

Statistical inference is the science of estimating probabilities using data assumed to be representative of what is to be estimated, which would be the case if the sample of data was chosen at random. If that is the case, then each numerical value of the sample is seen as a manifestation of the value of a random variable. S_n and $\bar{X} = \frac{S_n}{n}$ or any function of the random sample values is called a summary statistic. The derivation of distributions for summary statistics is called sampling distribution theory. In particular, some random functions of Gaussian random variables play a very important role in statistical inference, for example, the F statistic, the chi-square statistic and so on.

From a probability perspective, a random sample of data is assumed to be a set of independent and identically distributed random variables. The probability model for one of those random variables is considered to be the model prevalent in the population at large. The parameters of that model are assumed unknown, and the random sample helps estimate them by using the sample summary statistics. For example, the average of the random sample is used by statisticians to test hypotheses about the mean of a population's distribution. The distribution of the sample average is called sampling distribution. The following example describes the type of statistical reasoning involved in using the central limit theorem in statistics.

Example 9.5.13

A physician studies a randomly selected group of 25 patients and gives them a drug that could cause vasoconstriction. The physician conducting the study is trying to determine whether there are adverse effects on systolic blood pressure due to taking the drug. The physician finds that after taking the drug, the average blood pressure of the 25 patients is 124 mm Hg. The physician then asks: What is the probability that this or a higher blood pressure would happen if these patients had not taken the drug—that is, if the patients had remained like all patients of their kind, who have a mean systolic blood pressure of 120 and standard deviation 10?

That probability is 0.023. Why?

The statistician then interprets the result as follows: 2.3% of random samples of 25 patients not taking the drug would have had an average systolic blood pressure of 124 or more without taking the drug, just by chance. That is a very small number of samples, hence it is rare to find an average of 124 in a random sample of 25. But the physician found it in the sample at hand. Therefore, it seems that the drug has some effect on systolic blood pressure.

The physician used the property of the distribution of $\overline{X}_{25}$, assumed to be approximately Gaussian with expected value 120 and standard deviation $\frac{10}{\sqrt{25}}$,by the CLT. Statisticians call the distribution of $\overline{X}$ the sampling distribution of $\overline{X}$. The standard deviation of $\overline{X}$ is called the standard error.

Box 9.3

Misconceptions

A common misconception is that as n goes to infinity the distribution of X follows the Gaussian distribution. The Central Limit theorem says nothing about the distribution of one random variable. It is only a statement about the distribution of the sum S_n and the sample $\frac{S_n}{n}$. The CLT does not assume either that the population is large.

Example 9.5.14

A claim was made that 60% of the adult population thinks that there is too much violence on television and a random sample of size $n = 100$ was drawn from this population to check whether that is indeed true. A Poll was conducted. The people in the sample were asked: do you think there is too much violence on TV? YES or NO? 56% of the people in the sample said YES.

The sampling distribution of the $\hat{p}$ assuming that the claim of 60% is accurate can be assumed to be Gaussian with $E(\hat{p}) = p = 0.6$ and standard deviation $\frac{\sqrt{0.6(0.4)}}{\sqrt{100}} = 0.0489$. So

$$P(\hat{p} < 0.56) = P\left(z < \frac{0.56 - 0.6}{0.0489}\right) = P(z < -0.8179) = 1 - P(z < -0.844193) = 0.2067.$$

So approximately 20% random samples of 100 adults taken from a population where 60% think there is too much violence would have given 56% or less by chance. That is a large percentage of samples. Therefore seeing a 56% is no statistical evidence against the claim that 60% think there is too much violence. The 56% is a result of chance variation.

Example 9.5.15

Go to the Reese's pieces samples applet

http://statweb.calpoly.edu/chance/applets/Reeses/ReesesPieces.html

To run it, you need to click on the handle on the drawn candy dispenser, as if you were buying the candy from the machine. Get the distribution of $\hat{p}$ for sample size of n = 50 and p = 0.03. Draw 1000 samples to get a better approximation. Is the normal a good approximation to the distribution of $\hat{p}$ in this case?

You will see that the normal curve does not fit the distribution of $\hat{p}$ that is centered around 0.03 well. The sample size is not large enough. That can be seen by the fact that the normal curve centered at 0.03 gives negative proportions (there is blue in the negative range), but we never get in the simulation a negative proportion (all the black dots are in the positive range), as they should be. Proportions are between 0 and 1. The parent distribution is too skewed (a Bernoulli with p = 0.03 is very skewed).

The standard deviation of $\hat{p}$ is 0.0241. This means that if we go two standard deviations to the left of 0.03 we hit negative numbers.

9.5.6 Proof of the CLT

The CLT plays a very important role in almost every aspect of life.

There are several ways to prove the CLT, but some of them require more background than what is assumed in this book. Since we have already talked throughout the book about the moment generating function of a random variable, we will refer the reader to a proof by moment generating functions in Rice (2007). Before we do that, we need to learn a couple of properties of the moment generating function.

- Property 1: The moment generating function of a sum of independent and identically distributed random variables is the product of the moment generating functions.
- Property 2: The moment generating function of the product of a constant times a random variable is the moment generating function of the random variable evaluated at the constant times t.

9.5.7 Exercises

Exercise 1. Suppose that $X_1, X_2, \ldots\ldots\ldots X_{200}$ is a set of independent and identically distributed Gamma random variables with parameters $\alpha = 4$, $\lambda = 3$. Describe the normal distribution that would correspond to the sum of those 200 random variables if the Central Limit Theorem holds.

Exercise 2. A lot acceptance sampling plan for large lots calls for sampling 50 items and accepting the lot if the number of nonconformances is no more than 5. Find the approximate probability of acceptance if the true proportion of noncomformances in the lot is 10%.

Exercise 3. (This exercise is from Mosteller, Rourke and Thomas (1967, 333).) In crossing two pink flowers of a certain variety the resulting flowers are white, red, or pink, and the probabilities that attach to these various outcomes are $\frac{1}{4}, \frac{1}{4}, \frac{1}{2}$, respectively. If 300 flowers are obtained by crossing pink flowers of this variety, what is the probability that 90 or more of these flowers are white?

Exercise 4. Government statistics in the United Kingdom suggest that 20% of individuals live below the poverty level (Purdam, Royston and Whitham (2017)). What percentage of individuals in a random sample of 100 individuals are within two standard deviations of that proportion?

Exercise 5. According to Plewis (2014), Punjab produces 11.84% of India's cotton (2012 figures). Twenty-six percent of farmers in Punjab produce cotton (Bt cotton). Thirteen percent of 100,000 farmers (15+) commit suicide. Fifty-six percent of farmers producing cotton produce genetically modified cotton.

Maharashtra produces 20.42% of cotton in Maharashtra. Twenty percent of farmers produced cotton. There is a forty-six percent suicide rate per 100,000. Fifty-six percent of farmers producing cotton produce genetically modified cotton.

What is the probability that, in a random sample of 100 Punjab farmers, less than 40 produce genetically modified cotton?

Exercise 6. According to the Department of Motor Vehicles (DMV), the entity in charge of providing driving licenses in the United States, it is illegal to drive with a blood alcohol content of 0.08% or more if you are 21 or older. In the DMV's guidelines to determine when a person is driving under the influence, which can be found at https://www.dmv.ca.gov/portal/dmv/detail/pubs/hdbk/actions_drink, it is indicated that fewer than five percent of the population weighing 100 pounds will exceed the 0.33 alcohol level. Assume this is accurate. If the Highway Patrol stops in a random day 200 unrelated cars where the individual weighs one hundred pounds, on a Friday night, what is the probability that six percent or more individuals stopped exceeds the 0.33 alcohol level?

Exercise 7 (This exercise is from Kinney (2002, 75).) A bridge crossing a river can support at most 85,000 lbs. Suppose that the weights of automobiles using the bridge have mean weight 3,200 lbs and standard deviation 400 lbs. How many automobiles can use the bridge simultaneously so that the probability that the bridge is not damaged is at least 0.99?

Exercise 8. Resistors of a certain type have resistances that are exponentially distributed with parameter $\lambda = 0.04$. An operator connects 50 independent resistors in series, which causes total resistance in the circuit to be the sum of individual resistances. Find the probability that the total resistance is less than 1245.

9.6 When the expectation is itself a random variable

We have been saying throughout the book that the parameters of the probability mass functions and density functions studied are constant, and in all the chapters and exercises done so far they have been constant. However, there are situations where that is not the case. Bayesian Statistics for example assumes that the parameters are themselves random variables. But without getting into Bayesian statistics, there are some theorems regarding conditional expectations that help us compute expected values and variance of random variables, when solving the problems otherwise would be very difficult.

These theorems are

$$E(X) = E[E(X|Y)]$$
$$Var(X) = E[Var(X|Y)] + Var[E(X|Y)]$$

Example 9.6.1

A quality control plan for an assembly line involves sampling n = 10 finished items per day and counting *Y*, the number of defective items. If *p* denotes the probability of observing a defective item, then *Y* has a binomial distribution, when the number of items produced by the line is large. However, p varies from day to day and is assumed to have a uniform distribution on the interval from 0 to ¼. What is the expected value of Y for any given day.

$$E(E(Y \mid P)) = E(nP) = nE(p) = n(1/8) = 10(1/8) = 5/4.$$

9.7 Other generating functions

We have seen the moment generating function in earlier chapters. There is also the probability generating function, *G*(*t*). This is defined as

$$G_X(t) = E(t^X).$$

A property of this generating function is that the probability generating function of the sum of two independent random variables is the product of the probability generating functions of the two random variables. Let X and Y be two independent random variables. Then

$$G_{X+Y}(t) = E(t^{X+Y}) = E(t^X)E(t^Y) = G_X(t)G_Y(t).$$

Example 9.7.1.

(This example is from Newman (1998).) Let's recall the discrete pmf for the roll of a six sided die in Table 1.2 in Chapter 1.

X	**1**	**2**	**3**	**4**	**5**	**6**
$P(X = x)$	1/6	1/6	1/6	1/6	1/6	1/6

Then

$$G_X(t) = E(t^X) = \frac{1}{6}t + \frac{1}{6}t^2 + \frac{1}{6}t^3 + \frac{1}{6}t^4 + \frac{1}{6}t^5 + \frac{1}{6}t^6$$

As we can see, this looks like a polynomial, whose coefficients are the probabilities and the powers are the values of the random variable.

Suppose we roll the die twice and we are interested in the sum of the two rolls. Then

$$G_{X+Y}(t) = E(t^{X+Y}) = \left(\frac{1}{6}t + \frac{1}{6}t^2 + \frac{1}{6}t^3 + \frac{1}{6}t^4 + \frac{1}{6}t^5 + \frac{1}{6}t^6\right)^2$$

$$= \frac{1}{36}t^2 + \frac{2}{36}t^3 + \frac{3}{36}t^4 + \frac{4}{36}t^5 + \frac{5}{36}t^6 + \frac{6}{36}t^7 + \frac{5}{36}t^8 + \frac{4}{36}t^9 + \frac{3}{36}t^{10} + \frac{2}{36}t^{11} + \frac{1}{36}t^{12}$$

As we can see, the coefficients of this new polynomial contain the probabilities of the values of the sums given in the *exponents of the polynomial.*

9.8 Mini quiz

Question 1. Which of the following statements is NOT true according to the Central Limit Theorem? Select all that apply.

a. The mean of a distribution of sample means is equal to the population mean divided by the square root of the sample size.
b. The larger the sample size, the more the distribution of the sample means resembles the shape of the original distribution of one random variable.
c. The mean of the distribution of sample means for samples of size n = 15 will be the same as the mean of the distribution for samples of size n = 100.
d. The larger the sample size, the more the distribution of sample means will resemble a normal distribution.
e. An increase in n will produce a distribution of sample means with a smaller standard deviation.

Question 2. The distribution of income (in tens of thousands) of females in a small population can be modeled by a gamma distribution with mean 4 and standard deviation 8 dollars. A simulation is done, where each trial consists of drawing a random sample of 500 women from this population and computing the average salary of the women in this population. 1000 trials are done. Which of the following statements is true? (you may want to do this question after doing the simulation in Section 9.9)

a. The distribution of the sample obtained in one trial should be close to normal
b. If we plotted the distribution of the sample obtained in each of the 1000 trials, we would have 1000 distributions that look like the normal.
c. The mean of the 500 random variables in each sample is close to 2000.
d. The distribution of the 1000 averages is exactly gamma.
e. The distribution of the 1000 averages is close to normal.

Question 3. Blood pressure in a population of very at risk people has expected value of 195 and a standard deviation of 20. Suppose you take a random sample of 100 of these people. There would be a 68% chance that the average blood pressure would be between

a. 155 to 235
b. 193 to 197
c. 175 to 215
d. 191 to 199

Question 4. An airline knows that over the long run, 90% of passengers who reserve seats show up for their flight. On a particular flight with 300 seats, the airline accepts 324 reservations. Assuming that passengers show up independently of each other, what is the chance that the flight will be overbooked?

a. 0.91
b. 0.455
c. 0.05297
d. 0.1

Question 5. The service times for customers coming through a checkout counter in a retail store are independent random variables, with a mean of 1.5 minutes and a variance of 1.0 minute. Approximate the probability that 100 customers can be serviced in less than 2 hours of total service time.

a. 0.4987
b. 0.5
c. 0.0013
d. 0.23

Question 6. The amount of money college students spend each semester on textbooks is normally distributed with an expected value of \$195 and a standard deviation of \$20. Suppose you take a random sample of 100 college students from this population. There would be a 68% chance that the average amount spent on textbooks would be:

a. \$155 to \$235
b. \$191 to \$199
c. \$193 to \$197
d. \$175 to \$215
e. \$ 235 to \$155

Question 7. The median age of residents of the United States is 31 years. If a survey of 100 randomly selected United States residents is taken, find the approximate probability that at least 60 of them will be under 31 years of age.

a. 0.02
b. 0.5
c. 0.471
d. 5

Question 8. Chebyshev's and Markov's theorems give

a. exact probabilities that a random variable is in an interval in the real line
b. a bound for the probability that a random variable is in an interval in the real line
c. the expected value of a random variable
d. the variance of a random variable

Question 9. Let *X*, *Y* have the joint pdf

$$f(x,y) = 6y, 0 \leq y \leq x \leq 1$$

For this example, it is true that (circle all that applies)

a. $E(E(Y|X)) = E(Y)$
b. $Var(E(Y|X)) < Var(Y)$
c. X and Y are independent
d. The marginal density function of X is exponential

Question 10. An online computer system is proposed. The manufacturer gives the information that the mean response time is 10 seconds. Estimate the probability that the response time will be more than 20 seconds.

a. Less than 1/2
b. More than 1/2
c. 1/3
d. Less than 2

9.9 R code

9.9.1 Monte Carlo integration

Read Section 9.3.1 before starting this simulation.

```
n=1000;
x=runif(n, min=0, max=1) #draw 1000 uniform(0,1) r.n.
I = mean(cos(2*pi*x*x)) # compute the arithmetic mean of f(x)
I # see the mean computed
n=10000;
x=runif(n, min=0, max=1) # draw 10000 uniform(0,1) r.n.
I = mean(cos(2*pi*x*x)) # compute the arithmetic mean of f(x)
I # see the mean computed
n=100000; # n increases to 100000
x=runif(n, min=0, max=1) # draw 100000 uniform(0,1) r.n.
I = mean(cos(2*pi*x*x))
I
```

9.9.2 Random sampling from a population of women workers

It is impossible to verify that the central limit theorem holds in the random sampling that we do in reality. But we can explore whether it works by simulation. To this end, we will be concerned about the whole population of full-time year-round white female workers between 16 and 65 years of age, in a small town in the Midwest (Chatterjee, Handcock, and Simonoff (1995)). The characteristic under study for these women is their income from wages (as measured in thousands of dollars). We happen to have access to census information on the income in the population of all these women, and we are going to just read it using R. The first few lines of the population look like this:

Income
11.652
23.015
5.604
6.710
7.293
8.918
14.176
11.363

.....

.....

with each line representing the Income of a particular woman in the population.

1. Accessing the population

```
population=read.table("http://pages.stern.nyu.edu/~jsimonof/Casebook/
Data/Tab/census1.TAB",header=T)
attach(population) # Makes the observations available
population[1:8, ] # to see the first 8 rows of Income. Do not copy
paste this.
N= length(Income) # to find out how many incomes are observed in
this population
mu= mean(income) # Population mean is the expected value mu
sigma=sqrt(((N-1)/N)*var(income)) # population standard deviation is
sigma
N; mu; sigma
hist(income)
```

Question 1. Describe the population distribution.

$N =$

$\mu =$

$\sigma =$

Distribution is:

2. Start simulation.

One trial: Draw a random sample of women from this population and analyze their income. We do now 4 trials. Follow steps indicated below.

2.1. Trial 1. Draw a random sample of 300 women from the population and look at the histogram of their incomes and the sample mean. We will use density histograms, because we will be comparing histograms. Density histograms, like probability models for quantitative variables, have proportion represented by the area under the curve. The area under the curve represents the proportion of women.

```
par(mfrow=c(4,2)) # a window will pop up. The following 8 graphs
will go here
sample1Income=sample(Income,300) # get random sample of n=300, look
at income
xbar1=mean(sample1Income) #find the mean of the 300 incomes
hist(sample1Income,xlim=c(0,221 ),prob=T) #do histogram of sample
income
boxplot(sample1Income,ylim=c(0,221))
```

2.2. Trial 2. Now draw another random sample of 300 women from the same population and repeat the analysis you did above.

```
sample2Income=sample(Income,300)
xbar2= mean(sample2Income)
hist(sample2Income,xlim=c(0,221),prob=T)
boxplot(sample2Income,ylim=c(0,221))
```

2.3. Trial 3. Now draw another random sample of 300 women from the same population and repeat the analysis you did above.

```
sample3Income=sample(Income,300)
xbar3= mean(sample3Income)
hist(sample3Income,xlim=c(0,221),prob=T )
boxplot(sample3Income,ylim=c(0,221))
```

2.4. Trial 4. One more time: draw another random sample of 300 women from the same population and repeat the analysis you did above.

```
sample4Income=sample(Income,300)
xbar4=mean(sample4Income)
hist(sample4Income,xlim=c(0,221),prob=T)
boxplot(sample4Income,ylim=c(0,221))
xbar1; xbar2; xbar3;xbar4
dev.off() #to close the graph window we had. Don't type until you
  have copy pasted your graph
```

Question 2. Does each of the four histograms resemble the population distribution? Summarize their shape.

Question 3. Look at all your four boxplots. Are there outliers in any of them? What can you conclude about where the majority of the salaries are in each sample if you excluded the outliers (if you got any)?

Question 4. Let's see now what are the sample means $\left(\bar{X}_1, \bar{X}_2, \bar{X}_3, \bar{X}_4\right)$ that we obtained and compare them. Write them down and describe whether they are very different or not. Can you explain what you found?

Do many more trials

If we continued drawing random samples of 300 from the population, we would end up with a lot of different values of $\bar{X}$ (The number of possible random samples of size 300 is $\binom{5000}{300}$. Doing that by hand, however, is, as you have noticed, a little too tedious. So we are going to use a program that will continue the exercise we did in part 2 by taking samples of 300 over and over and computing their average income $\overline{X}$ for each. We will do that 1000 times only, so we will just approximate the sampling distribution of $\overline{X}$. We will end up with 1000 sample means $\overline{X}_1, \overline{X}_2, \overline{X}_3, \overline{X}_4 \ldots\ldots \overline{X}_{1000}$. You may want to write this function below in a separate file, because errors typing can be fatal now.

```
samples=matrix(rep(0,300000),ncol=300) #this is just space to put the
  samples
xbars=matrix(rep(0,1000),ncol=1) # this is space to put each sample
mean
for(i in 1:1000){ #repeat what we did above 1000 times.
samples[i,] = sample(Income,300) # each time take a random sample
  of 300 salaries
xbars[i]=mean(samples[i,]) # each time compute the sample mean
}
E.xbar=mean(xbars)
Sigma.xbar=sqrt((999/1000)*var(xbars))
E.xbar; Sigma.xbar
hist(xbars)
```

Question 5. Summarize the sampling distribution of xbar.
Relate the mean and the standard deviation of the sampling distribution to the mean and standard deviation of the distribution of Income in the population (question 1). In particular is each of the following true?

$$\mu_{\bar{x}} = \mu_{Population}$$

$$\sigma_{\bar{x}} = \frac{\sigma_{Population}}{\sqrt{n}}$$

Comment on the shape of the distribution that would be expected if the Central Limit Theorem holds.

9.10 Chapter Exercises

Exercise 1. Let $X_1, X_2, \ldots\ldots, X_{50}$ be 50 independent and identically distributed exponential random variables with expected value 2. (i) Compute $P(X_i < 3)$. (ii) Compute $P\left(\sum_{i=1}^{50} 10X_i < 30\right)$. (iii) Compute $P((X_1 < 3) \cap (X_2 < 3) \cap \ldots\ldots \cap (X_{50} > 3))$. (iv) Let $Y = 3e^{tX_i} + 4X_i$. Find $E(Y)$.

Exercise 2. Let X be a continuous random variable with the following density function:

$$f(x) = \frac{1}{2}x,\ 0 \le x \le 2.$$

Let X_1, X_2,....., X_{50} be 100 independent and identically distributed random variables each with that density function. Find (i) $E\left(\sum_{i=1}^{100}\frac{X_i}{4}\right)$ and $Var\left(\sum_{i=1}^{100}\frac{X_i}{4}\right)$.

Exercise 3. Consider an unfair six-sided die that has the following probabilities for each number.

x	1	2	3	4	5	6
P(X = x)	0.1	0.1	0.4	0.2	0.1	0.1

What is the probability that in 200 tosses of such a die we would get more than 120 odd numbers? Show work.

Exercise 4. Consider Exercise 10 in Section 7.2.1. Assume that for this particular matter, $\alpha = 0.5$. Calculate the probability that the average angle of 100 emitted electrons is larger than 0.5.

Exercise 5. The monthly salary of women that are in the labor force in a large town, *Y*, follows a gamma distribution with expected value 500 and variance 125. The city is planning to obtain a random sample of 100 women to obtain some information. The city is planning to ask the 100 women about their salary. (i) What is the probability that the average salary of the 100 women in the sample is larger than 1,500? (ii) If, in this town, 20% of the women have a monthly salary larger than 3,000, what is the probability that in the random sample of 100 women more than 20% of them make a salary larger than 3,000? Show work. (iii) Would $W = \frac{4}{100}\Sigma_y y_i$ be an unbiased estimator of the average salary of the women in the large town? (Note: "unbiased" means that the expected value of *W* equals the mean of the population of women.) (iv)What is the joint distribution of the 100 random variables with the same density function as in this problem?

Exercise 6. In 1,000 flips of a supposedly fair coin, heads came up 560 times and tails 440 times. What is the probability that a number of heads that large or larger occurs if the coin is fair?

Exercise 7. Demonstration of the law of large numbers with simulation with R. Run the following code, and report in the answer to this problem only what is being asked.

1. Give the 20 rolls of a fair six-sided die, the approximate (empirical) discrete distribution from those rolls and a plot of that distribution. Does it look like the probability mass function for the roll of a fair six-sided die? How far are you from it?

```
first20=sample(1:6, 20, rep=T) #roll the fair die 20 times
first20 # see the rolls you got
table(first20)/20
X=1:6
plot(table(first20)/20,xlab="X",ylab="empirical probability (20
rolls)", ylim=c(0,1),type="h")
```

2. Roll another 20 times, and now append the new numbers you got here to those in (a) to have 40 numbers

```
plus20=sample(1:6,20,rep=T)
first40=append(first20,plus20)
first40 #See your numbers
table(first40)/40
X=1:6
plot(table(first40)/40,xlab="X",ylab="empirical probability (40
rolls)", ylim=c(0,1),type="h")
```

3. Keep rolling: roll another 60 times, and append these new numbers to the ones you already got to have 100 numbers

```
more60=sample(1:6,60,rep=T)
first100=append(first40,more60)
first100 # see your numbers.
table(first100)/100
plot(table(first100)/100,xlab="X",ylab="empirical probability
(100 rolls)",
ylim=c(0,1),type="h")
```

4. Now roll an additional 1000 times, append these 1000 new numbers to the 100 you already have obtained and find the distribution and plot of the 1100 numbers.

5. Explain what is happening in a, b, c, d. Do you think the law of large numbers is at work? Why? How is this behavior different from what you would observe if you were illustrating the Central Limit Theorem with a simulation. Explain the difference. (no need to simulate).

What if the die had not been fair? What would be the final distribution you would observe after rolling 20, 100 and 1000 times? Rewrite the code. Most of the code will be the same except the first command For example,

```
sample(1:6,prob=c(0.5,0.05,0.2,0.05,0.1,0.1)
```

Repeat (i-iv) for the unfair die.

9.11 Chapter References

Chatterjee, Samprit, Mark S. Handcock, and Jeffrey S. Simonoff. 1995. *A Casebook for a First Course in Statistics and Data Analysis*. John Wiley & Sons, Inc.

Dinov, Ivo, Nicolas Christou and Juana Sanchez. (2008). "Central Limit Theorem: New SOCR Applet and Demonstration Activity." *Journal of Statistics Education* Volume 16, Number 2 (2008) http://www.amstat.org/publications/jse/v16n2/dinov.html

Freedman, David, Robert Pisani, and Roger Purves. 1998. *Statistics*. Third Edition. W.W. Norton & Company.

Grinstead, Charles M., and J. Laurie Snell. 1997. *Introduction to Probability*. Second revised edition. American Mathematical Society.

Kinney, John J. 2002. *Statistics for Science and Engineering*. Addison-Wesley.

Lanier, Jaron. 2018. *Ten Arguments for Deleting Your Social Media Accounts Right Now*. New York: Henry Holt and Company.

Mosteller, Frederic, Robert E. K. Rourke, and George B. Thomas. 1967. *Probability and Statistics*. Addison-Wesley.

Newman, Donald J. 1998. *Analytic Number Theory*. Springer Verlag,

Parzen, Emanuel. 1960. *Modern Probability Theory and Its Applications*. New York. John Wiley and Sons, Inc.

Pittman, Jim. 1993. *Probability*. Springer Texts in Statistics.

Plewis, Ian. 2014. "Indian farmer suicides. Is GM cotton to blame?" *Significance* 11, no. 1 (February): 14–18.

Purdam, Kingsley, Sam Royston, and Graham Whitham. Measuring the "Poverty Penalty" in the UK. *Significance*, August 2017, p 34 to 37.

Rice, John A. 2007. *Mathematical Statistics and Data Analysis*. Thomson Brooks/Cole

Chapter 10

How All of the Above Gets Used in Unsuspected Applications

▸ Thermodynamics, quantum mechanics, modern communication technology, genetics, social issues. After reading this book, what have you concluded they have in common?

10.1 Random numbers and clinical trials

- Eight people who suffer from migraine headaches volunteer to take part in a medical study of the effect of a new drug on migraine headaches. The names of the volunteers are:

1. Chang	2. Donley	3. Elfring	4. Miller	5. Reed	6. Ting
7. Toman	8. Whittinghill	9. Chib	10. Amir	11. Mason	12. Bradley

Your job is to allocate half of these people to the experimental group taking the drug and the other half to the control group that will not take the drug. How would you do it?

This is done on a daily basis by all clinical trials out there to determine the effects of drugs on people. Clinical trials and intervention studies must allocate subjects at random to treatment and control groups. The simplest way of allocation is by selecting at random six people to be in the treatment group. The remaining ones will be in the control group. The point for us is that random numbers are used for the allocation, namely the following command in R would do the job.

```
sample(1:12, 6, replace=F)
```

10.2 What model fits your data?

▶ The following list of 24 test scores has an average of approximately 50 and a standard deviation of approximately 10.

29, 36, 37, 39, 41, 44, 47, 48, 49, 50, 50, 52, 52, 53, 54, 56, 58, 59, 62, 64, 65.

How many scores are within one standard deviation of the mean? Is that the number of scores that you would have gotten using the normal model with the same mean and the same standard deviation?

The number of scores between 40 and 60, within one standard deviation in the data, is 16. The normal model predicts $0.68*24 = 16.32$, so approximately correct.

Model fitting to data is one of the day-to-day activities of statisticians and data scientists. One possible method goes as follows:

a. Data are available. For example, consider the baby boom data set presented in Table 10.1 below. This data set can be downloaded from http://ww2.amstat.org/publications/jse/datasets/babyboom.dat.txt

   The data set has the following variables:

   - Time of birth recorded on the 24-hour clock (column 1)
   - Sex of the child (1 = girl, 2 = boy) (column 2)
   - Birth weight in grams (column 3)
   - Number of minutes after midnight of each birth (column 4)

b. The Poisson distribution could be fit to the number of births per hour and the empirical proportion of births found in the data each hour could be compared to the theoretical number of births per hour predicted by a Poisson model, using the average number of births per hour of the data as the proxy for the parameter of the theoretical Poisson. Dunn (1999) did this. The results are found in Table 10.2.

c. The statisticians are not happy with just a table like that. They need to determine some criteria to accept that table as indication that the Poisson model is a good fit. To this end, statisticians design test statistics. These test statistics are summaries that themselves are random variables and have sampling distributions. One test statistic is the chi-Square goodness of fit statistic, which follows a chi-square density function.

Table 10.2 does not give the chi-square statistic. The reader should try to compute it after studying Section 10.5. But bear in mind that the criterion for statistically determining whether the Poisson fits a data set is probability based.

Table 10.1

0005	1	3837	5
0104	1	3334	64
0118	2	3554	78
0155	2	3838	115
0257	2	3625	177
0405	1	2208	245
0407	1	1745	247
0422	2	2846	262
0431	2	3166	271
0708	2	3520	428
0735	2	3380	455
0812	2	3294	492
0814	1	2576	494
0909	1	3208	549
1035	2	3521	635
1049	1	3746	649
1053	1	3523	653
1133	2	2902	693
1209	2	2635	729
1256	2	3920	776
1305	2	3690	785
1406	1	3430	846
1407	1	3480	847
1433	1	3116	873
1446	1	3428	886
1514	2	3783	914
1631	2	3345	991
1657	2	3034	1017
1742	1	2184	1062
1807	2	3300	1087
1825	1	2383	1105
1854	2	3428	1134
1909	2	4162	1149
1947	2	3630	1187
1949	2	3406	1189
1951	2	3402	1191
2010	1	3500	1210
2037	2	3736	1237
2051	2	3370	1251
2104	2	2121	1264
2123	2	3150	1283
2217	1	3866	1337
2327	1	3542	1407
2355	1	3278	1435

Table 10.2

Births per hour	Tally (hours)	Empirical Probability	Theoretical Probability (with lambda = 44/24 = 1.83 births per hour)
0	3	$3/24 = 0.125$	$\frac{1.83^0 e^{-1.83}}{0!} = 0.160$
1	8	$8/24 = 0.333$	$\frac{1.83^0 e^{-1.83}}{0!} = 0.293$
2	6	0.250	0.269
3	4	0.167	0.164
4	3	0.125	0.075
5+	0	0.000	0.039
Total	24	1	1

10.3 Communications

An area where randomness prevails is in communications networks, such as the internet, radio, etc. Randomness is the essence of communication.

Example 10.3.1

(This exercise is based on Grami (2016, chapter 4).) In a binary symmetric communication (BSC) channel, the input bits transmitted over the channel are either 0 or 1 with probabilities p and $1 - p$, respectively. Due to channel noise, errors are made.

If a channel is assumed to be symmetric, the probability of receiving 1 when 0 is transmitted is the same as the probability of receiving 0 when 1 is transmitted.

The conditional probabilities of error are assumed to be each ε. Determine the probability of error, also known as the bit error rate, as well as the a posteriori probabilities.

Priori probability of transmitting bits are:

$$P(0 \text{ transmitted}) = p; \qquad P(1 \text{ transmitted}) = 1 - p$$

Posterior probability of transmitting bits (are called transition probabilities):

$$P(1 \text{ received} \mid 0 \text{ transmitted}) = P(0 \text{ received} \mid 1 \text{ transmitted}) = \varepsilon$$

Average probability of error is:

$$P(\text{error}) = P(1 \text{ received, } 0 \text{ transmitted}) + P(0 \text{ received, } 1 \text{ transmitted}) =$$
$$P(1 \text{ received} \mid 0 \text{ transmitted})P(0 \text{ transmitted}) + P(0 \text{ received} \mid 1 \text{ transmitted})P(1 \text{ transmitted})$$

$$= \varepsilon p + \varepsilon(1 - p) = \varepsilon$$

The posteriori probabilities of interest are the probabilities:

$$P(\text{sent} = 1 \mid \text{received} = 1) \text{ and } P(\text{sent} = 0 \mid \text{received } 0).$$

$$P(\text{sent} = 1 \mid \text{received} = 1) = (1 - p - \varepsilon + \varepsilon p)/(1 - p - \varepsilon + 2\varepsilon p)$$

$$P(\text{sent} = 0 \mid \text{received} = 0) = (p - \varepsilon p)/(p + \varepsilon - 2\varepsilon p)$$

The following interesting observations regarding BSC channels can be made:

For $\varepsilon = 0$, i.e., when the channel is ideal, both a posteriori probabilities are one.
For $\varepsilon = \frac{1}{2}$, the a posteriori probabilities are the same as the a priori probabilities
For $\varepsilon = \frac{1}{1}$, when the channel is most destructive, both a posteriori probabilities are zero.

It is very insightful to note than in the absence of a channel, the optimum receiver, which minimizes the average probability of error, P(error), would always decide in favor of the bit whose a priori probability was the greatest. Moreover, if $P(error) > 1/2$, that is, more often than not an error is made, an inverter can then be employed to reduce the bit error rate to $1 - P(error) > 1/2$, simply by turning a 1 into 0 and a 0 into 1.

10.3.1 Exercises

Exercise 1. In a digital commutation system, bits are transmitted over a channel in which the bit error rate is assumed to be 0.0001. The transmitter sends each bit five times, and a decoder takes a majority vote of the received bits to determine what the transmitted bit was. Determine the probability that the receiver will make an incorrect decision.

Exercise 2. The normal distribution represents the distribution of thermal noise in signal transmission. For example, in a communication system, the noise level is modeled as a Gaussian random variable with mean 0, and $\sigma^2 = 0.0001$. What is the probability that the noise level is larger than 0.01?

10.4 Probability of finding an electron at a given point

Before quantum mechanics, it was thought that the accuracy of any measurement was limited only by the accuracy of the instruments used. However, Karl Heisenberg showed that no matter how accurate the instruments used, quantum mechanics limits the accuracy when two properties are measured at the same time. For the moving electron, the properties or variables are considered in pairs: momentum and position are one pair, and energy and time are another. The conclusion that came from Heisenberg's theory was that the more accurate the measurement of the position of an electron or any particle for that matter, the more inaccurate the measurement of momentum, and vice versa. In the most extreme case, 100% accuracy of one variable would result in 0% accuracy in the other.

Heisenberg challenged the theory that for every cause in nature there is a resulting effect. In "classical physics," this means that the future motion of a particle can be exactly predicted by knowing its present position and momentum and all of the forces acting upon it (according to Newton's laws). However, Heisenberg declared that because one cannot know the precise position and momentum of a particle at a given time, its future can not be determined. One can not calculate the precise future motion of a particle, but only a range of possibilities for the future motion of the particle. This range of probabilities can be calculated from Schrödinger equation, which describes the probability of finding the electron at a certain point.

This uncertainty leads to many strange phenomena. For example, in a quantum mechanical world, one cannot predict where a particle will be with 100% certainty. One can only speak in terms of probabilities. For example, if you calculate that an atom will be at some location with 99% probability, there will be a 1% probability it will be somewhere else (in fact, there will be a small but finite probability that it will be found as far away as across the universe).

https://history.aip.org/history/exhibits/heisenberg/p08a.htm
http://www.umich.edu/~chem461/QMChap7.pdf

Example 10.4.1 Hypothetical problem

When two hypothetical atoms (called X atoms) are randomly brought close together, they may be able to form a bond. Each atom has one electron somewhere near its nucleus, with the electron position determined by the Schrödinger density. For the bond to form, an atom with an electron in the outer layer (3–4 AU) must meet an atom with an electron within 1 AU from its nucleus. If this occurs, the atom with the electron within 1 AU will have enough energy to attract the electron 3–4 AU (which is weakly attracted to its own nucleus) from its nucleus. This results in the atom which gave up its electron gaining a positive charge and the atom which received the electron gaining a negative charge. These two atoms will electrostatically attract each other and form a bond.

We may ask: How many times would it take to bring the X atoms near each other before they bond, meaning that both atoms, upon encountering each other, have the correct electron configuration in order to bond? Note: if the atoms do not bond, you pull them away and bring them near again. The desired configuration is as indicated in Figure 10.1.

https://www.chemistry.mcmaster.ca/esam/Chapter_3/section_2.html

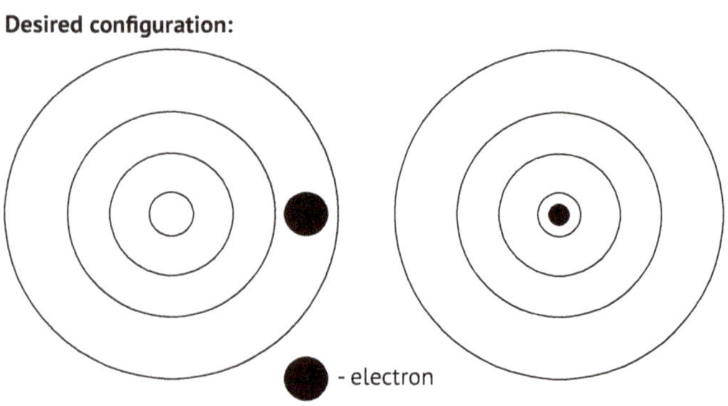

Figure 10.1 Desired configuration: an atom with electron in outer layer and an atom with electron in nucleus.

The following simulation was created, and the whole problem discussed in this section was brought to my attention, by a student in a Freshman seminar conducted by the author at UCLA in 2004.

Simulation (Park 2004)

Step 1. The probability model used for this simulation will be the electron charge density graph of Schrodinger. This graph shows the probabilities of the electron being in a specific region in reference to the X's nucleus. This is the data or probabilities that are gained from this probability model

- Probability of finding electron within 1 AU from nucleus: 32%
- Probability of finding electron within 1–2 AU from nucleus: 42%
- Probability of finding electron within 2–3 AU from nucleus: 19%
- Probability of finding electron within 3–4 AU from nucleus: 6%

We simulate where the electron is located by using R's random number generator. Have the program generate random integers between 1 and 100. Numbers between 0 and 32 will be within 1 AU, between 32 and 74 will be within 2 AU, etc. You would have to generate two numbers at once, each number representing the electron position of one atom.

Step 2. One trial will consist of repeatedly generating the 2 numbers until one number is between 0 and 32 and the other is between 94 and 100 (must occur at the same time)-the correct configuration for the atoms to exchange electrons and bond.

Step 3. The quantity to keep track of is how many sets of numbers the random number generator had to generate until one number was between 0 and 32 and the other between 94 and 100.

Step 4. Repeat steps 2 and 3 many times. The simulation ends when ten trial successes or bonds are made, which will take a sufficient number of number generations in order to do. This will make the calculated probability of the two atoms bonding a more accurate figure due to the high number of trials.

Step 5. Student says: the proportion of successful trials can be used to estimate the probability of two atoms encountering each other and having the correct electron configuration in order to bond, based on the number of trials performed.

Without doing the simulation (which would take a long time), one can calculate the probability: $(32/100)(6/100)(2) = 0.0192(2) = 0.0384$.

You multiply, not add, the two probabilities because they must occur at the same time. Then you multiply by 2 because the position of the 3–4 AU electron and within 1 AU electron can be on either atom for them to bond. Therefore you would expect the atoms to bond about every 4 out of 100 times they're put near each other, or 4% of the time.

10.5 Statistical tests of hypotheses in general

A weather model predicts the following for a certain month (a month = 31 days) in your area:

- Fifteen days will have no rain
- Ten days will have rain, but less than one inch
- Six days will have more than one inch of rain.

The rainfall during the month is looked back on after the month is over, and the actual results are as follows:

- Thirteen days had no rain
- Eleven days had rain, but less than one inch
- Seven days had more than one inch of rain

Of interest is knowing how good is this model. In other words, did the model predict well what the actual rainfall would be?

Statisticians also use the chi-square test to answer this question. The test uses the probability distribution for a chi-square random variable. Here is what they do.

First, they calculate a sample summary statistic, which we do in Table 10.3.

Table 10.3

Outcome	Observed (O)	Expected (E)	$(O-E)^2$	$\frac{(O-E)^2}{E}$
No rain	13	15	4	0.266
Rain but < 1 inch	11	10	1	0.1
>1 inch rain	7	6	1	0.166

When we add the numbers in the last column, the total is 0.532. This is the value of the chi-square statistic with 2 degrees of freedom (3 categories − 1 = 2).

$$\chi_2^2 = 0.532.$$

The next thing is to look at

$$P(\chi_2^2 > 0.532).$$

And we find from the tables that this probability is larger than 0.25. This means it is not unlikely that by chance we would see the difference from the expected value that we observed. Thus, the statistician would conclude that the forecasting model is not systematically deviating from the actual weather, the discrepancies observed are something that one would expect just by chance variation.

10.6 Geography

Geographers know the coordinates of the points on earth that they are interested in studying. For example, medical geographers are interested in the spatial patterns of mortality from various diseases. If certain areas contain more than expected mortality, they claim to have found a pattern, nonrandomness, or something that must have an identifiable cause. Then they seek the cause.

To conduct this type of analysis, geographers must design metrics and find the probability distributions of these metrics. The probability distributions that they use must convey distance and location somehow, and they must convey the correlation of the observations.

The task of geographical data analysis is not too different from that of time series data analysis. Time series data is data that has been collected over time. Both in geographical statistics and in time series analysis, the relevant models are multivariate, like those we studied in Chapters 6 and 8 in this book. The correlations among the different observations must be modeled, and that requires multivariate modeling.

10.7 Chapter References

Dunn, Peter K. (1999). A Simple Dataset for Demonstrating Common Distributions. *Journal of Statistics Education*, Vol 7 Number 3. American Statistics Education.

Grami, Ali. 2016. *Introduction to Digital Communications.* Cambridge: Academic Press.

Park, Dalnam. 2004. *Final project for Stats 19.* Reproduced with permission of the author.

www.ingramcontent.com/pod-product-compliance
Ingram Content Group UK Ltd.
Pitfield, Milton Keynes, MK11 3LW, UK
UKHW052007041026
15590UKWH00008B/2200